T0229166

Principles of Transistor Circuits

Principles of Transistor Circuits

Ninth Edition

Introduction to the Design of Amplifiers, Receivers and Digital Circuits

S. W. Amos, BSc, CEng, MIEE

M. R. James, BSc, CEng, MIEE

Newnes

OXFORD AUCKLAND BOSTON JOHANNESBURG MELBOURNE NEW DELHI

Newnes
An imprint of Elsevier Science
Linacre House, Jordan Hill, Oxford OX2 8DP
200 Wheeler Road, Burlington, MA 01803

First published by Iliffe Books Ltd 1959
Second edition 1961
Third edition 1965
Fourth edition 1969
Fifth edition 1975
Sixth edition 1981
Seventh edition 1990
Eighth edition 1994
Ninth edition 2000
Transferred to digital printing 2003

Copyright © 2000, S. W. Amos and M. R. James. All rights reserved.

The right of S. W. Amos and M. R. James to be identified as the authors
of this work has been asserted in accordance with the Copyright, Designs
and Patents Act 1988

No part of this publication may be reproduced in any material form (including
photocopying or storing in any medium by electronic means and whether
or not transiently or incidentally to some other use of this publication) without
the written permission of the copyright holder except in accordance with the
provisions of the Copyright, Designs and Patents Act 1988 or under the terms of
a licence issued by the Copyright Licensing Agency Ltd, 90 Tottenham Court Road,
London, England W1T 4LP. Applications for the copyright holder's written
permission to reproduce any part of this publication should be addressed
to the publisher

British Library Cataloguing in Publication Data
A catalogue record for this book is available from the British Library

Library of Congress Cataloguing in Publication Data
A catalogue record for this book is available from the Library of Congress

ISBN 0 7506 4427 3

For information on all Newnes publications
visit our website at www.newnespress.com

Printed and bound in Great Britain by Antony Rowe Ltd, Eastbourne

Contents

Preface to the ninth edition

This ninth edition was introduced to bring the material up-to-date and to render all of the diagrams to the same standard. Some of the information from previous editions has been left out; either because it was obsolete or because it is not relevant to modern electronics. Most students are taught discrete component circuit analysis and design with silicon npn transistors as the main active devices. Although a flexibility of approach is important (i.e. to be able to use both npn and pnp devices of any semiconductor type), the redrawn diagrams have been changed to conform to the npn silicon arrangement so that the learning process does not involve unfamiliar configurations. Some of the abbreviations have been modernised, and the gate turn off thyristor introduced along with optically coupled devices. Much of the section on digital techniques has been reworked to reflect current practice.

S. W. Amos
M. R. James

Chapter 1

Semiconductors and junction diodes

Introduction

The 1950s marked the beginning of a revolution in electronics. It started with the invention by William Shockley of the transistor, a minute three-terminal device which could switch, amplify and oscillate yet needed only a few microwatts of power; it was also robust and virtually everlasting. Inevitably the transistor replaced the electron tube (valve) in all except very high power applications.

The pace of the revolution was accelerated a decade later by the development of the integrated circuit or i.c. (popularly known as the silicon chip) in which transistors and other components are manufactured and interconnected by the planar process (see Appendix A) to form amplifiers, signal stores and other functional units on a single silicon slice. The miniaturisation now possible is such that several million transistors can be accommodated on an i.c. less than $1\,\mathrm{cm}^2$.

The applications of i.c.s seem boundless. They feature in activities as diverse as satellite communication and control of model railways. They are widely used in audio, video and radio equipment and they made possible the computers and microprocessors now universally employed in commerce and industry. Perhaps their most familiar applications are in digital watches, calculators and toys.

This book describes the properties of the various types of transistor and shows how they can be used in the design of electronic circuits. The principles described apply to circuits employing discrete transistors and those embodied in i.c.s. To explain the properties of transistors it is useful to begin with an account of the physics of semiconductors because all transistors, irrespective of type, depend on semiconducting material for their action.

Mechanism of semiconduction

As the name suggests a semiconducting material is one with a conductivity lying between that of an insulator and that of a conductor: that is to say one for which the resistivity lies between, say 10^{12} Ω-cm (a value typical of glass) and 10^{-6} Ω-cm (approximately the value for copper). Typical values for the resistivity of a semiconducting material lie between 1 and 100 Ω-cm.

Such a value of resistivity could, of course, be obtained by mixing a conductor and an insulator in suitable proportions but the resulting material would not be a semiconductor. Another essential feature of a semiconducting material is that its electrical resistance decreases with increase in temperature over a particular temperature range which is characteristic of the semiconductor. This behaviour contrasts with that of elemental metallic conductors for which the resistance increases with rise in temperature. This is illustrated in Fig. 1.1, which gives curves for a conductor and a semiconductor. The resistance of the conductor increases linearly, whereas that of the semiconductor decreases exponentially, as temperature rises. Over the significant temperature range the relationship between resistance and temperature for a semiconductor could be written

$$R_t = ae^{bT}$$

where R_t is the resistance at an absolute temperature T, a and b are constants characteristic of the semiconductor material and e is the base of the natural logarithms, i.e. 2.81828 . . . The two curves in Fig. 1.1 are not to the same vertical scale of resistance.

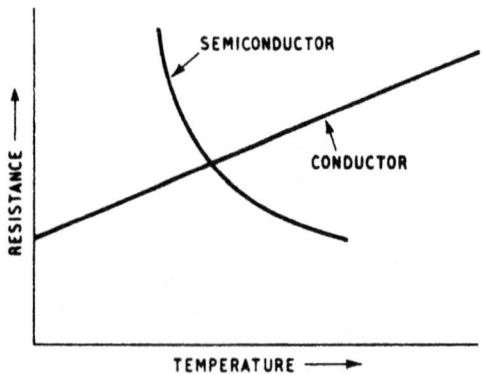

Fig. 1.1. Resistance–temperature relationship for a conductor and a semiconductor

All semiconducting materials exhibit the temperature dependence discussed in the paragraphs above in the pure state: the addition of impurities raises the temperature at which the material exhibits this behaviour, i.e. the region of negative temperature coefficient.

The element most widely used in transistor manufacture is silicon. It has largely replaced germanium which was also used in early transistors. When pure both elements have very poor conductivity and are of little direct use in transistor manufacture. But by the addition of a very small but controlled quantity of a particular type of impurity the conductivity can be increased and the material made suitable for use in transistors.

The behaviour of semiconductors can be explained in terms of atomic theory. The atom is assumed to have a central nucleus which carries most of the mass of the atom and has a positive charge. A number of electrons carrying a negative charge revolve around the nucleus. The total number of electrons revolving around a particular nucleus is sufficient to offset the positive nuclear charge, leaving the atom electrically neutral. The number of electrons associated with a given nucleus is equal to the atomic number of the element. The electrons revolve in a number of orbits and, for the purpose of this discussion, the orbits may be regarded as concentric, the nucleus being at the centre, as shown in Fig. 1.2. This diagram is greatly simplified; the orbits are in practice neither concentric nor co-planar.

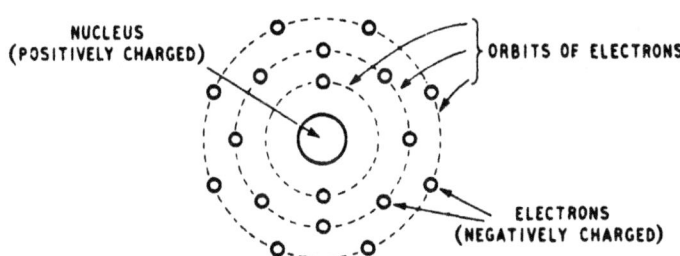

Fig. 1.2. Simplified diagram of structure of atom: for simplicity, electron orbits are shown as circular and co-planar

The first orbit (sometimes called a ring or a shell) is complete when it contains 2 electrons, and an atom with a single complete shell is that of the inert gas, helium. The second ring is complete when it has 8 electrons, and the atom with the first 2 rings complete is that of the inert gas, neon. The third ring is stable when it has 8 or 18 electrons, and the atom having 2, 8 and 8 electrons in the 1st, 2nd and 3rd rings is that of the inert gas, argon. All the inert gases have their outermost shells stable. It is difficult to remove any electrons from a stable ring or to insert others into it. Atoms combine by virtue of the electrons in the outermost rings: for example an atom with one electron in the outermost ring will willingly combine with another whose outermost ring requires one electron for completion.

The inert gases, having their outer shells stable, cannot combine with other atoms or with each other. The number of electrons in the outermost ring or the number of electrons required to make the outermost ring complete has a bearing on the chemical valency of the element and the outermost ring is often called the *valence ring*.

Now consider the copper atom: it has 4 rings of electrons, the first 3 being complete and the 4th containing 1 electron, compared with the 32 needed for completion. Similarly the silver atom has 5 rings, 4 stable and the 5th also containing 1 out of 50 needed for completion. The atoms of both elements thus contain a single electron and this is loosely bound to the nucleus. It can be removed with little effort and is termed a *free electron*. A small e.m.f. applied to a collection of these atoms can set up a stream of free electrons, i.e. an electric current through the metal. Elements in which such free electrons are available are good electrical conductors.

It might be thought that an atom with 17 electrons in the outermost orbit would be an even better conductor, but this is not so. If one electron is added to such an orbit it becomes complete and a great effort is needed to remove it again.

The arrangement of orbital electrons in a silicon atom is pictured in Fig. 1.3. There are three rings, the first containing 2 electrons, the second 8 and the third 4. The total number of electrons is 14, the atomic number of silicon. For comparison the germanium atom has four rings

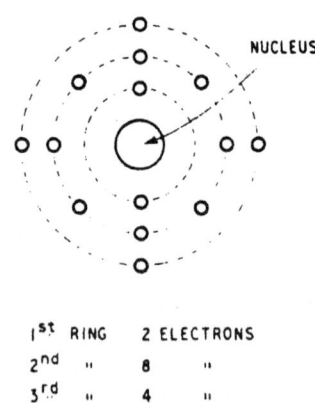

1st RING	2	ELECTRONS
2nd "	8	"
3rd "	4	"

Fig. 1.3. Structure of silicon atom

containing 2, 8, 18 and 4 electrons. These total 32, the atomic number for germanium. A significant feature of both atomic structures is that the outermost ring contains 4 electrons, a property of elements belonging to Group IV of the Periodic Table.

Covalent bonds

It might be thought that some of the 4 electrons in the valence ring of the silicon atom could easily be displaced and that these elements would therefore be good conductors. In fact, crystals of pure silicon are very poor conductors. To understand this we must consider the relationships between the valence electrons of neighbouring atoms when these are arranged in a regular geometric pattern as in a crystal. The valence electrons of each atom form bonds, termed *covalent bonds*, with those of neighbouring atoms as suggested in Fig. 1.4. It is difficult to portray a three-dimensional phenomenon in a two-dimensional diagram, but the diagram does show the valence electrons oscillating between two neighbouring atoms. The atoms behave in some respects as though each outer ring had 8 electrons and was stable. There are no free electrons and such a crystal is therefore an insulator: this is true of pure silicon at a very low temperature.

At room temperatures, however, silicon crystals do have a small conductivity even when they are as pure as modern chemical methods can make them. This is partly due to the presence of minute traces of impurities (the way in which these increase conductivity is explained later) and partly because thermal agitation enables some valence

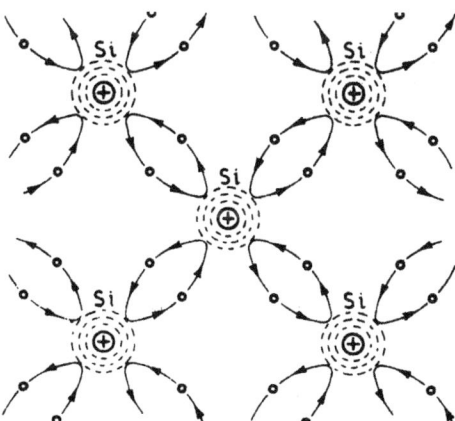

Fig. 1.4. Illustrating covalent bonds in a crystal of pure silicon: for simplicity only electrons in the valence rings are shown

electrons to escape from their covalent bonds and thus become available as charge carriers. They are able to do this by virtue of their kinetic energy which, at normal temperatures, is sufficient to allow a very small number to break these bonds. If their kinetic energy is increased by the addition of light or by increase in temperature, more valence electrons escape and the conductivity increases. If the temperature is raised sufficiently conductivity becomes so great that it swamps semi-conductor behaviour. This sets an upper limit to the temperature at which semiconductor devices can operate normally. For silicon devices the limit is sometimes quoted as 150°C.

Donor impurities

Suppose an atom of a Group-V element such as arsenic is introduced into a crystal of pure silicon. The atom enters into the lattice structure, taking the place of a silicon atom. Now the arsenic atom has 5 electrons in its outermost orbit and 4 of these form covalent bonds with the electrons of neighbouring atoms as shown in Fig. 1.5. The remaining (5th) electron is left unattached; it is a free electron which can be made to move through the crystal by an e.m.f., leaving a positively charged ion. These added electrons give the crystal much better conductivity than pure silicon and the added element is termed a *donor* because it gives free electrons to the crystal. Silicon so treated with a Group-V element is termed *n-type* because negatively charged particles are available to carry charge through the crystal. It is significant that the

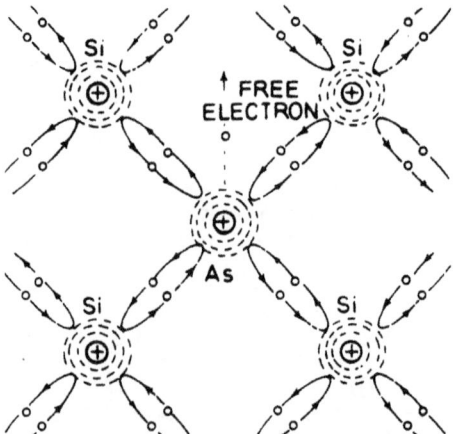

Fig. 1.5. Illustrating covalent bonds in the neighbourhood of an atom of a Group-V element introduced into a crystal of pure silicon. For simplicity only electrons in the valence rings are shown

addition of the arsenic or some other Group-V element was necessary to give this improvement in conductivity. The added element is often called an impurity and in the language of the chemist it undoubtedly is. However, the word is unfortunate in this context because it suggests that the pentavalent element is unwanted; in fact, it is essential.

When a battery is connected across a crystal of n-type semiconductor the free electrons are attracted towards the battery positive terminal and repelled from the negative terminal. These forces cause a drift of electrons through the crystal from the negative to the positive terminal: for every electron leaving the crystal to enter the positive terminal another must be liberated from the negative terminal to enter the crystal. The stream of electrons through the crystal constitutes an electric current. If the voltage applied to the crystal is varied the current varies also in direct proportion, and if the battery connections are reversed the direction of the current through the crystal also reverses but it does not change in amplitude; that is to say the crystal is a *linear* conductor.

Acceptor impurities

Now suppose an atom of a Group-III element such as boron is introduced into a crystal of pure silicon. It enters the lattice structure, taking the place of a silicon atom, and the 3 electrons in the valence ring of the boron atom form covalent bonds with the valence electrons of the neighbouring silicon atoms. To make up the number of covalent bonds to 4, each boron atom competes with a neighbouring atom and may leave this deficient of one electron as shown in Fig. 1.6. A group of covalent bonds, which

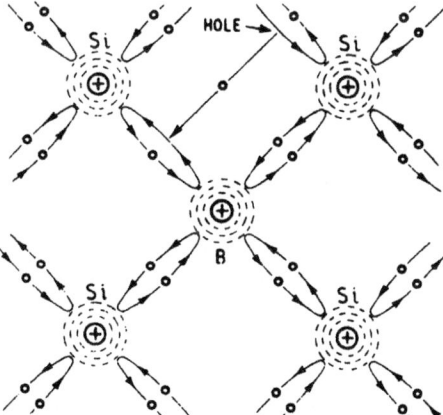

Fig. 1.6. Illustrating covalent bonds in the neighbourhood of an atom of a Group-III element, introduced into a crystal of pure silicon. For simplicity, only electrons in the valence rings are shown

is deficient of one electron, behaves in much the same way as a positively charged particle with a charge equal in magnitude to that of an electron. Such a particle is called a *hole* in semiconductor theory, and we may say that the introduction of the Group-III impurity gives rise to holes in a crystal of pure silicon. These can carry charge through the crystal and, because these charge carriers have a positive sign, silicon treated with a Group-III impurity is termed p-type. Such an impurity is termed an *acceptor* impurity because it takes electrons from the silicon atoms. Thus the introduction of the Group-III element into a crystal lattice of pure silicon also increases the conductivity considerably and, when a battery is connected across a crystal of p-type silicon, a current can flow through it in the following manner.

The holes have an effective positive charge, and are therefore attracted towards the negative terminal of the battery and repulsed by the positive terminal. They therefore drift through the crystal from the positive to the negative terminal. Each time a hole reaches the negative terminal, an electron is emitted from this terminal into the hole in the crystal to neutralise it. At the same time an electron from a covalent bond enters the positive terminal to leave another hole in the crystal. This immediately moves towards the negative terminal, and thus a stream of holes flows through the crystal from the positive to the negative terminal. The battery thus loses a steady stream of electrons from the negative terminal and receives a similar stream at its positive terminal. It may be said that a stream of electrons has passed through the crystal from the negative to the positive terminal. A flow of holes is thus equivalent to a flow of electrons in the opposite direction.

If the battery voltage is varied the current also varies in direct proportion: thus p-type silicon is also a linear conductor.

It is astonishing how small the impurity concentration must be to make silicon suitable for use in transistors. A concentration of 1 part in 10^6 may be too large, and concentrations commonly used are of a few parts in 10^8. A concentration of 1 part in 10^8 increases the conductivity by 16 times. Before such a concentration can be introduced, the silicon must first be purified to such an extent that any impurities still remaining represent concentrations very much less than this. Purification is one of the most difficult processes in the manufacture of transistors. The addition of the impurity is commonly termed *doping*.

Intrinsic and extrinsic semiconductors

If a semiconductor crystal contains no impurities, the only charge carriers present are those produced by thermal breakdown of the covalent bonds. The conducting properties are thus characteristic of the pure semiconductor. Such a crystal is termed an *intrinsic* semiconductor.

In general, however, the semiconductor crystals contain some Group-III and some Group-V impurities, i.e. some donors and some acceptors. Some free electrons fit into some holes and neutralise them but there are some residual charge carriers left. If these are mainly electrons they are termed *majority carriers* (the holes being *minority carriers*), and the material is n-type. If the residual charge carriers are mainly holes, these are majority carriers (the electrons being minority carriers) and the semiconductor is termed p-type. In an n-type or p-type crystal the impurities are chiefly responsible for the conduction, and the material is termed an *extrinsic* semiconductor.

Compound semiconductors

In a silicon crystal covalent bonds between the valence electrons of neighbouring atoms cause the atoms to behave as though the outermost electron orbits were complete. The crystal is therefore effectively a non-conductor until an impurity is introduced. A similar process can occur in a compound of a trivalent and a pentavalent element. Here, too, sharing of the valence electrons yields an effectively complete outer shell and the resulting insulating property can again be destroyed by the introduction of a suitable impurity.

There are a number of such compound semiconductors (known as III-V compounds from the columns of the Periodic Table) but the most widely used is gallium arsenide, GaAs. This has a number of advantages over silicon. For example, the mobility of electrons in GaAs is five times that in silicon, making GaAs transistors suitable for use at microwave frequencies and in computers where high-speed switching is required. Moreover intrinsic GaAs is a better insulator than silicon which helps in the manufacture of integrated circuits. Thirdly GaAs retains its semiconducting properties up to a higher temperature. GaAs is widely used in the manufacture of light-emitting diodes.

Other compound semiconductors contain divalent and hexavalent elements. An example of a II–VI compound is cadmium sulphide CdS.

PN junctions

As already mentioned, an n-type or p-type semiconductor is a linear conductor, but if a crystal of semiconductor has n-type conductivity at one end and p-type at the other end, as indicated in Fig. 1.7, the crystal so produced has asymmetrical conducting properties. That is to say, the current which flows in the crystal when an e.m.f. is applied between the ends depends on the polarity of the e.m.f., being small when the e.m.f. is in one direction and large when it is reversed. Crystals with such conductive properties have obvious applications as detectors or rectifiers.

It is not possible, however, to produce a structure of this type by placing a crystal of n-type semiconductor in contact with a crystal of p-type semiconductor. No matter how well the surfaces to be placed together are planed, or how perfect the contact between the two appears, the asymmetrical conductive properties are not properly obtained. The usual way of achieving a structure of this type is by treating one end of a single crystal of n-type semiconductor with a Group-III impurity so as to offset the n-type conductivity at this end and to produce p-type conductivity instead at this point. Alternatively, of course, one end of a p-type crystal could be treated with a Group-V impurity to give n-type conductivity at this end. The semiconducting device so obtained is termed a *junction diode*, and the non-linear conducting properties can be explained in the following way.

Behaviour of a pn junction

Fig. 1.7 represents the pattern of charges in a crystal containing a pn junction. The ringed signs represent charges due to the impurity atoms and are fixed in position in the crystal lattice: the unringed signs represent the charges of the free electrons and holes (majority carriers) which are liberated by the impurities. The n-region also contains a few holes and the p-region also a few free electrons: these are minority carriers which are liberated by thermal dissociation of the covalent bonds of the semiconducting element itself.

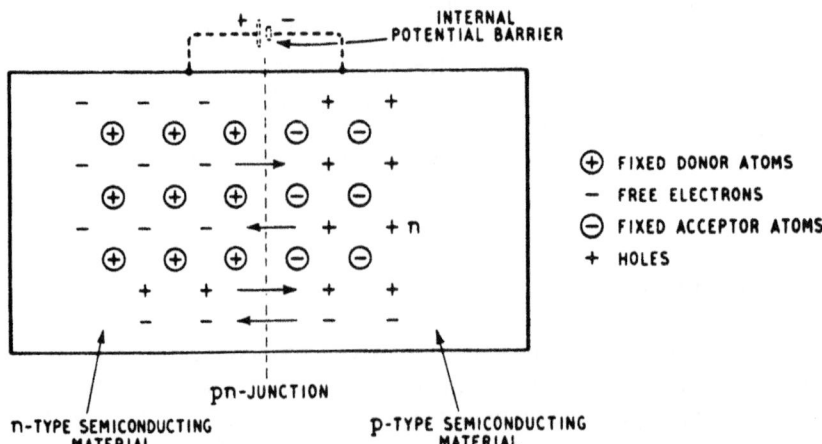

Fig. 1.7. Pattern of fixed and mobile charges in the region of a pn junction

Even when no external connections are made to the crystal, there is a tendency, due to diffusion, for the free electrons of the n-region to cross the junction into the p-region: similarly the holes in the p-region tend to diffuse into the n-region. However, the moment any of these majority carriers cross the junction, the electrical neutrality of the two regions is upset: the n-region loses electrons and gains holes, both causing it to become positively charged with respect to the p-region. Thus a potential difference is established across the junction and this discourages further majority carriers from crossing the junction: indeed only the few majority carriers with sufficient energy succeed in crossing. The potential difference is, however, in the right direction to encourage minority carriers to cross the junction and these cross readily in just sufficient numbers to balance the subsequent small flow of majority carriers. Thus the balance of charge is preserved even though the crystal has a potential barrier across the junction. In Fig. 1.7 the internal potential barrier is represented as an external battery and is shown in dashed lines.

The potential barrier tends to establish a carrier-free zone, known as a *depletion area*, at the junction. The depletion area is similar to the dielectric in a charged capacitor.

Reverse-bias conditions

Suppose now an external battery is connected across the junction, the negative terminal being connected to the p-region and the positive terminal to the n-region as shown in Fig. 1.8. This connection gives a

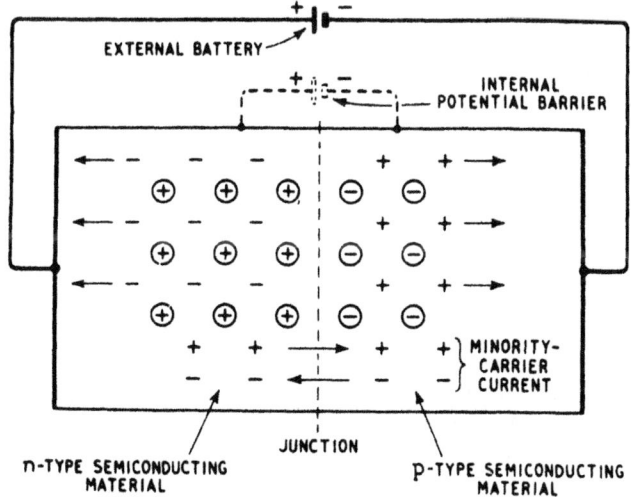

Fig. 1.8. Reverse-bias conditions in a pn junction

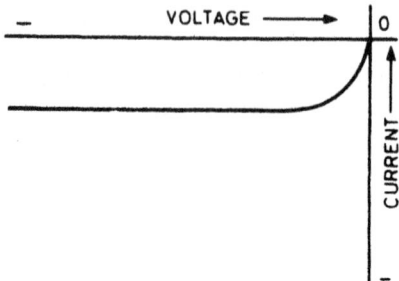

Fig. 1.9. Current–voltage relationship for a reverse-biased pn junction

reverse-biased junction. The external battery is in parallel with and aiding the fictitious battery, increasing the potential barrier across the junction and the width of the depletion area. Even the majority carriers with the greatest energy now find it almost impossible to cross the junction. On the other hand the minority carriers can cross the junction as easily as before and a steady stream of these flows across. When the minority carriers cross the junction they are attracted to the battery terminals and can then flow as a normal electric current in a conductor. Thus a current, carried by the minority carriers and known as the *reverse current,* flows across the junction. It is a small current because the number of minority carriers is small: it increases as the battery voltage is increased as shown in Fig. 1.9 but at a reverse voltage of less than 1 V becomes constant: this is the voltage at which the rate of flow of minority carriers becomes equal to the rate of production of carriers by thermal breakdown of covalent bonds. Increase in the temperature of the crystal produces more minority carriers and an increase in reverse current. A significant feature of the reverse-biased junction is that the width of the depletion area is controlled by the reverse bias, increasing as the bias increases.

Forward-bias conditions

If the external battery is connected as shown in Fig. 1.10 with the positive terminal connected to the p-region and the negative terminal to the n-region, the junction is said to be *forward-biased.* The external battery now opposes and reduces the potential barrier due to the fictitious battery and the majority carriers are now able to cross the junction more readily. The depletion area has now disappeared. Some of the holes and electrons recombine in the junction area so that the current flowing through the device (which can be very large) is carried by holes in the p-region and electrons in the n-region. If the

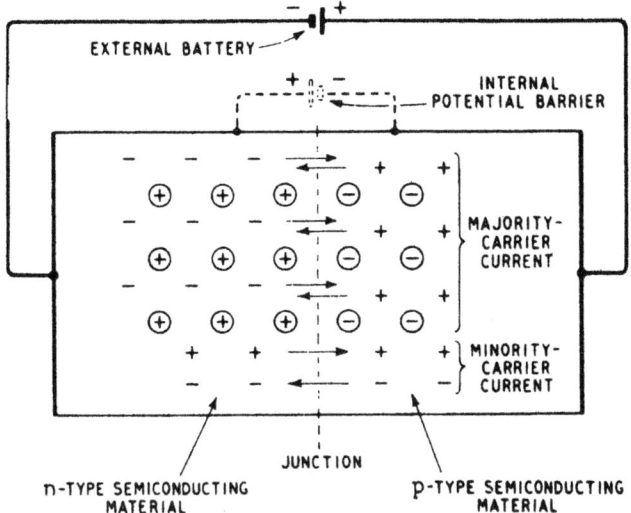

Fig. 1.10. Forward-bias conditions in a pn junction

semiconductor regions are equally doped the number of holes is equal to the number of electrons but the contribution to the forward current made by the holes and electrons depends on the degree of doping of the p- and n-regions. For example, if the doping of the p-region is much heavier than that of the n-region, the forward current will be carried mostly by holes.

The flow of minority carriers across the junction also continues as in reverse-bias conditions but at a reduced scale and these give rise to a second current also taken from the battery but in the opposite direction to that carried by the majority carriers. Except for very small external battery voltages, however, the minority-carrier current is very small compared with the majority-carrier current and can normally be neglected in comparison with it.

The relationship between current and forward-bias voltage is illustrated in Fig. 1.11. The curve has a small slope for small voltages because the internal potential barrier discourages movements of majority carriers across the junction. Increase in applied voltage tends to offset the internal barrier and current increases at a greater rate. Further increase in voltage almost completely offsets the barrier and gives a steeply-rising current. The curve is, in fact, closely exponential in form.

A pn junction thus has asymmetrical conducting properties, allowing current to pass freely in one direction but hardly at all in the reverse direction.

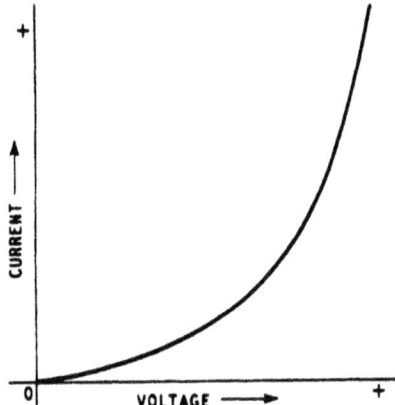

Fig. 1.11. Current–voltage relationship for a forward-biased pn junction

Junction diodes

Small-signal diodes, i.e. those used for detection, mixing and switching, require a small junction area to minimise capacitance and are usually encapsulated in a cylindrical glass or plastic envelope with coaxial leads which are soldered directly into circuit. Germanium was used in the 1950s and the diodes were manufactured by the alloy-junction process described in Appendix A but this was later superseded by a process using diffusion. Silicon junction diodes manufactured by the planar process were introduced in the 1960s. Germanium diodes have a much greater reverse current than silicon but conduction occurs at a lower forward voltage (0.2 V) compared with 0.7 V for silicon and they are preferred to silicon where this smaller voltage drop is important.

In rectifier diodes used for the production of d.c. supplies from a.c. sources the junction area must be large enough to carry the output current required and the chief problems are minimising the rise in junction temperature (a heat sink may be necessary to limit this) and the prevention of breakdown under the stress of the peak inverse voltage. Mains-voltage types are capable of handling currents of tens of amperes.

Fig. 1.12 shows several variations of the circuit symbol for a junction diode. The circular envelope is optional, and although symbol (a) with the continuous line is the recommended symbol, (b) and (c) are both acceptable. The connections are called the anode and cathode, although the circuit abbrieviations are A and K respectively. The anode contains the p-type semiconductor and the cathode contains the n-type semiconductor.

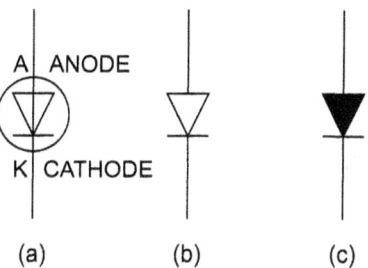

Fig. 1.12. Variations of the circuit symbol for a junction diode

Recovery time

When a conventional pn diode is forward biased some of the majority carriers crossing the junction are neutralised by combination with majority carriers of opposite polarity. Others remain (as minority carriers) and, when the applied voltage is reversed, return across the junction in the form of a substantial pulse of reverse current which takes a significant time to decay to the normal value of reverse current. This delay is a serious disadvantage in diodes required for operation at microwave frequencies or in high-speed switching.

Schottky diode ('hot-carrier' or 'hot-electron' diode)

A diode which overcomes this recovery-time difficulty is the Schottky diode which uses a metal-semiconductor contact instead of a pn junction. For example, in one form of construction a region of epitaxial* n-type GaAs is grown on a GaAs substrate and a metallic layer is deposited on this. Ohmic connections are made to the substrate and the metallic layer. Only one type of charge carrier is involved in operation of the diode. When the metal is biased positively electrons from the n-region are attracted to it to neutralise the charge so giving rise to the forward current. When the metal is negatively charged electrons are repelled and there is no reverse current. There is no p-layer in which electrons could be stored and the resulting diode is highly efficient at frequencies as high as 20 GHz.

Avalanche effect

When a pn junction is reverse-biased the current is carried solely by the minority carriers, and at a given temperature the number of minority carriers is fixed. Ideally, therefore, we would expect the reverse current

* See Appendix A.

for a pn junction to rise to a saturation value as the voltage is increased from zero and then to remain constant and independent of voltage, as shown in Fig. 1.9. In practice, when the reverse voltage reaches a particular value which can be 100 V or more the reverse current increases very sharply as shown in Fig. 1.13, an effect known as breakdown. The effect is reproducible, breakdown in a particular junction always occurring at the same value of reverse voltage. This is known as the *Avalanche effect* and reversed-biased diodes known as *Avalanche diodes* (sometimes called – perhaps incorrectly – *Zener diodes*) can be used as the basis of a voltage stabiliser circuit. The junction diodes used for this purpose are usually silicon types and examples of voltage stabilising circuits employing such diodes are given in Chapter 16.

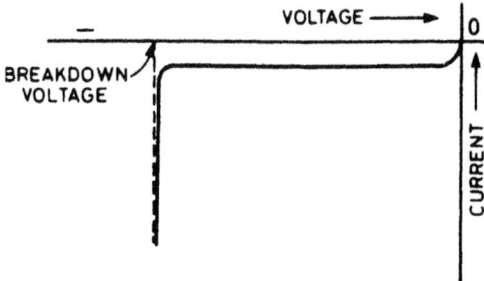

Fig. 1.13. Breakdown in a reverse-biased pn junction

The explanation of the Avalanche effect is thought to be as follows. The reverse voltage applied to a junction diode establishes an electric field across the junction and minority electrons entering it from the p-region are accelerated to the n-region as illustrated in Fig. 1.8. When this field exceeds a certain value some of these electrons collide with valence electrons of the atoms fixed in the crystal lattice and liberate them, thus creating further hole-electron pairs. Some new carriers are themselves accelerated by the electric field due to the reverse bias and in turn collide with other atoms, liberating still further holes and electrons. In this way the number of charge carriers increases very rapidly: the process is, in fact, regenerative. This multiplication in the number of charge carriers produces the sharp increase in reverse current shown in Fig. 1.13. Once the breakdown voltage is exceeded, a very large reverse current can flow and unless precautions are taken to limit this current the junction can be damaged by the heat generated in it. Voltage stabilising circuits using Avalanche diodes must therefore include protective measures to avoid damage due to this cause.

Capacitance diode (varactor diode)

As pointed out above, the application of reverse bias to a pn junction discourages majority carriers from crossing the junction, and produces a depletion area, the width of which can be controlled by the magnitude of the reverse bias. Such a structure is similar to that of a charged capacitor and, in fact, a reverse-biased junction diode has the nature of a capacitance shunted by a high resistance. The value of the capacitance is dependent on the reverse-bias voltage and can be varied over wide limits by alteration in the bias voltage. This is illustrated in the curve of Fig. 1.14: the capacitance is inversely proportional to the applied voltage. A voltage-sensitive capacitance such as this has a number of useful applications: it can be used as a frequency modulator, as a means of remote tuning in receivers or for automatic frequency control (a.f.c.) purposes in receivers. An example of one of these applications of the reverse-biased junction diode is given in Chapter 16.

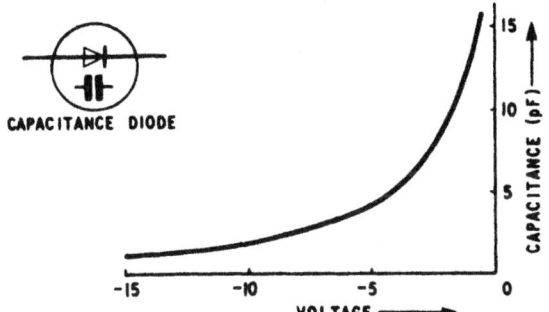

Fig. 1.14. Symbol and typical capacitance–voltage characteristic for a capacitance diode

Zener effect

Some reverse-biased junction diodes exhibit breakdown at a very low voltage, say below 5 V. In such examples breakdown is thought to be due, not to Avalanche effect, but to Zener effect which does not involve ionisation by collision. Zener breakdown is attributed to spontaneous generation of hole-electron pairs within the junction region from the inner electron shells. Normally this region is carrier-free but the intense field established across the region by the reverse bias can produce carriers which are then accelerated away from the junction by the field, so producing a reverse current. The graphical symbol for a Zener diode is given in Fig. 1.15.

Fig. 1.15. Graphical symbol for a Zener diode

Voltage reference diode

The breakdown voltage of a reverse-biased junction diode can be placed within the range of a few volts to several hundred volts but for stabiliser and voltage reference applications it is unusual to employ a diode with a breakdown voltage exceeding a few tens of volts. Some of the reasons for this are given below.

The breakdown voltage varies with temperature, the coefficient of variation being negative for diodes with breakdown voltages less than approximately 5.3 V and positive for diodes with breakdown voltages exceeding approximately 6.0 V. Diodes with breakdown voltages between these two limits have very small coefficients of variation and are thus well suited for use in voltage stabilisers. However, for voltage reference purposes, the slope resistance of the breakdown character-istic must be very small and the slope resistance is less for diodes with breakdown voltages exceeding 6 V than for those with lower breakdown voltages. Where variations in temperature are likely to occur it is probably best to use a diode with a breakdown voltage between 5.3 and 6.0 V for voltage reference purposes but if means are available for stabilising the temperature it is probably better to use a diode with a higher breakdown voltage to obtain a lower slope resistance. Diodes with breakdown voltages around 6.8 V have a temperature coefficient (2.5 mV/°C) which matches that of forward-biased diodes. Voltage reference diodes therefore often consist of two Zener diodes connected in series back-to-back as shown in Fig. 1.16. D1 is reverse-biased when voltage is applied and this has a positive temperature coefficient. D2 is forward-biased and has an equal negative temperature coefficient. The voltage across D2 is small compared with that across D1 and thus the voltage across the device is substantially the Zener voltage of D1 but independent of temperature.

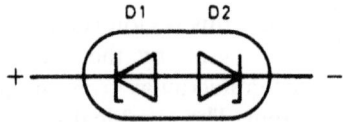

Fig. 1.16. Construction of a voltage reference diode

Voltage reference diodes are usually marketed with preferred values of breakdown voltage (4.7, 5.6, 6.8 V, etc.) and with tolerances of 5 per cent or 10 per cent.

Early voltage reference diodes were rated for only 30 mW dissipation but modern types can withstand a pulse power up to 2.5 kW for a duration of 1 ms provided the pulse is not repetitive. Large diodes are used to protect radars, communications systems and delicate instruments from large electrical transients from nearby electrical equipment.

Backward diode

If the breakdown voltage of a germanium diode is made very low, the region of low slope resistance virtually begins at the origin. Such a junction has a reverse resistance lower than the forward resistance and can be used as a diode which, by contrast with normal diodes, has low resistance when the p-region is biased negatively relative to the n-region. Such backward diodes, manufactured with low capacitance, make highly efficient detectors up to 40 GHz. The current–voltage characteristic for a backward diode is shown in Fig. 1.17 and for comparison the curve for a normal junction diode is also included.

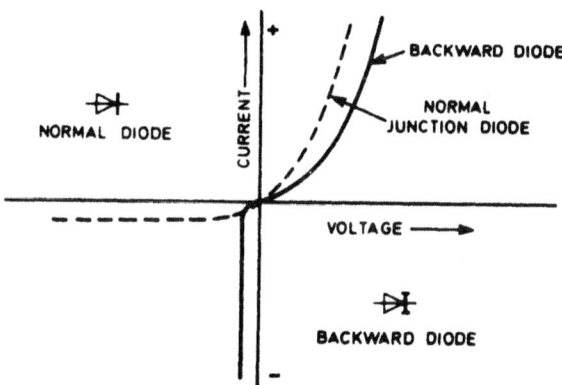

Fig. 1.17. Characteristic curves for a backward diode and normal junction diode. Inset shows graphical symbols used in circuit diagrams

Gunn-effect diode

In certain semiconductors, notably GaAs, electrons can exist in a high-mass low-velocity state as well as their normal low-mass high-velocity state and they can be forced into the high-mass state by a steady electric field of sufficient strength. In this state they form clusters or domains which cross the field at a constant rate causing current to flow as a series of pulses. This is the Gunn effect and one form of diode which makes use of it consists of an epitaxial layer of n-type GaAs grown on a GaAs substrate. A potential of a few volts applied between ohmic contacts to the n-layer and substrate produces the electric field which causes clusters. The frequency of the current pulses so generated depends on the transit time through the n-layer and hence on its thickness. If the diode is mounted in a suitably tuned cavity resonator, the current pulses cause oscillation by shock excitation and r.f. power up to 1 W at frequencies between 10 and 30 GHz is obtainable.

Pin diode

As its name suggests, this is a junction diode with a region of intrinsic semiconductor between the n- and p-regions. When such a diode is reverse-biased the intrinsic layer is depleted of carriers and the diode behaves as a capacitor. When it is forward-biased carriers are injected into the intrinsic region to give a forward resistance which varies linearly between, say, 1 ohm and 10 kilo-ohms with the current through the device. This property makes the diode useful as a modulator or switch in microwave systems and at frequencies between 1 MHz and 20 GHz.

Light-emitting diodes (LEDs)

When a pn junction is forward-biased, electrons are driven into the p-region and holes into the n-region as shown in Fig. 1.10. Some of these charge carriers combine in the junction area and, in some of the combinations, energy is given out in the form of light. By using an alloy of gallium, arsenic and phosphorus as the semiconducting material, the emitted light can be made in shades of red, yellow and green. These can be used as indicators or as seven-segment numeric displays. A combination of doping levels and materials can create LEDs that emit in the infra-red part of the spectrum, which is invisible to human eyes. This is the type of LED used in remote controls for domestic equipment.

When conducting, an LED has typically a forward voltage drop in the region of 2 V and gives a reasonable light output when passing 10 mA (although high-intensity LEDs can draw as much as 25 mA, and there are high-efficiency LED indicators that are useful for battery-powered equipment requiring only 1 mA). However, LEDs do not have a very high reverse voltage capability. Some can be permanently damaged by reverse voltages as low as 5 V.

Blue-emitting LEDs use more exotic semiconductors such as silicon carbide, and are more expensive and less efficient than red and green emitters. They have forward voltages in the region of 5 V.

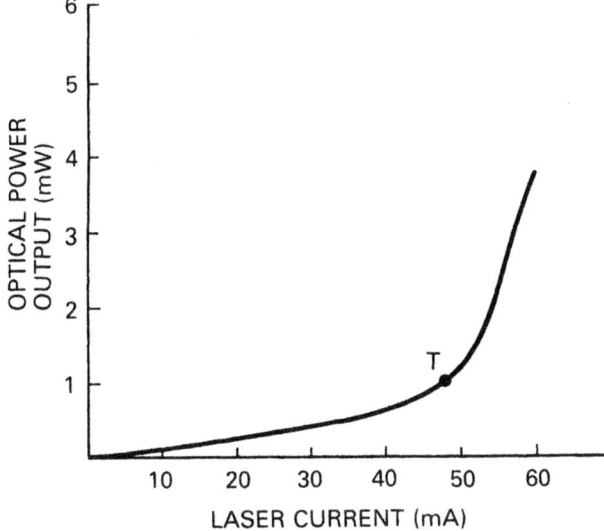

Fig. 1.18. Laser diode operating curve: note the abrupt threshold region at point T

Laser diodes

When high-energy photons pass through a material some are absorbed by atoms, which acquire a higher energy level as a result. Normally the excited atoms quickly return to their normal state by re-emitting a photon. In some semiconductors the high energy level can be sustained so that a photon released from one atom can stimulate the release of another from an adjacent atom, and so on. The process is triggered by normal LED action and builds up to a point governed by the current available from the diode supply. In an injection-laser LED a pn junction is formed in a GaAs crystal, the end faces of which (parallel to the junction plane) are polished so that they act as semi-transparent mirrors and feed back into the junction region some of the emitted light. The photons bounce back and forth in what amounts to an optical cavity resonator, and above a certain threshold bias current they are released in a continuous stream of coherent light from the end of the crystal; a typical operating curve is shown in Fig. 1.18. The most popular application for laser diodes is in audio compact disc (CD) players, where they provide a very narrow infra-red light beam to read the information from micro-pits on the disc surface.

Chapter 2

Basic principles of transistors

Bipolar transistors

Introduction

Chapter 1 showed that a junction between n-type and p-type materials has asymmetrical conducting properties enabling it to be used for rectification. A bipolar transistor includes two such junctions arranged as shown in Fig. 2.1. Fig. 2.1(a) illustrates one basic type consisting of a layer of n-type material sandwiched between two layers of p-type material: such a transistor is referred to as a *pnp-type*.

A second type, illustrated in Fig. 2.1(b), has a layer of p-type material sandwiched between two layers of n-type semiconducting material: such a transistor is referred to as an *npn-type*.

In both types, for successful operation, the central layer must be thin. However, it is not possible to construct bipolar transistors by placing suitably treated layers of semiconducting material in contact. One method which is employed is to start with a single crystal of, say, n-type germanium and to treat it so as to produce regions of p-type conductivity on either side of the remaining region of n-type conductivity.

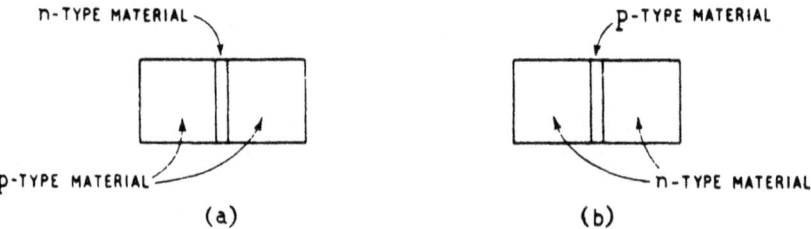

Fig. 2.1. Theoretical diagrams illustrating the structure of (a) a pnp and (b) an npn transistor

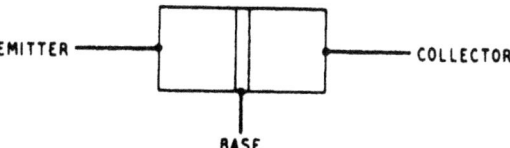

Fig. 2.2. Electrical connections to a bipolar transistor

Electrical connections are made to each of the three different regions as suggested in Fig. 2.2. The thin central layer is known as the *base* of the transistor, one of the remaining two layers is known as the *emitter* and the remaining (third) layer is known as the *collector.* The transistor may be symmetrical and either of the outer layers may then be used as emitter: the operating conditions determine which of the outer layers behaves as emitter, because in normal operation the emitter-base junction is forward-biased whilst the base-collector junction is reverse-biased. In practice most bipolar transistors are unsymmetrical with the collector junction larger than the emitter junction and it is essential to adhere to the emitter and collector connections prescribed by the manufacturer.

The symbols used for bipolar transistors in circuit diagrams are given in Fig. 2.3. The symbol shown at (a), in which the emitter arrow is directed towards the base, is used for a pnp transistor and the symbol shown at (b), in which the emitter arrow is directed away from the base, is used for an npn transistor.

An account of the principal methods used in the manufacture of transistors is given in Appendix A.

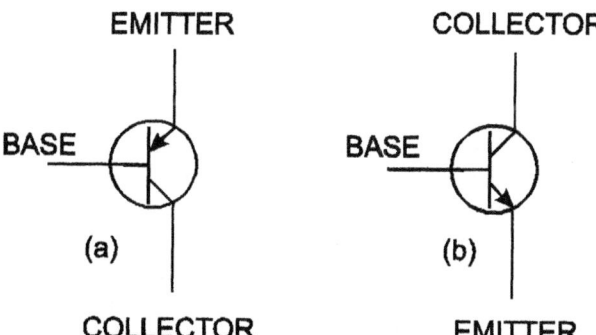

Fig. 2.3. Circuit diagram symbols for (a) pnp and (b) npn bipolar transistors

Operation of a pnp transistor

Fig. 2.4 illustrates the polarity of the potentials which are necessary in a pnp-transistor amplifying circuit. The emitter is biased slightly positively with respect to the base: this is an example of forward bias and the external battery opposes the internal potential barrier associated with the emitter-base junction. A considerable current therefore flows across this junction and this is carried by holes from the p-type emitter (which move to the right into the base) and by electrons from the n-type base (which move to the left into the emitter). However, because the impurity concentration in the emitter is normally considerably greater than that of the base (this is adjusted during manufacture), the holes carrying the emitter-base current greatly outnumber the electrons and we can say with little error that the current flowing across the emitter-base junction is carried by holes moving from emitter to base. Because holes and electrons play a part in the action, this type of transistor is known as *bipolar*.

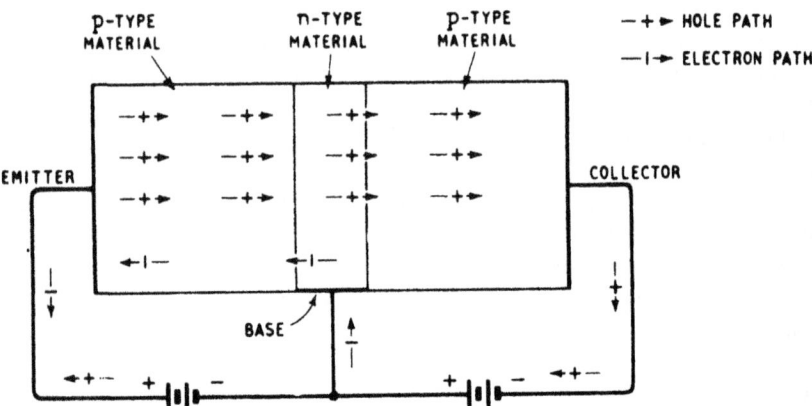

Fig. 2.4. Hole and electron paths in a pnp transistor connected for amplification

The collector is biased negatively with respect to the base: this is an example of reverse bias and the external battery aids the internal potential barrier associated with the base-collector junction. If the emitter-base junction were also reverse-biased, no holes would be injected into the base region from the emitter and only a very small current would flow across the base-collector junction. This is the reverse current (described in Chapter 1): it is a saturation current independent of the collector-base voltage. However, when the emitter-base junction is forward-biased, the injected holes have a marked effect on the collector current. Because the base is a particularly thin layer most of the injected holes cross the base by diffusion and on reaching the collector-base junction are swept into the collector region. The reverse bias of the base-collector junction ensures the collection of all the holes crossing this junction, whether these are present in the base region as a result of breakdown of covalent bonds by virtue of thermal agitation or are injected into it by the action of the emitter. A few of the holes which leave the emitter combine with electrons in the base and so cease to exist but the majority of the holes (commonly more than 95 per cent) succeed in reaching the collector. Thus the increase in collector current due to hole-injection by the emitter is nearly equal to the current flowing across the emitter-base junction. The balance of the emitter carriers (equal to, say, 5 per cent) is neutralised by electrons in the base region and to maintain charge neutrality more electrons flow into the base, constituting a base current.

Thus a small current flowing in the base controls a much larger collector current: this is the essence of transistor action and from what has been said above it is clear that to achieve high current gain we need a heavily doped emitter area and a very thin but lightly doped base region. In early alloy-junction transistors the base thickness exceeded 10^{-4} cm but in more modern planar types it is less than 10^{-5} cm.

The collector current, even though it may be considerably increased by forward bias of the emitter-base junction, is still independent of the collector voltage. This is another way of saying that the output resistance of the transistor is extremely high: it can in fact be several megohms. The input resistance is approximately that of a forward-biased junction diode and is commonly of the order of $25\,\Omega$. A small change in the input (emitter) current of the transistor is faithfully reproduced in the output (collector) current but, of course, at a slightly smaller amplitude. Clearly such an amplifier has no current gain but because the output resistance is many times the input resistance it can give voltage gain. To illustrate this suppose a 1-mV signal source is connected to the 25-Ω input. This gives rise to an emitter current of $1/25$ mA, i.e. $40\,\mu$A. The collector current is slightly less than this but as an approximation suppose the output current is also $40\,\mu$A. A common value of load resistance is $5\,\mathrm{k}\Omega$ and for this value the output voltage is given by $5,000 \times 40 \times 10^{-6}$, i.e. $200\,\mathrm{mV}$, equivalent to a voltage gain of 200.

Bias supplies for a pnp transistor

Fig. 2.5 shows a pnp transistor connected to supplies as required in one form of amplifying circuit. For forward bias of the emitter-base base junction, the emitter is made positive with respect to the base; for reverse bias of the base-collector junction, the collector is made negative with respect to the base. Fig. 2.5 shows separate batteries used to provide these two bias supplies and it is significant that the batteries are connected in series, the positive terminal of one being connected to the negative terminal of the other. The base voltage in fact lies between that of the collector and the emitter and thus a single battery can be used to provide the two bias supplies by connecting it between emitter and collector, the base being returned to a tapping point on the battery or to a potential divider connected across the battery. The potential divider technique (Fig. 2.6) is often used in transistor circuits and a pnp transistor operating with the emitter circuit earthed requires a negative collector voltage. The arrow in the transistor symbol shows the direction of conventional current flow, i.e. is in the opposite direction to that of electron flow through the transistor.

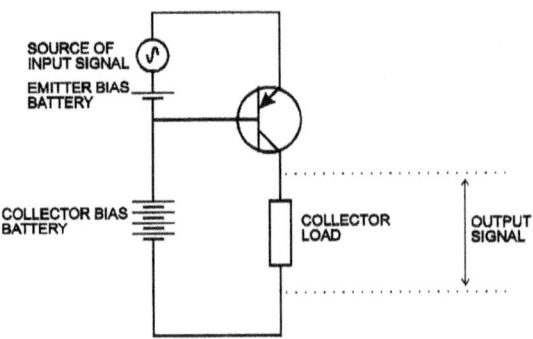

Fig. 2.5. Basic circuit for using a pnp transistor as an amplifier

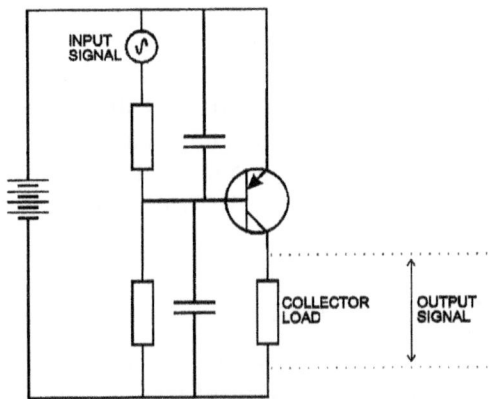

Fig. 2.6. The circuit of Fig. 2.5 using a single battery and a potential divider providing base bias

Operation of an npn transistor

The action of an npn junction transistor is similar to that of a pnp type just described but the bias polarities and directions of current flow are reversed. Thus the charge carriers are predominantly electrons and the collector bias voltage for an earthed-emitter circuit must be positive. Originally, pnp transistors were easier to manufacture and so were the most commonly encountered variant of the bipolar transistor. As manufacturing technology developed, npn transistors were used more often because they had a better performance. The mobility of holes in the semiconductor matrix is lower than that of electrons, and thus npn transistors operate at higher frequencies.

Common-base, common-emitter and common-collector amplifiers

So far we have described amplifying circuits in which the emitter current determines the collector current: it is, however, more usual in transistor circuits to employ the external base current to control the collector or emitter current. Used in this way the transistor is a current amplifier because the collector (and emitter) current can easily be 100 times the controlling (base) current and variations in the input current are faithfully portrayed by much larger variations in the output current.

Thus we can distinguish three ways in which the transistor may be used as an amplifier:

(a) with emitter current controlling collector current,
(b) with base current controlling collector current,
(c) with base current controlling emitter current.

It is significant that in all these modes of use, operation of the transistor is given in terms of input and output current. This is an inevitable consequence of the physics of the bipolar transistor: such transistors are *current-controlled* devices: by contrast field-effect transistors are voltage-controlled devices.

Corresponding to the three modes of operation listed above there are three fundamental transistor amplifying circuits: these are shown in Fig. 2.7. At signal frequencies the impedance of the collector voltage supply is assumed negligibly small and thus we can say for circuit (a) that the input is applied between emitter and base and that the output is effectively generated between collector and base. Thus the base connection is common to the input and output circuits: this amplifier is therefore known as the *common-base* type.

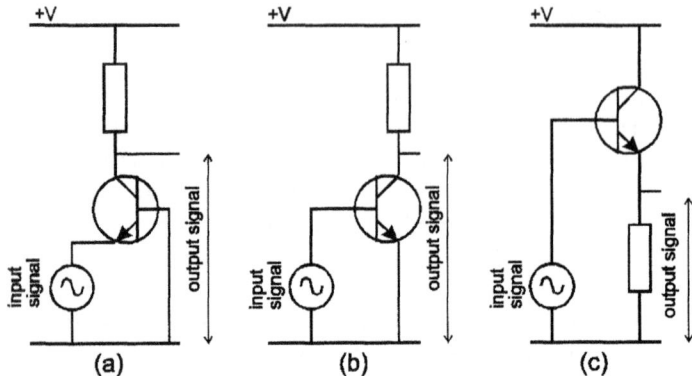

Fig. 2.7. The three basic forms of transistor amplifier; (a) common-base, (b) common-emitter and (c) common-collector (emitter follower). For simplicity base d.c. bias is omitted

Fig. 2.5 illustrates one significant feature of the common-base circuit, namely that the base acts as a screen between the input and output circuits. The elimination of capacitive coupling between them makes stable v.h.f. and u.h.f. amplification possible.

In (b) the input is again applied between base and emitter but the output is effectively generated between collector and emitter. This is therefore the *common-emitter* amplifier, probably the most used of all transistor amplifying circuits.

In (c) the input is effectively between base and collector, the output being generated between emitter and collector. This is the *common-collector* circuit but it is better known as the *emitter follower*.

Current amplification factor

In a common-base amplifier the ratio of a small change in collector current i_c to the small change in emitter current i_e which gives rise to it is known as the current amplification factor α. It is measured with short-circuited output. Thus we have

$$\alpha = \frac{i_c}{i_e} \tag{2.1}$$

As we have seen i_c is very nearly equal to i_e. Thus α is nearly equal to unity and is seldom less than 0.95. In approximate calculations α is often taken as unity.

In a common-emitter amplifier the ratio of a small change in collector current i_c to the small change in base current i_b which gives rise to it is represented by β. It is also measured with short-circuited output and indicates the maximum possible current gain of the transistor. Thus

$$\beta = \frac{i_c}{i_b} \qquad (2.2)$$

We have seen how the emitter current of a transistor effectively divides into two components in the base region. Most of it passes into the collector region and emerges as external collector current. The remaining small fraction forms an external base current. This division applies equally to steady and signal-frequency currents. Thus we have

$$i_e = i_c + i_b \qquad (2.3)$$

From Eqns 2.1, 2.2 and 2.3 we can deduce a relationship between α and β thus

$$\beta = \frac{i_c}{i_b} = \frac{i_c}{i_e - i_c}$$

But $i_c = \alpha i_e$

$$\therefore \beta = \frac{\alpha i_e}{i_e - \alpha i_e}$$

$$= \frac{\alpha}{1 - \alpha} \qquad (2.4)$$

As α is nearly equal to unity there is little error in taking β as given by

$$\beta = \frac{1}{1 - \alpha} \qquad (2.5)$$

Thus for a transistor for which $\alpha = 0.98$

$$\beta = \frac{1}{1 - 0.98} = \frac{1}{0.02} = 50$$

In practice, values of β lie between 20 and 500. β is one of the properties of a transistor normally quoted in manufacturer's literature: here it is usually known as h_{fe}. This is one of a range of parameters called the 'h' parameters. h_{fe} is the forward current gain in common emitter configuration. As explained further in Appendix B there are two variants: h_{fe} and h_{FE}. In common with all notation, the upper-case subscripts refer to d.c. values and the lower-case refer to a.c. values. In this instance, the current gain is different. It also depends on the transistor bias current, the temperature and, for a.c. signals, the frequency.

Collector current–collector voltage characteristics

Fig. 2.8(a) illustrates the way in which the collector current of a bipolar transistor varies with collector voltage for given values of the emitter current. The characteristics are straight, horizontal and equidistant. The fact that the curves are horizontal shows that collector current is independent of collector voltage: in other words, the collector a.c. resistance is very high. The regular spacing implies low distortion if the device is used as an analogue amplifier. Fig. 2.8(b) illustrates the way in which the collector current of a bipolar transistor varies with collector voltage for given values of base current. These characteristics are of the same general shape as those of Fig. 2.8(a).

A number of parameters of the transistor can be obtained from these characteristics. For example the slope of the curves is not so low as for the common-base connection showing that the collector a.c. resistance is smaller: it is in fact approximately $30\,\text{k}\Omega$. To deduce β consider the intercepts made by the characteristics on the vertical line drawn through $V_c = 4\,\text{V}$. When the base current is $40\,\mu\text{A}$ (point A) the collector current is $2.5\,\text{mA}$ and when I_b is $50\,\mu\text{A}$ (point B) I_c is $3.5\,\text{mA}$. A change of base current of $10\,\mu\text{A}$ thus causes a change in collector current of $1\,\text{mA}$: this corresponds to a value of β of 100.

Collector current–base voltage characteristics

Fig. 2.9 gives the I_c–V_{be} characteristics for a silicon transistor. This shows that the relationship between base voltage and collector current is not linear. It also shows that collector current does not start until the base voltage exceeds about $0.7\,\text{V}$. This was pointed out in Chapter 1 in respect of silicon junction diodes.

Equivalent circuit of a transistor

For calculating the performance of transistor circuits, it is useful to regard the transistor as a three-terminal network which is specified in terms of its input resistance, output resistance and current gain – all fundamental properties which can readily be measured. The properties of such a network are expressed in a number of ways notably as z parameters, y parameters or h parameters. The basic equations for these

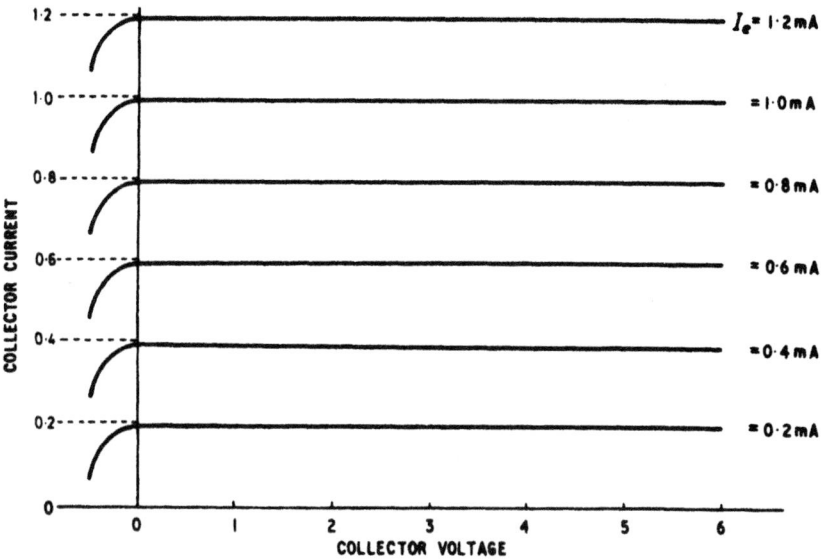

Fig. 2.8(a). A set of $I_c - V_c$ characteristics for a common-base amplifier

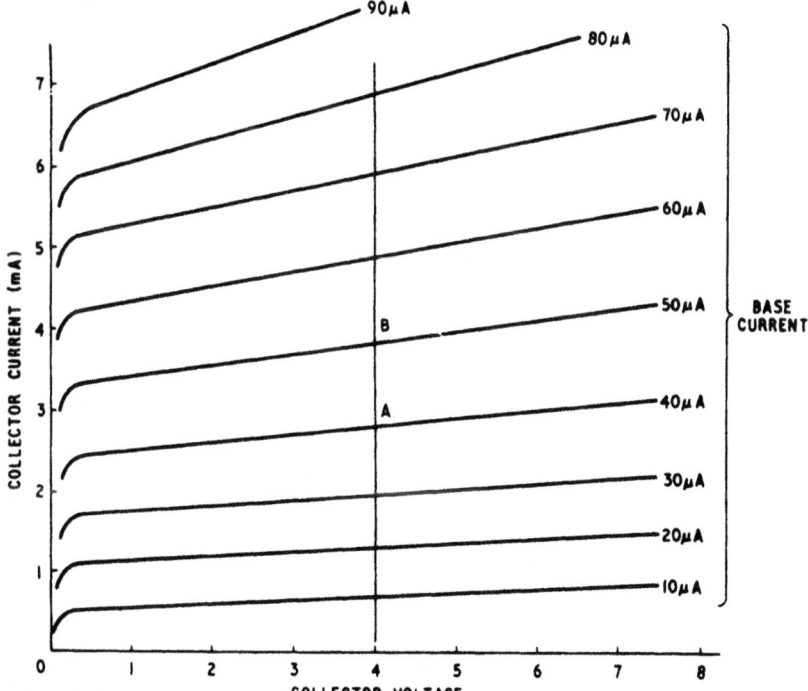

Fig. 2.8(b). Typical collector current–collector voltage characteristics for common-emitter connection

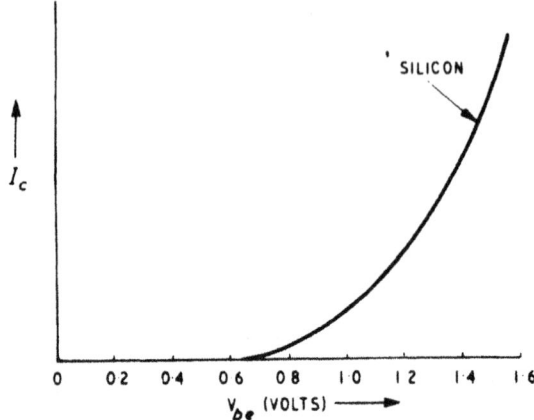

Fig. 2.9. I_c–V_{be} curves for a silicon transistor

parameters are given in Appendix B and this shows that these parameters, in spite of their variety, express the same four properties of the network, namely its input resistance, output resistance, forward gain and reverse gain.

One of the disadvantages of this method of expressing transistor properties is that the values of the fundamental properties which apply to the common-base connection do not apply to the common-emitter connection or to the common-collector connection and three sets of values are therefore required in a complete expression of a transistor's properties. Moreover the numerical values of these properties vary with emitter current and frequency and can be regarded as constant only over a narrow frequency and emitter-current range.

An alternative approach to the problem of calculating transistor performance is to deduce an equivalent network which has a behaviour similar to that of the transistor. The constants of such a network can often be directly related to the physical construction of the transistor but they cannot be directly measured: they can, however, be deduced from measurements on the transistor. If the network is truly equivalent it will hold at all frequencies and by applying Kirchhoff's laws or other network theorems to this equivalent circuit we can calculate the performance of the circuit. Much useful work is possible by representing a transistor as a simple T-network of resistance as shown in Fig. 2.10.

The transistor cannot, however, be perfectly represented by three resistances only because such a network cannot generate power (as a transistor can) but can only dissipate power. In other words the network illustrated in Fig. 2.10 is a *passive* network and to be accurate the equivalent network must include a source of power, i.e. must be *active*.

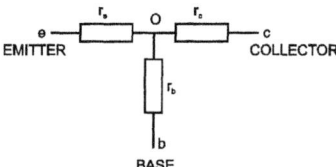

Fig. 2.10. A three-terminal passive network which can be used to build an equivalent circuit for a transistor

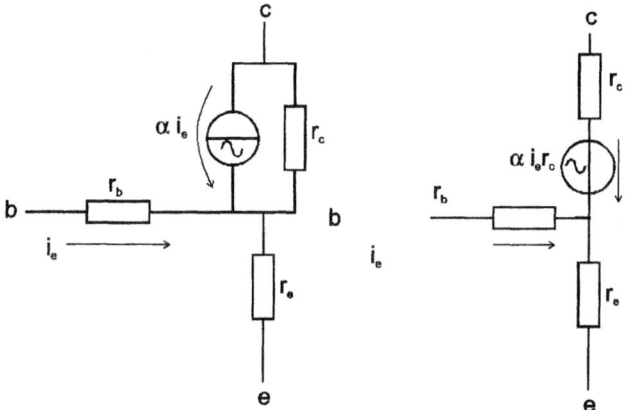

Fig. 2.11. A three-terminal active network which can be used as an equivalent circuit for a transistor (a) including a constant-current generator, and (b) a constant-voltage generator

The source of power could be shown as a constant-current source connected in parallel with the collector resistance r_c as in Fig. 2.11(a) and the current so supplied is, as we have already seen, equal to αi_e, where α is the current amplification factor of the transistor and i_e is the current in the emitter circuit, i.e. in the emitter resistance r_e. Alternatively the source of power can be represented as a constant-voltage generator connected in series with r_c as shown in Fig. 2.11(b).

Such a voltage has precisely the same effect as the constant-current generator, provided the voltage is given the correct value, and the value required is equal to $\alpha i_e r_c$ as can be shown by applying Thevenin's theorem to Fig. 2.11(a).

In some books on transistors the voltage generator is given as $r_m i_e$, where r_m is known as the *mutual* or *transfer resistance** and is equal to αr_c. The mutual resistance may be defined as the ratio of the e.m.f. in the collector circuit to the signal current in the emitter circuit which

---

*The word 'transistor' is, in fact, derived from 'transfer resistance'.

gives rise to it. This may be regarded as the dual of the mutual conductance g_m of a field-effect transistor which is defined as the ratio of the current in the drain circuit to the signal voltage in the gate circuit which causes it.

This network is quite satisfactory for calculating the performance of transistor amplifiers at low frequencies such as audio amplifiers because the reactances of the internal capacitances have, in general, negligible effects on performance.

At higher frequencies, however, and in particular at radio frequencies, it is necessary to include such capacitances in the T-network to obtain accurate answers.

Of the various internal capacitances within a bipolar transistor, that between the collector and the base has the greatest effect on the high-frequency performance.

In a transistor r.f. amplifier the capacitance between collector and base provides feedback from the output to the input circuit. This is illustrated in Fig. 2.12(a) in which the capacitance c_{bc} is shown connected directly between collector and base terminals. However, a better approximation to the performance of a transistor at high frequencies is obtained by assuming that the collector capacitance is returned to a tapping point b' on the base resistance as shown in Fig. 2.12(b).

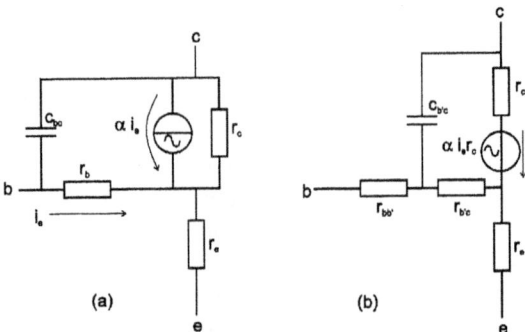

(a) (b)

Fig. 2.12. T-section equivalent circuit of a transistor showing collector capacitance, (a) returned to base input terminal and (b) returned to a tapping point on the base resistance

This modifies the feedback which now occurs via $c_{b'c}$ and $r_{bb'}$ in series and the high-frequency performance of the transistor is now dependent on the time constant $r_{bb'} c_{b'c}$ which is probably the most important characteristic of a transistor intended for high-frequency use.

Values of r_e, r_b and r_c

The networks of Fig. 2.11 can be used in analyses of transistor circuits. The resistances represent differential, i.e. a.c. quantities and, for a given value of mean emitter current I_e, are constant provided that the variations in voltage or current which occur during operation of the transistor are small.

The value of r_e for all bipolar transistors, irrespective of size and type, is given by

$$r_e = \frac{kT}{eI_e}$$

where k = Boltzmann's constant, i.e. 1.374×10^{-23} J/°C

e = charge on the electron, i.e. 1.59×10^{-19} C

and T = absolute temperature.

Thus r_e is directly proportional to the absolute temperature and inversely proportional to the emitter current. The coefficient kT/e has the dimensions of a voltage and if we substitute numerical values for k and e this voltage is 25 mV at a temperature of 20°C. Thus we have

$$r_e \ (\Omega) = \frac{25 \ (\text{mV})}{I_e \ (\text{mA})}$$

If $I_e = 1$ mA, r_e is 25 Ω (at 20°C).

The value of r_c is also inversely proportional to I_e in simple transistors but the value depends on the impurity grading in the base region and is very high for diffused-base transistors.

The value of r_b is largely independent of I_e and is a function of the size and geometry of the transistor. For small silicon transistors it is unlikely to exceed 100 Ω but for a large power transistor r_b may be as low as 5 Ω.

The values of r_e, r_b and r_c cannot be measured directly because it is impossible to obtain a connection to the point O (in Fig. 2.12) but the values can be deduced from measurements made at the transistor terminals.

Although the bipolar transistor is a current-controlled device there are occasions (e.g. when it is used as a voltage amplifier) when it is useful to be able to express the performance in terms of the signal voltage applied between base and emitter. For such calculations we need to know the mutual conductance g_m of the transistor: this is the ratio of a small change in collector current to the change in base-emitter voltage which causes it.

The mutual conductance of a transistor is given by

$$g_m = \frac{\alpha_o}{r_e}$$

where α_o is the current amplification factor at low frequencies. α_o is very nearly equal to unity and thus we may say

$$g_m \approx \frac{1}{r_e}$$

We know, from the previous page that r_e is inversely proportional to the mean emitter current I_e according to the relationship

$$r_e \ (\Omega) = \frac{25 \ (\text{mV})}{I_e \ (\text{mA})}$$

Eliminating r_e between the last two expressions we have

$$g_m \ (\text{A/V}) \approx \frac{I_e \ (\text{mA})}{25 \ (\text{mV})}$$

which is perhaps more conveniently expressed

$$g_m \ (\text{mA/V}) \approx 40 I_e \ (\text{mA})$$

Thus for an emitter current of 1 mA the mutual conductance is 40 mA/V. This relationship will be used in Chapter 4 to aid the design of amplifiers.

Frequency f_T

Modern circuitry makes extensive use of silicon planar transistors as common-emitter amplifiers and the high-frequency performance is usually expressed in terms of the parameter f_T, the transition frequency. This is defined as the frequency at which the modulus of β (the current gain of the common-emitter amplifier) has fallen to unity. It thus measures the highest frequency at which the transistor can be used as an amplifier: it also gives the gain-bandwidth product for the transistor.

Field-effect transistors

Introduction

Field-effect transistors operate on principles quite different from those of bipolar transistors. They consist essentially of a channel of semiconducting material, the charge-carrier density of which is controlled by the input

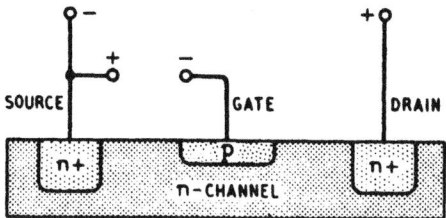

Fig. 2.13. Structure of an n-channel *jfet* with bias polarities indicated

signal. The input signal thus determines the conductivity of the channel and hence the current which flows through it from the supply. The connections from the ends of the channel to the supply are known as the *source* and *drain* terminals and they correspond to the emitter and collector terminals of a bipolar transistor. The control terminal is known as the *gate*. The potential on the gate can control the channel conductivity via a reverse-biased pn junction: transistors using this principle are termed *junction-gate field-effect transistors* (jfets). Alternatively, the gate potential can control the channel conductivity via a capacitance link the gate being physically isolated from the conducting layer. Transistors using this principle could be called *insulated-gate field-effect transistors* igfets, but are more commonly referred to as mosfets, which stands for *metal-oxide semiconductor fets*. Fets in general can be regarded as voltage-controlled variable resistors.

There is only one type of charge carrier in an fet. namely electrons in an n-channel device and holes in a p-channel device. Thus fets can be termed unipolar* transistors: those previously described in this chapter are, of course, bipolar. An important consequence of the single type of charge carrier is that a fet. can introduce less noise than bipolar types. Because there are no minority carriers, fets are also free of the carrier-storage effects which limit the switching times of bipolar transistors.

Junction-gate Field-effect transistors (jfets)

The structure of an n-channel jfet is illustrated in Fig. 2.13. This shows a slice of high-resistance n-type silicon with ohmic contacts formed by highly doped (n+) regions near the two ends: these provide the source and drain connections to the external circuit. A region of p-type conductivity is formed between the ohmic contacts: this forms the gate connection and is reverse-biased, i.e. negatively biased with respect to the source connection. When the source and drain terminals are connected to a supply, current flows longitudinally through the slice, being carried by the free electrons of the n-type material. However, near

---

* Strictly this should be monopolar if we are using Greek prefixes.

the p-region there is a depletion area, i.e. an area free of charge carriers and if the reverse bias of the gate is increased the depletion area spreads, confining the longitudinal current to a small cross-sectional area of the slice and reducing its amplitude. If the reverse bias is increased sufficiently the depletion area spreads to the whole cross-section of the slice and cuts off the current completely. The gate-source voltage required to do this is known as the pinch-off voltage. By connecting a signal source in series with the gate bias the effective width of the channel can be modulated and the signal waveform is impressed on the current. If the current is passed through a suitable external impedance, the voltage generated across it is a magnified version of that of the signal source.

This can be shown graphically as in the transconductance graph of Fig. 2.14. This shows that when the gate voltage is the same as the source, a fixed amount of current flows through the fet. This is called the saturated drain-source current, or I_{DSS}. As the gate voltage is made more negative (for an n-channel fet) the drain current is reduced until it is cut off completely. This is the pinch-off voltage V_P already mentioned. Typical values for I_{DSS} are 4 to 10 mA, while V_P is −2 to −5 V.

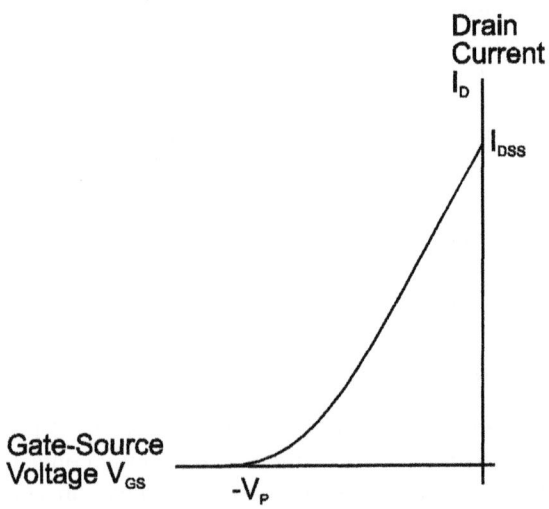

Fig. 2.14. Fet transconductance characteristic

The input resistance of a jfet is that of a reverse-biased pn junction and can be very high: a typical value is $10^{10}\,\Omega$ with a shunt capacitance of say 5 pF. The characteristics (Fig. 2.15) have the same general shape as those of a bipolar transistor but, of course, the control parameter here is the gate voltage not current. If the gate terminal of the transistor

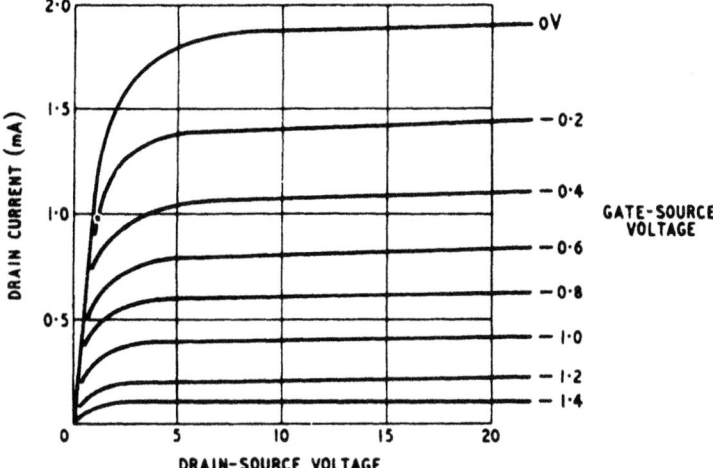

Fig. 2.15. Typical characteristics for a field-effect transistor

illustrated in Fig. 2.13 is biased positively with respect to the source, the pn junction is forward biased giving a very low input resistance.

There is a complementary type of jfet with a p-type channel and n-type gate: this requires a positive gate-source voltage to cut off the channel current. The graphical symbols for both types of jfet are given in Fig. 2.16.

The jfet takes a significant drain current with zero gate bias and, for an n-channel device, a negative bias is required to cut the current off. When the fet is used as an amplifier the gate bias is normally between zero and cut-off and lies outside the range of the drain-source voltage; for example typical voltages are $V_g = -1$ V, $V_s = 0$ and $V_d = +20$ V. Devices with this property are said to operate in the *depletion mode*. By contrast, the base bias voltage for a bipolar transistor lies between that of the emitter and the collector.

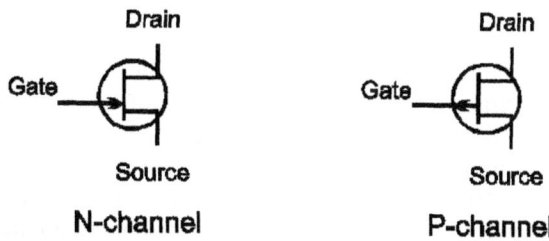

Fig. 2.16. Graphical symbols for jfets

Insulated-gate field-effect transistors (mosfets)

A cross-section of a mosfet is shown in Fig. 2.17. It consists of a base (substrate) layer of p-type silicon into the surface of which two closely spaced stripes of n-type conductivity are diffused. Ohmic connections are made to the n-regions to give source and drain connections, and to the p-type substrate to give a base connection. The device is sealed by a silicon dioxide coating and a thin layer of aluminium between the n-regions provides a gate connection. The device can be manufactured by the techniques used for planar transistors described in Appendix A.

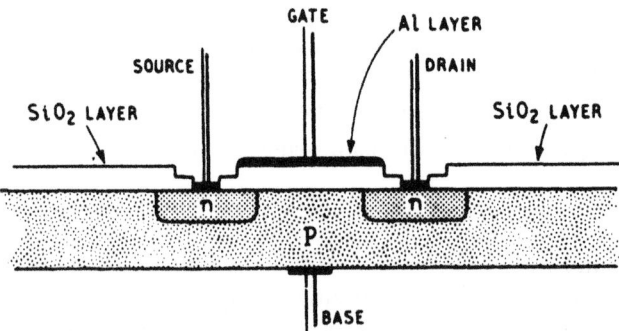

Fig. 2.17. Structure of enhancement-type n-channel mosfet

The substrate, the silicon dioxide layer and the aluminium skin form a parallel-plate capacitor. The voltage applied between gate and base controls the conductivity of the surface area of the substrate between the n-regions and hence the current which flows between source and drain when these are connected to a supply. If the gate-base voltage is zero the only drain current is the leakage current of one of the pn junctions and this, in a silicon device, is very small indeed. If the gate is made positive with respect to the base, positive charge carriers are repelled into the body of the substrate and negative charge carriers are attracted to its surface either from thermal breakdown of the p-material or from the n-regions. In this way a layer of mobile charge carriers is induced on the surface of the substrate and this makes ohmic contact with the diffused n-regions. The induced layer provides the conducting channel between the source and drain terminals and permits drain current to flow. The induced layer is known as an *inversion layer* because it changes its conductivity from p-type to n-type as the gate voltage is increased positively from zero. Increase in gate voltage increases the number of charge carriers in the induced layer, increasing channel conductivity and drain current. The gate voltage thus controls the drain current and the characteristics have a form similar to those of a jfet. A significant feature of the behaviour of this type of mosfet is that drain current is zero in the absence of a gate voltage and that a forward (positive for an n-channel device) gate bias is necessary to give working values of drain current: this type of operation is known as the *enhancement mode*.

For an enhancement mosfet to obtain a working value of drain current, the gate voltage must lie between the source and drain potentials. This compares with bipolar transistors and contrasts with depletion devices such as jfets.

It is, however, possible in the manufacture of a mosfet to produce a thin n-type layer on the surface of the p-type substrate. This provides a conducting channel between source and drain and ensures that a useful drain current flows even at zero gate-base voltage. Negative gate bias reduces channel conductivity and drain current and in the limit will change the conductivity of the channel to p-type, reducing drain current to zero: this is the depletion mode of operation again. Positive voltages on the gate increase channel conductivity and drain current as in enhancement-mode operation. Mosfets of this type can thus operate in depletion and enhancement modes.

Complementary mosfets with an n-type substrate and p-type channel are also available, giving a total of four basic types of mosfet: the graphical symbols are shown in Fig. 2.18. The non-conductivity of the enhancement type for zero gate bias is indicated in the symbol by breaks in the rectangle representing the channel.

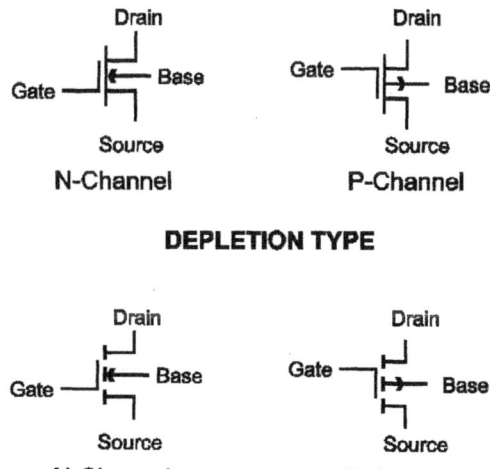

Fig. 2.18. Graphical symbols for mosfets

Input resistances greater than 10^{12} Ω have been achieved in mosfets and input capacitances may be as low as 1 pF.

The connection to the substrate provides a second input terminal (the base) to a mosfet and the potential applied to it with respect to the source controls the drain current in the same way as a potential on the gate of a jfet. The base terminal is not so sensitive a control electrode as the gate terminal but is used in certain types of circuit. In many mosfet applications the base terminal is connected to the source terminal internally or externally.

The silicon dioxide dielectric of the mosfet is very thin and can easily be broken down by transient voltages on the gate terminal. A practical precaution normally taken in handling such transistors is to keep the leads short-circuited whilst the transistor is being connected into circuit.

Short-channel fets

Mosfets can be manufactured by the planar process and they can be made so small that thousands can be accommodated on a single silicon chip. They are therefore widely used for digital signal storage in computers and similar equipment: typical circuits are described later. In mosfets used for switching, the length of the conducting channel is important because it controls the switching speed. The length of the channel determines the drain-source resistance when the transistor is conducting and in the type of construction shown in Fig. 2.17 the channel is of considerable length, occupying the whole of the distance between the two n-regions. Much shorter channels are needed in switching transistors and the manufacturers have adopted a number of different geometries to achieve them. In the double-diffused metal-oxide-semiconductor transistor (DMOST) successive diffusions by opposite types of impurity through the same opening in the silicon dioxide layer are used to produce a very short channel the length of which can be closely controlled during manufacture (Fig. 2.19). The channel occupies only a small fraction of the distance

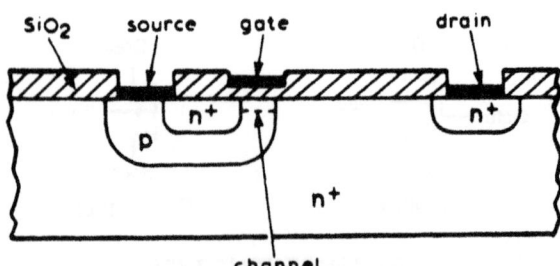

Fig. 2.19. Simplified construction of DMOS transistor

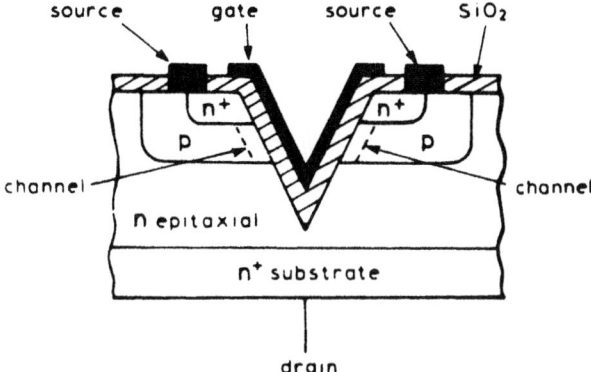

Fig. 2.20. Simplified construction of VMOS transistor

between the n-regions, the remainder being a drift space. In the T-metal-oxide-semiconductor transistor (TMOST) the current flow is normal to the plane of the layers as in bipolar transistors, the source being on top and the drain underneath. The source consists of a number of sites, each associated with a short channel, the parallel connection of which yields the desired low resistance and switching times. A somewhat similar approach is used in the vertical-groove metal-oxide-semiconductor transistor (VMOS) designed to deliver substantial output power, e.g. in audio amplifiers. In this type of transistor a groove-shaped gate projects into the n+, p and epitaxial layers giving very short channel lengths as shown in Fig. 2.20.

The fet as a linear resistor

The internal drain-source path of an fet differs in nature from the corresponding path of a bipolar transistor. It has no rectifying properties and conducts irrespective of the polarity of the applied drain-source voltage. Thus the I_d–V_{ds} characteristics occupy two quadrants of the current–voltage graph as shown in Fig. 2.21. The characteristics for low voltages are almost straight and all pass through the origin, showing the fet. to be a good approximation to a linear conductor. Its resistance is, of course, measured by the slope of the characteristics and varies greatly depending on the gate voltage. In practice a range from a few hundred ohms to as many megohms can be achieved. To obtain this near-linear behaviour, however, applied voltages must be confined to a small range centred on zero. Suitable inputs are alternating signals with an amplitude small compared with 1 V. The circuit diagram of an attenuator using this principle is given on page 366.

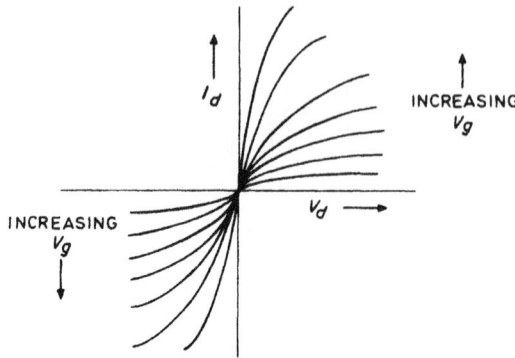

Fig. 2.21. $I_d V_d$ characteristics of a fet near the origin

Dual-gate mosfets

Dual-gate field effect transistors are of the same construction as those already described but have a second separate insulated gate. Initially intended for linear applications at v.h.f. as described on page 188, they have low feedback capacitance (c_{rs}) and high transconductance (y_{fs}) and so find their main uses in high-gain, high-frequency amplifiers. The transconductance can be controlled by varying the bias voltage on the gate which is not used for signal input, permitting a simple system of automatic gain control (a.g.c.).

The most common application of the dual-gate mosfet is as a mixer in r.f. circuits. An application of this will be shown in Chapter 12.

Insulated-gate bipolar transistor (igbts)

A relatively recently developed semiconductor type is the insulated-gate bipolar transistor. It combines the input characteristics of a mosfet with the output performance of a bipolar device and offers greater power gain than the bipolar type together with the higher voltage operation and lower losses of the mosfet. In effect it is an fet integrated with a bipolar transistor in a form of Darlington configuration: see page 131. The igbt is used for its very fast, high-power switching capability; 50 A in 50 ns is readily achievable.

Gallium arsenide fets

Electron mobility in gallium arsenide is about five times that in silicon, an important factor in s.h.f. (super high frequency) applications, mentioned in Chapter 1. The physical construction of a gasfet is relatively simple: for low-noise performance the epitaxial layer under the gate is etched into a recess, and the gate length made very short; this calls for use of an ion-implantation technique in manufacture (see Appendix A).

At s.h.f. the performance of a gasfet is superior to that of a silicon transistor by virtue of the lower parasitic capacitance due to the isolation effect of a layer of semi-insulating GaAs, the decrease of series resistance and the increase in transconductance g_m caused by the greater electron mobility inherent in GaAs devices.

HEMT

A development of the GaAs fet is the hemt (high electron-mobility transistor), now used almost exclusively in the front ends (low-noise blocks, l.n.b.s) of satellite receivers working in the 12-GHz band for direct-to-home broadcasts from space.

The hemt has a hetero-structure (junction of different semi-conductors) of aluminium-gallium-arsenide (AlGaAs) and GaAs, doubling the electron-mobility of a Schottky-barrier junction. Manufacture of a hemt uses a similar process to that of the GaAs fet, but because of its superior transconductance the hemt has a very good low-noise performance: the best yields in production have a noise factor of 0.6 dB, making it possible to construct a head-end (dish-mounted) frequency converter with an overall noise factor of less than 0.8 dB. This is important because the better the noise performance of the head-end, the smaller need be the receiving dish for a given field strength of the broadcast satellite.

Equivalent circuit for a fet

Fig. 2.22(a) gives the equivalent circuit of a fet working at very high frequency. At radio frequencies the representative diagram of Fig. 2.22(b) is adequate while at low frequencies and for d.c. applications the effects of internal capacitances can usually be ignored and the input resistance assumed to be infinite as shown in the further simplified diagram of Fig. 2.22(c).

Thyristors

These are four-layer devices of pnpn structure, the outer layers being known as anode and cathode, as indicated in Fig. 2.23. They are bistable, having very low resistance between anode and cathode in one state and very high resistance in the other state. Changes of state are very rapid, being accelerated by regenerative action within the device,

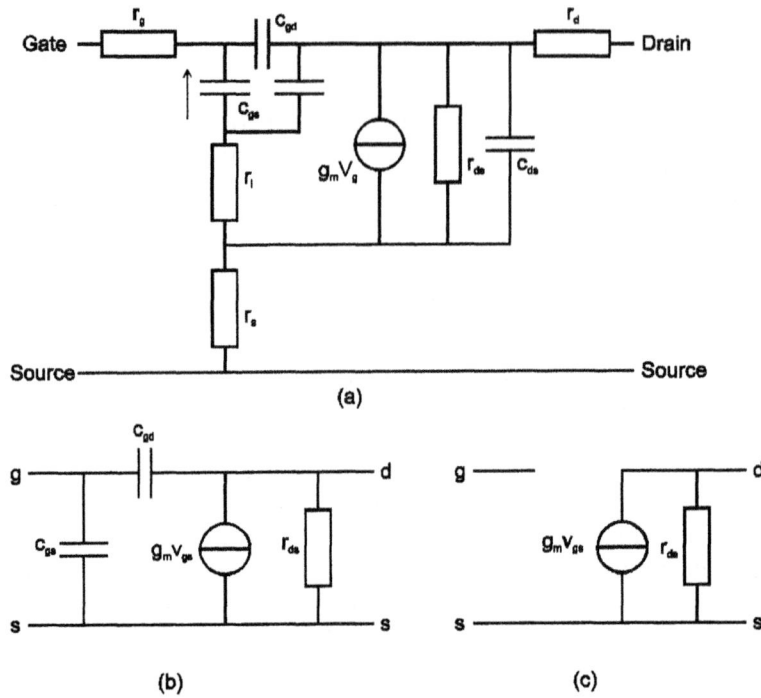

Fig. 2.22. Equivalent circuit for a fet for use at (a) very high frequencies; (b) radio frequencies; (c) low frequencies and d.c.

and very little power is required to initiate a change of state. Their chief applications are in switching and power control. They are sometimes known as semiconductor-controlled rectifiers or, more accurately, controlled semiconductor rectifiers. There are many different types of thyristor and they can be classified according to the number of external connections to the layers.

Diode (two-terminal) thyristor

The simplest type of thyristor has the two external connections indicated in Fig. 2.23. When the anode is made positive with respect to the cathode the two outer junctions are forward-biased and the centre junction is reverse-biased. Because the outer junctions are therefore of low resistance most of the applied voltage appears across the centre junction. As the applied voltage is increased the electric field across this junction increases and a value is reached at which it breaks down. The few charge carriers released by the intense field generate further carriers as a result of collisions and current therefore increases very rapidly to a value limited only by the applied voltage and the resistance of the

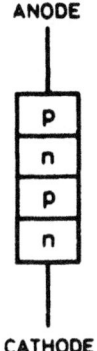

Fig. 2.23. Structure of a thyristor

external circuit: the device is now in its on-state. The applied voltage can now be reduced to a low value but the on-state persists until the current through the device has fallen below a critical value (known as the holding current) when the device becomes non-conductive again. If the anode is maintained at a voltage just below the critical value which causes breakdown, then a small superimposed voltage pulse can initiate the change to the on-state.

If the anode is made negative with respect to the cathode the two outer junctions are reverse-biased and the centre junction is forward-biased. The applied voltage is now shared between the outer junctions and the device has the properties of a reverse-biased pn junction, i.e. is non-conductive. The I–V characteristic of a diode thyristor has the form shown in Fig. 2.24. The device can be regarded as a diode with an exceptionally low forward resistance.

Such a thyristor is known as a *reverse-blocking diode thyristor* and two such devices connected in reverse parallel form a bidirectional diode thyristor sometimes called a *diac*. By suitable treatment of the

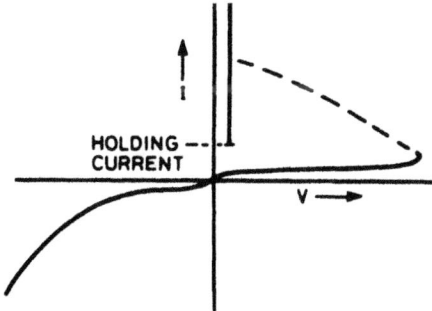

Fig. 2.24. I–V characteristic of a reverse-blocking diode thyristor

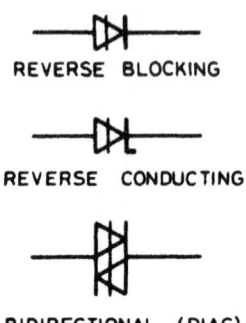

REVERSE BLOCKING

REVERSE CONDUCTING

BIDIRECTIONAL (DIAC)

Fig. 2.25. Graphical symbols for diode thyristors

outer junctions it is possible to arrange for these to break down at a very low value of negative anode voltage. These devices are known as *reverse-conducting diode thyristors* and the graphical symbols for all these types of diode thyristor are shown in Fig. 2.25. Diode thyristors are not so widely used as the triode types now to be described. The commonest form of diac breaks down at approximately 32 V. It is mostly used to trigger a thyristor or triac into conduction when, as part of an R–C delay circuit, it starts to conduct when the capacitor voltage builds up to the 32 V.

Triode (three-terminal) thyristors

These have an external connection, known as a *gate*, to the inner p- or n-region and the behaviour of the device is best understood by regarding it as made up of two direct-coupled complementary bipolar transistors as shown in Fig. 2.26(a) which shows a gate connection to the inner p-region. Redrawn in the form shown in Fig. 2.26(b) the circuit can be

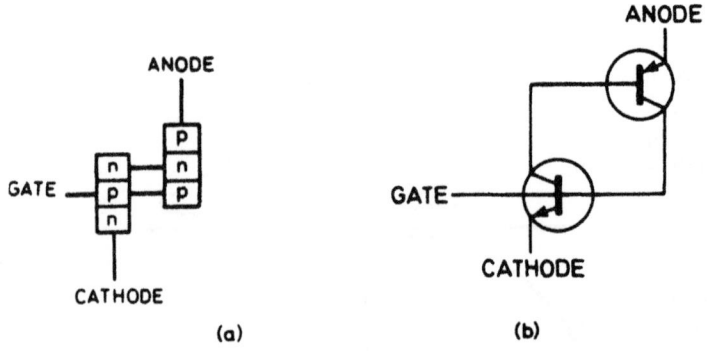

(a)   (b)

Fig. 2.26. Triode thyristor regarded as two direct-coupled complementary bipolar transistors

recognised as that of a complementary bistable multivibrator similar to that shown in Fig. 13.5 but without the four resistors. One of the features of the complementary circuit, as pointed out on page 260, is that in one of the stable states both transistors are cut off and in the other both are conductive.

When the anode is made negative with respect to the cathode the outer two junctions are reverse-biased: these are the base-emitter junctions of the constituent bipolar transistors and thus both transistors are cut off. Very little current can flow through the device, which is therefore in its off-state.

When the anode is made positive with respect to the cathode the outer two junctions of the device are forward biased whilst the centre junction (which is the base-collector junction for both transistors) is reverse-biased. These are the normal conditions for the two transistors to operate as amplifiers. If a positive voltage is applied to the inner p-region, so injecting a base current i_b, this increases the forward bias of the npn transistor with the result that current crosses the centre junction to give rise to a collector current of $\beta_1 i_b$ in the inner n-region, where β_1 is the current gain of the npn transistor. The inner n-region is, however, the base region of the pnp transistor and this increase in its forward bias gives rise to a collector current of $\beta_1 \beta_2 i_b$ in the inner p-region, where β_2 is the current gain of the pnp transistor. The inner p-region is, however, the base region of the npn transistor. In this way positive feedback occurs and if $\beta_1 \beta_2$ exceeds unity the feedback is regenerative, causing a very rapid build-up of current, so putting the device into the on-state. The current through the thyristor when regeneration starts is known as the latching current and it is greater than the holding current. The current gain of a bipolar transistor increases with increase in forward bias and thus regeneration can always be produced by increasing the external voltage applied to the inner p-region. Once regeneration has started, the external voltage which initiated it can be removed but the thyristor will remain in the on-state until the current through it has fallen below the holding value. Thus the thyristor can be triggered into conduction by a short-duration pulse applied to the gate. A thyristor can also usually be turned on by applying the anode–cathode voltage rapidly. This is a known and often unwelcome characteristic although, since many applications involve a.c., the build-up of voltage does not exceed the dV/dt limit. In d.c. applications, the build-up of the d.c. voltage to the anode is usually designed to be slow enough to avoid inadvertant switch on.

The advantage of the gate control is that it enables the anode breakdown voltage of the thyristor to be reduced considerably – down to a few volts in fact. This is illustrated in the characteristics of a triode thyristor shown in Fig. 2.27. Typically a pulse of 100 mA at 5 V on the

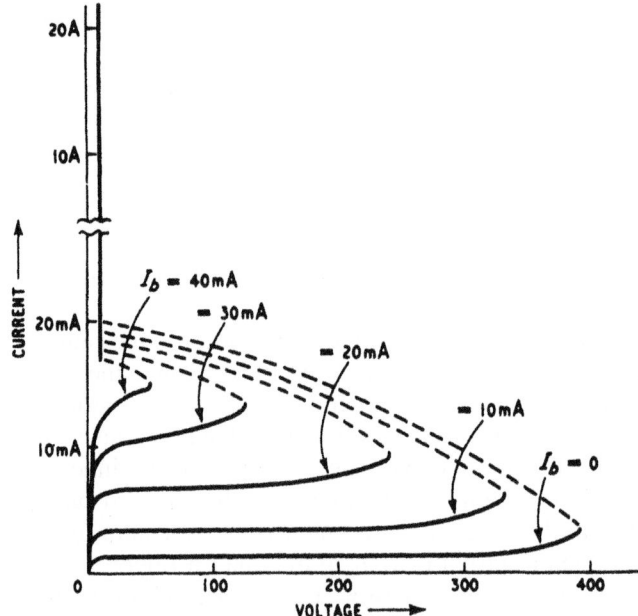

Fig. 2.27. Anode characteristics of a triode thyristor

gate can control a current of 50 A at 250 V and the voltage drop across the thyristor can be less than 2 V. Modern thyristors can carry currents of up to 1,000 A at voltages up to 2.5 kV, and thus find applications, for instance, in electric traction and lift control.

From the symmetry of the device it is clear that the thyristor could alternatively be triggered into conduction by a negative signal applied to the inner n-region. Thus there are two basic types of reverse-blocking triode thyristor, one with a p-gate (cathode gate) and the other with an

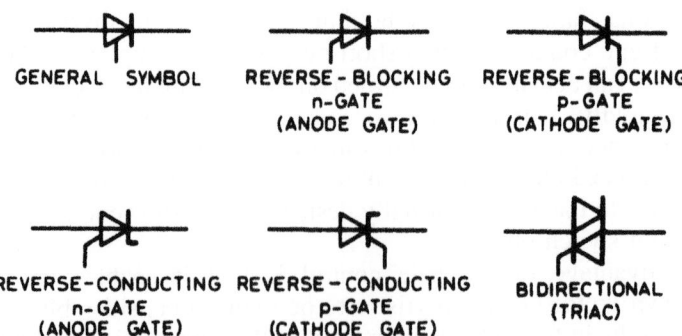

Fig. 2.28. Graphical symbols for triode thyristors

n-gate (anode gate). There are corresponding reverse-conducting triode thyristors. A useful bidirectional thyristor, known as a triac, can be made by connecting two reverse-blocking triode thyristors in reverse parallel with a common gate connection: this can be triggered into conduction by positive or negative voltages applied to the gate terminal. The graphical symbols for the various types of triode thyristor are given in Fig. 2.28.

Tetrode (four-terminal) thyristor

If a thyristor has separate gate connections to both the inner regions, as shown in Fig. 2.29, then it can be triggered into conduction by a positive signal applied to the p-gate or a negative signal applied to the n-gate. It also has the property that it can be switched from the on-state to the off-state by applying a positive signal to the n-gate or a negative signal to the p-gate. Such a thyristor is usually known as a *silicon-controlled switch*.

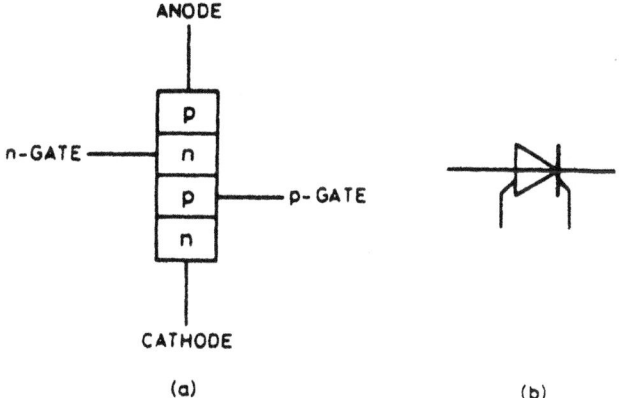

(a)   (b)

Fig. 2.29. Structure (a) and graphical symbol (b) for a tetrode thyristor (silicon-controlled switch)

Gate turn-off thyristor (GTO)

The conventional thyristor is turned off by reducing the anode current to zero. The structure of the gate turn-off thyristor has been altered as shown in Fig. 2.30 so that the device can be turned off by removing current from the gate. The p_n layer at the anode has highly doped n+ areas 'interdigitated' with the usual p_n region.

To turn on, current is pulsed into the gate in the normal way, but it needs to be maintained for longer than in a conventional thyristor to

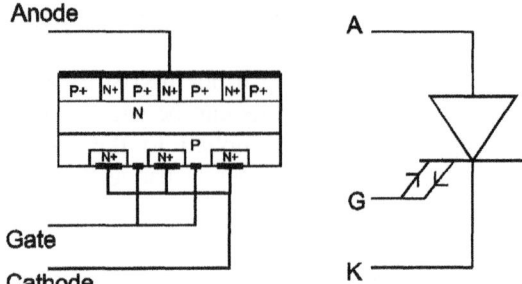

Fig. 2.30. Structure and circuit symbol for a gate turn-off thyristor

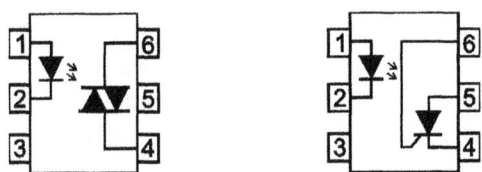

Fig. 2.31. An optically-coupled triac and thyristor

ensure that latching takes place. To turn the thyristor off, a negative (≈ -10 V) voltage must be applied between the gate and cathode. The gate turn-off pulse circuit must be capable of absorbing a short (1 µs) pulse of current near to the value of the anode current to ensure that turn-off occurs.

Optically-coupled triac and thyristors

These devices feature a low-power triac or thyristor which is triggered by an LED included in the same package. The main advantage is that the triggering circuit is electrically isolated from the power device. This means that low-voltage circuitry can be used to operate the triac or thyristor, which can be connected to high-voltage outputs. Quite often, this output can itself be the trigger for higher-current devices. The devices shown in Fig. 2.31 have a typical reverse blocking voltage of around 600 V with a forward current capability of around 100 mA.

Chapter 3

Common-base and common-gate amplifiers

Common-base amplifiers

Introduction

It was shown in the previous chapter that the common-base amplifier has a current gain of just less than unity, that its input resistance is low, output resistance high and that it can give substantial voltage gain. In this chapter we shall show how these resistances and the voltage gain depend on the values of r_e, r_b and r_c and on the value of the source and load resistance.

A simplified circuit of a transistor common-base amplifier is given in Fig. 3.1(a) in which R_s represents the internal resistance of the signal source and R_l represents the collector load for the amplifier. For the sake of simplicity, methods of applying bias are not indicated.

Fig. 3.1(b) represents the same circuit in which the transistor is represented by its equivalent T-section network. The directions of the currents in the two meshes of this network can be determined by considering the physical processes occurring in a transistor. From Chapter 2 we know that the emitter current i_e is equal to the sum of the

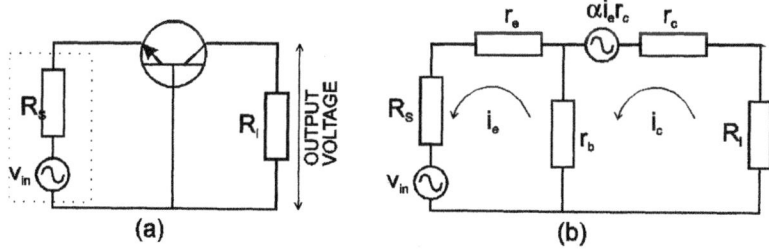

Fig. 3.1. Simplified circuit of a common-base amplifier (a), and its electrical equivalent (b)

collector current i_c and the base current i_b. If, therefore, i_e and i_c are given anti-clockwise directions as shown in Fig. 3.1(b), then the current in r_b is $(i_e - i_c)$ which is equal to i_b: all conventions are therefore satisfied in this diagram.

Input resistance

Applying Kirchhoff's laws to the circuit of Fig. 3.1(b) we have

$$v_{in} = i_e(R_s + r_e + r_b) - i_c r_b \tag{3.1}$$

$$0 = i_c(r_b + r_c + R_l) - i_e r_b - \alpha i_e r_c \tag{3.2}$$

To obtain an expression for the input resistance we can eliminate i_c between these two equations to obtain a relationship between i_e and v_{in}.

From Eqn 3.2

$$i_c = \frac{r_b + \alpha r_c}{r_b + r_c + R_l} i_e$$

Substituting for i_c in Eqn 3.1

$$v_{in} = i_e(R_s + r_e + r_b) - \frac{r_b(r_b + \alpha r_c)}{r_b + r_c + R_l} i_e$$

$$\therefore i_e = \frac{v_{in}}{R_s + r_e + r_b - \dfrac{r_b(r_b + \alpha r_c)}{r_b + r_c + R_l}} \tag{3.3}$$

If we represent the input resistance of the transistor by r_i the circuit of Fig. 3.1(b) takes the form shown in Fig. 3.2. For this circuit we have

$$i_e = \frac{v_{in}}{R_s + r_i} \tag{3.4}$$

Comparing Eqns 3.3 and 3.4

$$r_i = r_e + r_b - \frac{r_b(r_b + \alpha r_c)}{r_b + r_c + R_i} \tag{3.5}$$

$$= r_e + r_b \cdot \frac{R_l + r_c(1 - \alpha)}{r_b + r_c + R_l} \tag{3.6}$$

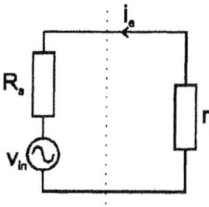

Fig. 3.2. Circuit illustrating the meaning of input resistance

For a given transistor operating under constant d.c. conditions, r_b, r_e, r_c and α are constant, and r_i therefore depends solely on R_l, increasing as R_l increases.

The range of input resistance can be calculated from Eqn 3.5. First consider Eqn 3.5 when R_l is made vanishingly small

$$r_i = r_e + r_b - \frac{r_b(r_b + \alpha r_c)}{r_b + r_c}$$

Now for bipolar transistors r_c and αr_c are both large compared with r_b and the input resistance is approximately given by

$$r_i = r_e + r_b - r_b \cdot \frac{\alpha r_c}{r_c}$$

$$= r_e + r_b (1 - \alpha)$$

From Eqn 2.5

$$(1 - \alpha) = \frac{1}{\beta}$$

$$\therefore r_i = r_e + \frac{r_b}{\beta} \tag{3.7}$$

and this gives the value of the input resistance of the common-base amplifier when the output terminals are short-circuited.

Frequently r_b/β can be neglected in comparison with r_e and thus we can say

$$r_i \approx r_e$$

As we have seen r_e is inversely proportional to emitter current. It follows that the input resistance for the common-base amplifier (for low values of collector load resistance) is inversely proportional to emitter current and thus to collector current. This has important consequences. For example suppose a series of steady potentials is applied between base and emitter as in plotting the $I_c - V_{be}$ curves for the transistor. Each time the voltage is increased, the emitter current increases and r_e decreases with the result that the final increase in I_e is more than proportional to the voltage increase. Similarly if V_{be} is decreased I_e decreases and r_e increases, the effect being that the final decrease in I_e is less than proportional to the change in voltage. Thus the $I_c - V_{be}$ curve is far from linear: indeed it approximates to exponential form as shown in Fig. 2.9.

Changes in input resistance also occur when emitter current varies during signal amplification. If the input signal is from a high-resistance source then variations in input resistance have little effect on input current and amplification is linear. If however the transistor is used as a voltage amplifier and the input signal is from a low-resistance source, variations in input resistance can cause severe distortion; this is to be expected from a device with a characteristic as non-linear as that shown in Fig. 2.9. To minimise this distortion as required in an analogue amplifier the effects of the input-resistance variation must be eliminated and one way of achieving this is to ensure that the internal resistance of the signal source is large compared with the average value of the input resistance of the transistor. If this is impossible and the source resistance is small, it can effectively be increased by connecting an external resistance in series with the signal source. By making R_s large we reduce gain, of course, but we ensure that the transistor is fed with a current that is a substantially undistorted copy of the voltage to be amplified: thus we still use the transistor as a current amplifier even though the input and output signals are in the form of voltages.

Now consider the input resistance when R_l is made very large. The final term in Eqn 3.5 vanishes leaving

$$r_i = r_e + r_b \tag{3.8}$$

which gives the value of the input resistance when the output terminals are open-circuited.

Input resistance for a bipolar transistor as common-base amplifier

To obtain an estimate of practical values of input resistance for a bipolar transistor we can substitute $r_e = 25 \ \Omega$, $r_b = 300 \ \Omega$ and $\beta = 50$ and from Eqn 3.7 we find that the input resistance for short-circuited output terminals is given by

$$r_i = 25 + \frac{300}{50}$$

$$= 31 \, \Omega$$

From Eqn 3.8 the input resistance for open-circuited output terminals is given by

$$r_i = 25 + 300$$

$$= 325 \, \Omega$$

Thus the input resistance of a bipolar transistor used as a common-base amplifier varies between a minimum value for short-circuited output terminals to a maximum value for open-circuited terminals. The dependence of input resistance on output load is illustrated in Fig. 3.3.

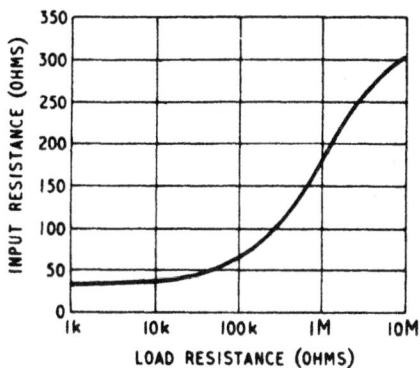

Fig. 3.3. Variation of input resistance with load resistance for a common-base amplifier

Output resistance

To obtain an expression for the output resistance of the common-base amplifier we can eliminate i_e between Eqns 3.1 and 3.2 to obtain a relationship between i_c and v_{in}.
From Eqn 3.2

$$i_e = \frac{r_b + r_c + R_l}{r_b + \alpha r_c} \cdot i_c$$

Substituting for i_e in Eqn 3.1

$$v_{in} = \frac{(r_b + r_c + R_l)(R_s + r_e + r_b)}{r_b + \alpha r_c} i_c - i_c r_b$$

$$\therefore i_c = \frac{v_{in}}{\dfrac{(r_b + r_c + R_l)R_s + r_e + r_b}{r_b + \alpha r_c} - r_b}$$

$$= \frac{(r_b + \alpha r_c)v_{in}}{(r_b + r_c + R_l)(R_s + r_e + r_b) - r_b(r_b + \alpha r_c)}$$

$$\therefore i_c = \frac{\dfrac{r_b + \alpha r_c}{R_s + r_e + r_b}v_{in}}{R_l + r_b + r_c - \dfrac{r_b(r_b + \alpha r_c)}{R_s + r_e + r_b}} \tag{3.9}$$

If we represent the output resistance of the transistor amplifier by r_o the circuit of Fig. 3.1(b) has the form shown in Fig. 3.4. For this circuit we have

$$i_c = \frac{v}{R_l + r_o} \tag{3.10}$$

Comparing Eqns 3.9 and 3.10 we have that v, the e.m.f. effectively acting in the output circuit, is given by

$$v = \frac{r_b + \alpha r_c}{R_s + r_e + r_b}v_{in} .$$

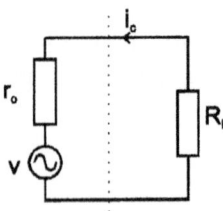

Fig. 3.4. Circuit illustrating the meaning of output resistance

Making the normally-justified assumptions that r_b is negligible compared with αr_c that α can be taken as unity and that $(r_e + r_b)$ can be neglected in comparison with R_s we have

$$v \approx \frac{r_c}{R_s} \cdot v_{in}$$

Eqn 3.10 thus becomes

$$i_c = \frac{\dfrac{r_c}{R_s} \cdot v_{in}}{R_l + r_o}$$

Comparing Eqns 3.9 and 3.10, we have

$$r_o = r_b + r_c - \frac{r_b(r_b + \alpha r_c)}{R_s + r_e + r_b} \tag{3.11}$$

$$= r_c + r_b \cdot \frac{R_s + r_e - r_c \alpha}{R_s + r_e + r_b} \tag{3.12}$$

For a given transistor operating with fixed d.c. conditions, r_b, r_e, r_c and α are fixed and r_o then depends on R_s, increasing as R_s is increased. The range over which r_o varies can be estimated as follows. First let $R_s = 0$. Eqn 3.12 then becomes

$$r_o = r_c + r_b \cdot \frac{r_e - r_c \alpha}{r_e + r_b}$$

which, if the relatively small term $r_b r_e/(r_e + r_b)$ is neglected, gives

$$r_o = r_c \cdot \frac{r_e + r_b(1 - \alpha)}{r_e + r_b} \tag{3.13}$$

Now let R_s approach infinity. The final term in Eqn 3.11 then vanishes, leaving the output resistance as

$$r_o = r_c + r_b \tag{3.14}$$

$$\approx r_c$$

because r_b is small compared with r_c.

Output resistance of a bipolar transistor as common-base amplifier

To obtain a numerical estimate of the range of output resistance for a bipolar transistor let $r_e = 25\,\Omega$, $r_b = 300\,\Omega$, $r_c = 1\,\text{M}\Omega$ and $\alpha = 0.98$. Substituting these values in Eqn 3.13 we find the output resistance for short-circuited input terminals is given by

$$r_o = 1{,}000{,}000 \times \left[\frac{25 + 300(1 - 0.98)}{25 + 300} \right] \Omega$$

$$= 100\,\text{k}\Omega \text{ approximately}$$

The output resistance for open-circuited input terminals is, from Eqn 3.14, given by

$$r_o = 1{,}000{,}000 + 300\,\Omega$$

$$= 1\,\text{M}\Omega \text{ approximately}$$

The variation of output resistance with source resistance for a common-base amplifier is illustrated in Fig. 3.5.

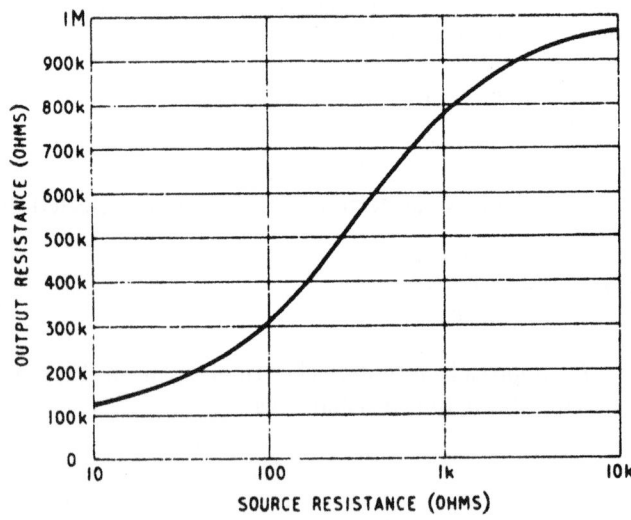

Fig. 3.5. Variation of output resistance with source resistance for a common-base amplifier

Voltage gain

From Fig. 3.1(b) we can see that the output voltage is given by $i_c R_l$. The voltage gain v_{out}/v_{in} is hence equal to $i_c R_l/v_{in}$. From Eqn 3.9 i_c is given by

$$i_c = \frac{(r_b + \alpha r_c)v_{in}}{(R_l + r_b + r_c)(R_s + r_e + r_b) - r_b(r_b + \alpha r_c)}$$

Hence

$$\frac{v_{out}}{v_{in}} = \frac{i_c R_l}{v_{in}} = \frac{(r_b + \alpha r_c)R_l}{(R_l + r_b + r_c)(R_s + r_e + r_b) - r_b(r_b + \alpha r_c)} \quad (3.15)$$

The dependence of the voltage gain on the value of R_l is illustrated in Fig. 3.6, which shows gain increasing with R_l, linearly for small values of R_l but becoming asymptotic to a limiting value of gain as R_l becomes large.

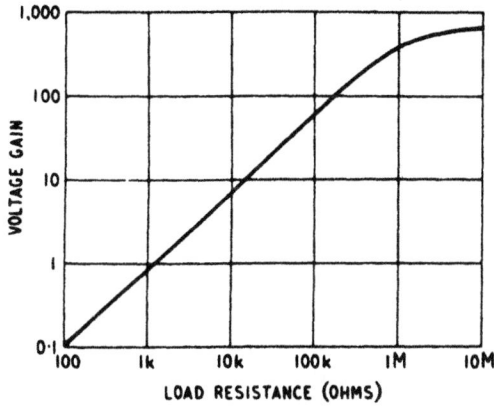

Fig. 3.6. Variation of voltage gain with load resistance for a common-base amplifier

For a bipolar transistor r_c is normally large compared with all other resistances in the expression, and it is possible to simplify Eqn 3.15 as follows:

$$\frac{v_{out}}{v_{in}} = \frac{\alpha r_c R_l}{r_c(R_s + r_e + r_b) - \alpha r_b r_c}$$

$$= \frac{\alpha R_l}{R_s + r_e + r_b(1 - \alpha)}$$

If α in the numerator is taken as unity, this may be written

$$\frac{v_{out}}{v_{in}} = \frac{R_l}{R_s + r_e + r_b/\beta}$$

From Eqn 3.7

$$r_i = r_e + \frac{r_b}{\beta}.$$

$$\therefore \frac{v_{out}}{v_{in}} = \frac{R_l}{R_s + r_i} \tag{3.16}$$

Now v_{in} is the voltage of the input-signal generator and in a practical circuit the terminals of v_{in} are not normally accessible. It is more useful therefore to express the voltage gain of the amplifier as the ratio of the signal voltage at the collector to that at the emitter because these two voltages can readily be measured. Reference to Fig. 3.1 shows that if we make R_s equal to zero, v_{in} equals v_{be}, the signal voltage at the emitter terminal. Thus to obtain this more practical expression for the voltage gain all that is necessary is to put $R_s = 0$ in the above expression and we have

$$\frac{v_{out}}{v_{be}} = \frac{R_l}{r_i}$$

This gives the voltage gain between collector and emitter terminals and R_s does not enter into it. As indicated earlier, however, R_s should be large compared with r_i to minimise distortion in a class-A amplifier.

In a typical common-base amplifier $R_l = 10\,\text{k}\Omega$ and $r_i = 40\,\Omega$ giving v_{out}/v_{be} as 250. If the source resistance R_s is taken as $400\,\Omega$ (ten times r_{in}) then v_{out}/v_{in} is 25.

To obtain an indication of the maximum voltage gain available from a bipolar transistor as a common-base amplifier suppose R_l is made large compared with r_c. With a few simplifications Eqn 3.15 reduces to the form

$$\frac{v_{out}}{v_{in}} = \frac{r_c}{R_s + r_e + r_b} \tag{3.17}$$

from which

$$\frac{v_{out}}{v_{be}} = \frac{r_c}{r_e + r_b}.$$

Substituting $r_c = 1\,\mathrm{M\Omega}$ $r_e = 25\,\Omega$ and $r_b = 300\,\Omega$ we have that the maximum value of v_{out}/v_{be} is given by $10^6/(25 + 300)$, approximately 3,000, a much larger value than is obtainable from an fet.

Applications of common-base amplifiers

The common-base amplifier has a very low input resistance, a very high output resistance, high voltage gain and unity current gain. It is widely used as an r.f. stage in v.h.f. and u.h.f. receivers, often as part of a cascode stage. Typical circuits are given in later chapters.

Common-gate amplifiers

Amplifier properties

Fig. 3.7(a) shows the basic form of the common-gate amplifier and Fig. 3.7(b) gives the (low-frequency) equivalent circuit for an fet connected to a signal source and to a load resistor. Because of the very high resistance of the gate circuit, the gate connection is shown as an open circuit in Fig. 3.7(b). The output of the constant-current generator $g_m v_{gs}$ divides between r_{ds} and $(R_s + R_l)$ but normally $(R_s + R_l)$ is small compared with r_{ds} and it can thus be assumed that the current $g_m v_{gs}$ flows wholly through R_s and R_l.

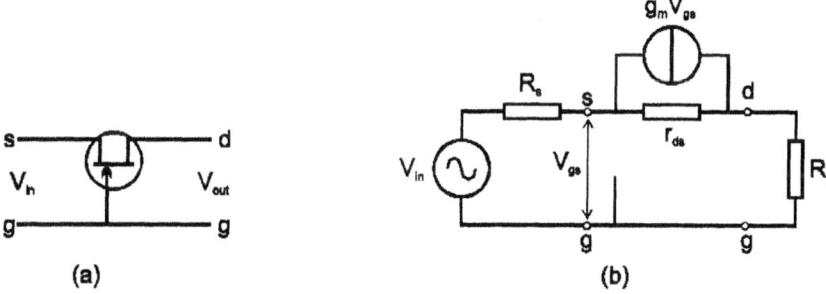

Fig. 3.7. Basic connections for a common-gate amplifier (a) and the electrical equivalent with source and load circuit (b)

The input resistance of the common-gate amplifier r_i is the ratio of the voltage across the source and gate terminals (v_{gs}) to the current flowing through these terminals ($g_m v_{gs}$) and is thus equal to $1/g_m$ – a very low resistance. If $g_m = 2\,\mathrm{mA/V}$, a typical value, r_i is $1/(2 \times 10^{-3})$, i.e. $500\,\Omega$.

The output resistance r_o can be seen from inspection of Fig. 3.7(b) to be $(r_{ds} + R_s)$ but as r_{ds} is normally large compared with R_s, r_o can usually be taken as approximately equal to r_{ds}.

The output voltage v_o of the common-gate amplifier can be seen from inspection of Fig. 3.7(b) to be iR_l where i is the current $(g_m v_{gs})$ in the circuit. Thus the voltage gain is given by

$$\frac{v_o}{v_{gs}} = g_m R_l$$

The current in the circuit $(g_m v_{gs})$ flows in the input and output circuits and the current gain of the common-gate amplifier is therefore unity.

Applications

The common-gate amplifier has output resistance and voltage gain similar to those of the common-source amplifier but its input resistance is low. This particular type of amplifier therefore lacks the most attractive feature of the fet, namely its very high input resistance, and is consequently not widely used. Its chief application is in wideband v.h.f. and u.h.f. amplifiers (often as part of a cascode circuit) where the low input resistance is no disadvantage.

Chapter 4

Common-emitter and common-source amplifiers

Common-emitter amplifiers

Introduction

We have already seen (page 29) that the common-emitter amplifier has a considerable current gain β. In the following analysis we shall derive expressions for the input resistance, output resistance and voltage gain for this circuit in terms of the parameters of the equivalent T-network, the source resistance R_s and the output load R_l

The basic circuit for a common-emitter amplifier is given in Fig. 4.1(a), in which R_s is the resistance of the signal source and R_l is the collector load resistance. Bias sources are omitted from this diagram for the sake of simplicity. Fig. 4.1(b) gives the equivalent circuit in which the transistor is represented by the T-network of resistances r_e, r_b and r_c introduced in Chapter 2. If the base current i_b is shown acting in an anti-clockwise direction and the collector current i_c in a clockwise direction

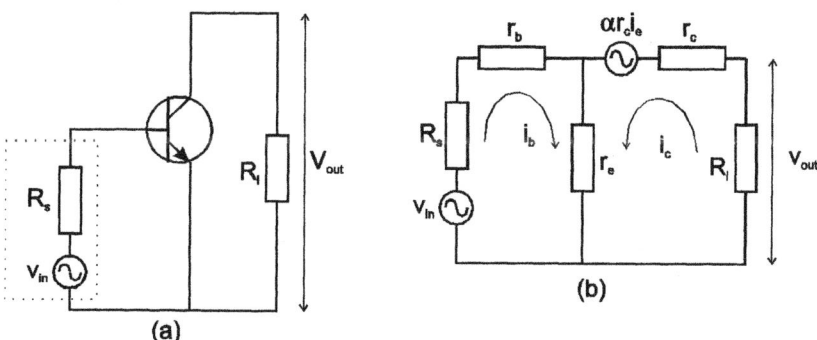

Fig. 4.1. Basic form of common-emitter amplifier (a), and its equivalent circuit (b)

as indicated in Fig. 4.1(b), then the current in the common-emitter resistance r_e is the sum $(i_b + i_c)$. From Chapter 2 we know that the sum of these two currents is the emitter current i_e: all conventions are therefore satisfied in this diagram.

Applying Kirchhoff's laws to the circuit of Fig. 4.1(b) we have

$$v_{in} = i_b (R_s + r_b + r_e) + i_c r_e \tag{4.1}$$

$$0 = i_c (R_l + r_c + r_e) + i_b r_e - \alpha r_c i_e \tag{4.2}$$

Now $i_e = i_b + i_c$ and Eqn 4.2 may therefore be written in the form

$$0 = i_c (R_l + r_c + r_e) + i_b r_e - \alpha r_c i_b - \alpha r_c i_c$$

$$= i_c [R_l + r_e + r_c (1 - \alpha)] + i_b (r_e - \alpha r_c) \tag{4.3}$$

Current gain

From Eqn 4.3 we can say

$$\beta = \frac{i_c}{i_b} = \frac{r_e - \alpha r_c}{R_l + r_e + r_c (1 - \alpha)}$$

Now r_e may be neglected in comparison with αr_c: normal values of R_l and r_e can be neglected in comparison with $r_c (1 - \alpha)$. We thus have

$$\beta = \frac{i_c}{i_b} = \frac{\alpha}{1 - \alpha}$$

$$\approx \frac{1}{1 - \alpha}$$

Values of β lie between 20 and 500 and tend to fall off at low and high values of emitter current.

Input resistance

As in the analysis of the common-base amplifier, we can obtain an expression for the input resistance of the amplifier by eliminating i_c between Eqns 4.1 and 4.3 to obtain a relationship between i_b and v_{in}. From Eqn 4.3

$$i_c = - \frac{r_e - \alpha r_c}{R_l + r_e + (1 - \alpha) r_c} \cdot i_b$$

Substituting for i_c in Eqn 4.1

$$v_{in} = i_b(R_s + r_b + r_e) - \frac{r_e(r_e - \alpha r_c)}{R_l + r_e + (1 - \alpha)r_c} \cdot i_b$$

$$\therefore i_b = \frac{v_{in}}{R_s + r_b + r_e - \dfrac{r_e(r_c - \alpha r_c)}{R_l + r_e + (1 - \alpha)r_c}}$$

Re-arranging

$$i_b = \frac{v_{in}}{R_s + r_b + r_e + \dfrac{r_e(\alpha r_c - r_e)}{R_l + r_e + (1 - \alpha)r_c}} \tag{4.4}$$

If the input resistance of the amplifier is represented by r_i the input current i_b can be expressed

$$i_b = \frac{v_{in}}{R_s + r_i} \tag{4.5}$$

Comparing this with Eqn 4.4 we have

$$r_i = r_b + r_e + \frac{r_e(\alpha r_c - r_e)}{R_l + r_e + (1 - \alpha)r_c} \tag{4.6}$$

which is a useful expression because it shows that r_i increases as R_l decreases. Eqn 4.6 can be simplified thus

$$r_i = r_b + \frac{r_e[R_l + r_e + (1 - \alpha)r_c] + r_e(\alpha r_c - r_e)}{R_l + r_e + (1 - \alpha)r_c}$$

$$= r_b + r_e \cdot \frac{R_l + r_c}{R_l + r_e + (1 - \alpha)r_c}$$

The denominator can be simplified by neglecting r_e in comparison with the other terms which are much larger in practice. We then have the result

$$r_i = r_b + r_e \cdot \frac{R_l + r_c}{R_l + (1 - \alpha)r_c} \tag{4.7}$$

Normally R_l is small compared with r_c and $(1 - \alpha)r_c$. The input resistance is then given by

$$r_i = r_b + r_e \cdot \frac{r_c}{r_c(1 - \alpha)} = r_b + \frac{r_e}{1 - \alpha}$$

$$= r_b + \beta r_e \qquad (4.8)$$

We have already shown that r_e is inversely proportional to the mean emitter current I_e and we can thus say

$$r_i = r_b + \beta \cdot \frac{25\ (\text{mV})}{I_e\ (\text{mA})}$$

For high values of β such as commonly encountered in silicon transistors the second term is large compared with the first. Thus

$$r_i \approx \beta \cdot \frac{25\ (\text{mV})}{I_e\ (\text{mA})} \approx \frac{\beta}{g_m}$$

This is a useful result which shows that the input resistance is approximately inversely proportional to emitter current and directly proportional to β. Thus for a given emitter current the input resistance gives a direct measure of β for the transistor.

The above expression also shows that the input resistance of the common-emitter amplifier varies with emitter current (as in the common-base amplifier) and can cause distortion in a voltage amplifier if the source resistance is not large compared with the average value of input resistance.

When R_l is large compared with r_c and $(1 - \alpha)r_c$ the input resistance has a value given by

$$r_i = r_b + r_e \cdot \frac{R_l}{R_l}$$

$$= r_b + r_e \qquad (4.9)$$

Input resistance of a bipolar transistor as common-emitter amplifier

To illustrate the range of values of input resistance likely to be encountered in practice in a common-emitter bipolar transistor amplifier, let us assume that $r_e = 25\ \Omega$, $r_b = 300\ \Omega$ and $\beta = 100$. Substituting in Eqn 4.8 to find the input resistance for short-circuited output terminals we have

$$r_i = 300 + 100 \times 25$$

$$= 2,800 \, \Omega$$

Substituting in Eqn 4.9 to obtain the input resistance for open-circuited output terminals we have

$$r_i = 300 + 25 \, \Omega$$

$$= 325 \, \Omega$$

These numerical examples show that the common-emitter amplifier has a higher input resistance than the common-base amplifier and that it decreases with increase in load resistance.

The variation of input resistance with output collector load for bipolar transistors is illustrated in Fig. 4.2.

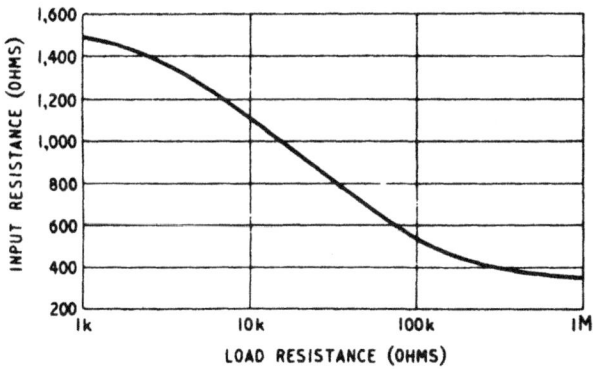

Fig. 4.2. Variation of input resistance with load resistance for a common-emitter amplifier

Output resistance

We can obtain an expression for the output resistance of the common-emitter amplifier by eliminating i_b between Eqns 4.1 and 4.3 to obtain a relationship between v_{in} and i_c. From Eqn 4.3 we have

$$i_b = -\frac{R_l + r_e + r_c \, (1 - \alpha)}{r_e - \alpha r_c} i_c$$

Substituting in Eqn 4.1

$$v_{in} = -\frac{R_l + r_e + r_c\,(1 - \alpha)}{r_e - \alpha r_c}\,(R_s + r_b + r_e)i_c + i_c r_e$$

$$\therefore i_c = \frac{v_{in}}{-\dfrac{[R_l + r_e + r_c\,(1 - \alpha)]\,(R_s + r_b + r_e)}{r_e - \alpha r_c} + r_e}$$

$$= \frac{\dfrac{r_e - \alpha r_c}{R_s + r_b + r_e}\cdot v_{in}}{-[R_l + r_e + r_c\,(1 - \alpha)] + \dfrac{r_e\,(r_e - \alpha r_c)}{R_s + r_b + r_e}}$$

This can be more conveniently written in the form

$$i_c = \frac{\dfrac{\alpha r_c - r_e}{R_s + r_b + r_e}\cdot v_{in}}{R_l + r_e + r_c\,(1 - \alpha) + \dfrac{r_e\,(\alpha r_c - r_e)}{R_s + r_b + r_e}} \tag{4.10}$$

In a simple circuit containing a signal source of voltage v and of internal resistance r_o feeding a load resistance R_l the current i_c is given by

$$i_c = \frac{v}{R_l + r_o} \tag{4.11}$$

Comparison between Eqns 4.10 and 4.11 shows that the effective voltage acting in the equivalent circuit is given by

$$v = \frac{\alpha r_c - r_e}{R_s + r_b + r_e}\cdot v_{in}$$

and it is normally justifiable to assume that r_e is negligible compared with αr_c, that α can be taken as unity and that $(r_b + r_e)$ can be neglected in comparison with R_s. Making these assumptions we have

$$v \approx \frac{r_c}{R_s}\cdot v_{in}$$

and Eqn 4.11 now becomes

$$i_c = \frac{\dfrac{r_c}{R_s} \cdot v_{in}}{R_l + r_o}$$

both results identical to those for the common-base amplifier. But r_o for the common-emitter amplifier (see expression (4.15), p. 72) is approximately $1/\beta$ that for the common-base amplifier whereas R_s for the common-emitter amplifier must be approximately β times that for the common-base amplifier to limit distortion in a class-A amplifier to an acceptable level. These two effects cancel and the net result is that i_c has approximately the same value for a given value of v_{in} for common-emitter and common-base amplifiers.

If we compare Eqns 4.10 and 4.11 we obtain the following expression for the output resistance of the common-emitter amplifier:

$$r_o = r_e + r_c\,(1 - \alpha) + \frac{r_e(\alpha r_c - r_e)}{R_s + r_b + r_e} \tag{4.12}$$

This expression shows that the output resistance depends on the source resistance (for a given transistor), decreasing as the source resistance is increased.

This expression can be simplified by combining the first term with the third; this gives the result

$$r_o = r_c\,(1 - \alpha) + r_e \cdot \frac{R_s + r_b + \alpha r_c}{R_s + r_b + r_e} \tag{4.13}$$

To find an expression for the output resistance for short-circuited input terminals, let R_s approach zero in Eqn 4.13. We then have

$$r_o = r_c\,(1 - \alpha) + r_e \cdot \frac{r_b + \alpha r_c}{r_b + r_e}$$

$$= r_c \cdot \frac{r_e + r_b\,(1 - \alpha)}{r_e + r_b}$$

$$= r_c \cdot \frac{r_e + r_b/\beta}{r_e + r_b} \tag{4.14}$$

if we neglect the term $r_e r_b/(r_e + r_b)$ which is small compared with the others.

To find an expression for the output resistance for open-circuited input terminals let R_s approach infinity in Eqn 4.12. We then have

$$r_o = r_e + r_c (1 - \alpha)$$

$$= r_e + r_e/\beta$$

and as r_e is small compared with r_c/β we can say

$$r_o = r_c/\beta \tag{4.15}$$

Output resistance of a bipolar transistor as common-emitter amplifier

We can determine the range of values of output resistance for a common-emitter bipolar transistor amplifier by substituting the typical practical values $r_e = 25\,\Omega$, $r_b = 300\,\Omega$, $r_c = 1\,\text{M}\Omega$ and $\beta = 50$ in Eqns 4.14 and 4.15.

From Eqn 4.14 the output resistance for short-circuited input terminals is given by

$$r_o = 1,000,000 \cdot \frac{25 + 300/50}{25 + 300}\ \Omega$$

$$= 95\,\text{k}\Omega$$

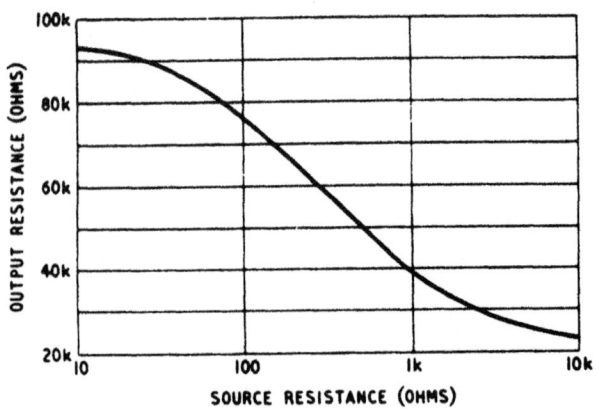

Fig. 4.3. Variation of output resistance with source resistance for a common-emitter amplifier

From Eqn 4.15 the output resistance for open-circuited input terminals is given by

$$r_o = \frac{1,000,000}{50} \ \Omega$$

$$= 20 \, \text{k}\Omega$$

The variation in output resistance with source resistance for bipolar transistors is illustrated in Fig. 4.3.

Voltage gain

From Fig. 4.1(b) we can see that the output voltage is given by $i_c R_l$. The voltage gain v_{out}/v_{in} is thus given by $i_c R_l/v_{in}$. From Eqn 4.10 i_c is given by

$$i_c = \frac{\dfrac{\alpha r_c - r_e}{R_s + r_b + r_e} \cdot v_{in}}{R_l + r_e + r_c \, (1 - \alpha) + \dfrac{r_e(\alpha r_c - r_e)}{R_s + r_b + r_e}}$$

Hence

$$\frac{v_{out}}{v_{in}} = \frac{\dfrac{\alpha r_c - r_e}{R_s + r_b + r_e} \cdot R_l}{R_l + r_e + r_c \, (1 - \alpha) + \dfrac{r_e(\alpha r_c - r_e)}{R_s + r_b + r_e}}$$

$$= \frac{(\alpha r_c - r_e) \, R_l}{[R_l + r_e + r_c \, (1 - \alpha)] \, (R_s + r_b + r_e) + r_e \, (\alpha r_c - r_e)} \tag{4.16}$$

Normally αr_c greatly exceeds r_e and this expression therefore gives a positive value for the voltage gain. But, to agree with the physics of the transistor, the collector current in Fig. 4.1(b) was assumed to be flowing in an anticlockwise direction whereas the base current was shown flowing in a clockwise direction. A positive value for the voltage gain thus implies that the output voltage is inverted with respect to the input voltage. Eqn 4.16 can be simplified by neglecting r_e in comparison with $R_l + r_c(1 - \alpha)$ in numerator and denominator.

We then have

$$\frac{v_{out}}{v_{in}} = \frac{\alpha r_c R_l}{[R_l + r_c(1-\alpha)](R_s + r_b + r_e) + \alpha r_e r_c}$$

For values of R_l and r_e small compared with $r_c(1-\alpha)$ Eqn 4.16 may be written

$$\frac{v_{out}}{v_{in}} = \frac{\alpha r_c R_l}{r_c(1-\alpha)(R_s + r_b + r_e) + \alpha r_e r_c}$$

$$= \frac{\alpha R_l}{R_s(1-\alpha) + r_b(1-\alpha) + r_e}$$

If α in the numerator is taken as unity, this may be written

$$\frac{v_{out}}{v_{in}} = \frac{\beta R_l}{R_s + r_b + \beta r_e}$$

From Eqn 4.8 we can say

$$\frac{v_{out}}{v_{in}} = \frac{\beta R_l}{R_s + r_i} \tag{4.17}$$

which compares with Eqn 3.16 in the previous chapter for the common-base amplifier.

As pointed out in the previous chapter it is more useful to express voltage gain as the ratio of v_{out} to v_{be} and that an expression for this ratio can be obtained by putting $R_s = 0$. Thus we have

$$\frac{v_{out}}{v_{be}} = \beta \cdot \frac{R_l}{r_i}$$

and from page 68 $\beta/r_i = g_m$

$$\therefore \frac{v_{out}}{v_{be}} = g_m R_l$$

a useful result which compares directly with that for an fet and is used later. It should always be remembered, of course, that R_s must be large compared with r_i for linear operation even though R_s does not enter into the expression for v_{out}/v_{be}.

We have seen that a typical value for the input resistance of a transistor with $\beta = 100$ is 2.8 kΩ. Such a transistor if used for voltage amplification will require R_s to be at least $30\,\text{k}\Omega$ to minimise distortion due to input-resistance variation. A typical value for R_l is $5\,\text{k}\Omega$ and hence the voltage gain is given by

$$\frac{v_{out}}{v_{in}} = 100 \times \frac{5}{30}$$

$$= 17.$$

The voltage gain v_{out}/v_{be} is given by

$$\beta \cdot \frac{R_l}{r_i} = 100 \times \frac{5,000}{2,800}$$

$$= 180 \text{ approximately.}$$

Both values of gain are of the same order as those for the common-base amplifier.

Consider now the voltage gain of the common-emitter amplifier for load resistor values which are very large compared with $r_c\,(1-\alpha)$. If r_e is neglected in comparison with r_c, Eqn 4.16 becomes

$$\frac{v_{out}}{v_{in}} = \frac{\alpha r_c R_l}{R_l(R_s + r_b + r_e) + \alpha r_e r_c}$$

The second term in the denominator can be neglected in comparison with the first and α may be taken as unity giving

$$\frac{v_{out}}{v_{in}} = \frac{r_c}{R_s + r_b + r_e} \tag{4.18}$$

which is identical with Eqn 3.17 for the common-base amplifier. Thus the voltage gain of a given transistor with a given large value of collector load resistance is approximately the same, no matter whether the transistor is connected up as a common-base or a common-emitter amplifier. The curve relating voltage gain with load resistance is similar to that for the common-base amplifier and is given in Fig. 4.4.

Applications of common-emitter amplifiers

The common-emitter amplifier has a high output resistance and a low input resistance but the ratio of the two is not so high as for the common-base amplifier. It also has high voltage gain and high current gain which accounts for it being the most widely used of the three

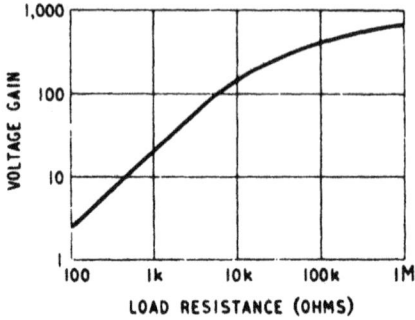

Fig. 4.4. Variation of voltage gain with load resistance for a common-emitter amplifier

fundamental bipolar transistor circuits. It forms the basis of most types of amplifier, oscillator and pulse generator. Numerous examples of its applications are given in the remaining chapters of this book.

Common-source amplifiers

Amplifier properties

The basic connections for the common-source amplifier are shown in Fig. 4.5(a). Fig. 4.5(b) shows the low-frequency equivalent circuit of the transistor with signal-source and load circuits.

The input resistance of the transistor, i.e. the resistance between gate and source terminals, is clearly infinite: there is hence no possibility of feeding any current into the input terminals and we are concerned only with the performance of the transistor as a voltage amplifier.

The output resistance, i.e. the resistance between the drain and source terminals, is simply r_{ds} the drain a.c. resistance, the constant-current generator being assumed to have infinite resistance.

Because of the infinite input resistance there is no input current and hence no loss of input signal across R_s. Thus $v_{gs} = v_{in}$ and the current

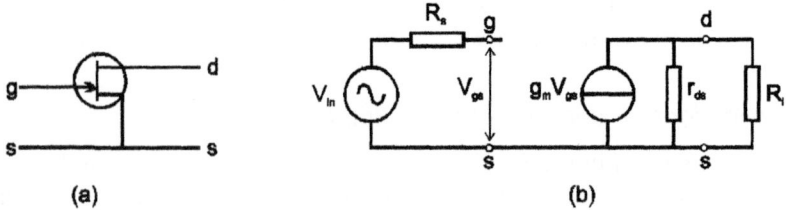

(a)   (b)

Fig. 4.5. Basic connections for a common-source amplifier (a) and the electrical equivalent with source and load circuit (b)

$g_m v_{in}$ divides between r_{ds} and R_l. The fraction i_{out} which enters R_l is given by

$$i_{out} = g_m v_{in} \cdot \frac{r_{ds}}{r_{ds} + R_l}$$

But $v_{out} = i_{out} R_l$

$$\therefore \frac{v_{out}}{v_{in}} = g_m \cdot \frac{r_{ds}}{r_{ds} + R_l}$$

Often R_l is small compared with r_{ds} and this can be simplified to

$$\frac{v_{out}}{v_{in}} = g_m R_l$$

Applications

The common-source amplifier has infinite input resistance and appreciable voltage gain. Both properties are useful and this is the most used of the fet linear circuits. Because of their low noise fets are often used in low-level a.f. applications, e.g. in microphone head amplifiers where the high input resistance makes such amplifiers particularly suitable for following capacitor and piezo-electric microphones.

Chapter 5

Common-collector and common-drain amplifiers (emitter and source followers)

Common-collector amplifiers

Introduction

The fundamental circuit of a common-collector amplifier is given in Fig. 5.1(a) in which R_s represents the resistance of the signal source and R_l represents the load resistance. For simplicity, bias sources are omitted from this diagram. In Fig. 5.1(b) the transistor is represented by an equivalent network of resistances r_e, r_b and r_c together with a voltage generator of e.m.f. $\alpha r_c i_e$ where i_e is the alternating current in the emitter circuit. If the base current i_b and emitter current i_e are both shown acting in clockwise directions as in Fig. 5.1(b), then the current in the common-collector resistance r_c is the difference $(i_e - i_b)$. From Chapter 2 we know that this difference is the collector current i_c: all conventions are therefore satisfied in this diagram.

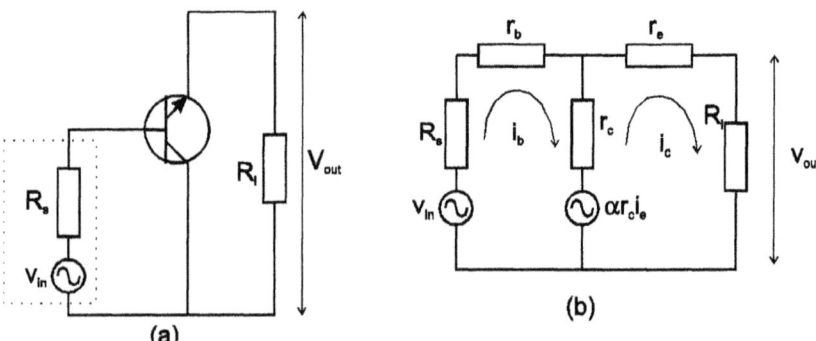

Fig. 5.1. The basic circuit for the common-collector transistor amplifier is given at (a), and the equivalent circuit at (b)

Current gain

Applying Kirchhoff's laws to the circuit of Fig. 5.1(b) we have

$$v_{in} = i_b(r_b + r_c + R_s) + \alpha r_c i_e - i_e r_c$$
$$= i_b(r_b + r_c + R_s) - i_e r_c(1 - \alpha) \tag{5.1}$$

$$0 = i_e(r_c + r_e + R_l) - \alpha r_c i_e - r_c i_b$$
$$= i_e[r_c(1 - \alpha) + r_e + R_l] - r_c i_b \tag{5.2}$$

From Eqn 5.2 we have

$$\frac{i_e}{i_b} = \frac{r_c}{r_c(1 - \alpha) + r_e + R_l}$$

Normally r_e and R_l are both small compared with $r_c(1 - \alpha)$ and may be neglected in comparison with it. Thus the current gain of the common-collector circuit is given approximately by

$$\frac{i_e}{i_b} = \frac{1}{1 - \alpha}$$

As we have seen this is the approximation often used for β, the current gain of the common-emitter circuit. Thus the current gains of the common-collector and common-emitter circuit are very nearly equal: because i_c is only very slightly smaller than i_e this is not an unexpected result.

Input resistance

We can obtain an expression for the input resistance if i_e is eliminated between Eqns 5.1 and 5.2 to obtain a relationship between v_{in} and i_b. From Eqn 5.2

$$i_e = i_b \cdot \frac{r_c}{r_c(1 - \alpha) + r_e + R_l}$$

Substituting for i_e in Eqn 5.1

$$v_{in} = i_b(r_b + r_c + R_s) - r_c(1 - \alpha) \cdot \frac{i_b r_c}{r_c(1 - \alpha) + r_e + R_l}$$

$$\therefore i_b = \frac{v_{in}}{r_b + r_c + R_s - \dfrac{r_c^2(1 - \alpha)}{r_c(1 - \alpha) + r_e + R_l}} \tag{5.3}$$

In a simple circuit containing a generator of resistance R_s and a load of resistance r_i the current is given by

$$i_b = \frac{v_{in}}{R_s + r_i} \qquad (5.4)$$

Comparison of Eqns 5.3 and 5.4 gives the input resistance as

$$r_i = r_b + r_c - \frac{r_c^2(1 - \alpha)}{r_c(1 - \alpha) + r_e + R_l} \qquad (5.5)$$

which shows that the input resistance of the common-collector amplifier depends on the parameters of the transistor (r_b, r_c, r_e and α) and on the load resistance R_l, Fig. 5.2.

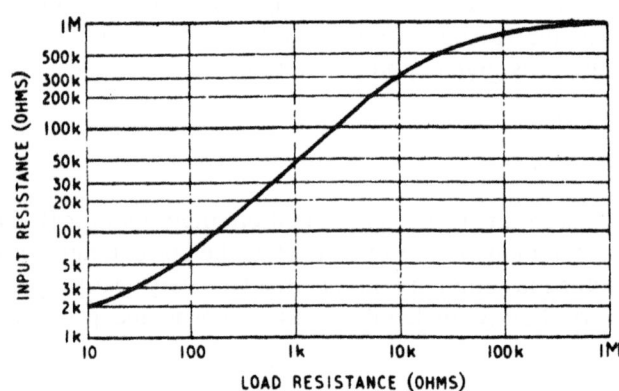

Fig. 5.2. Variation of input resistance with load resistance for a common-collector amplifier

This expression can be simplified slightly by combining the second and third terms. We then have

$$r_i = r_b + \frac{r_c^2(1 - \alpha) + r_e r_c + r_c R_l - r_c^2(1 - \alpha)}{r_c(1 - \alpha) + r_e + R_l}$$

$$= r_b + r_c \frac{r_e + R_l}{r_c(1 - \alpha) + r_e + R_l} \qquad (5.6)$$

Consider the limiting value of input resistance as R_l approaches zero and in addition neglect r_e in comparison with $r_c(1 - \alpha)$

$$r_i = r_b + \frac{r_e}{1 - \alpha} \tag{5.7}$$

$$= r_b + \beta r_e$$

which is, of course, the same result as for the common-emitter amplifier.

In practice values of R_l are likely to be large compared with r_e but small compared with r_c/β. If we neglect r_e in comparison with R_l in the numerator of Eqn 5.6 and R_l in comparison with $r_c(1 - \alpha)$ in the denominator we have

$$r_i = r_b + \beta R_l \tag{5.8}$$

$$\approx \beta R_l$$

a simple and useful result. Although it does not include r_e (as for the common-base and common-emitter amplifiers) the input resistance of the emitter follower is nevertheless dependent on emitter current because β normally varies with I_e. This does not mean however that the source resistance must be large compared with r_i when the emitter follower is used as a voltage amplifier. The circuit has a very large degree of negative feedback (sufficient, in fact, to reduce the voltage gain to less than unity) and this reduces distortion to a small value.

If R_l is large compared with $r_c(1 - \alpha)$ the input resistance is given by

$$r_i = r_b + r_c$$

but, of course, r_c is very large compared with r_b and thus we may say

$$r_i \approx r_c$$

Numerical examples

Consider a transistor with $r_b = 300\,\Omega$, $r_e = 25\,\Omega$, $r_c = 1\,\text{M}\Omega$ and $\beta = 50$. From Eqn 5.7 the input resistance for short-circuited output terminals is given by

$$r_i = 300 + 25 \times 50\,\Omega$$

$$= 1{,}550\,\Omega$$

which is the same value as for the common-emitter circuit. Suppose a load resistance of 1 kΩ is used in the emitter circuit. The input resistance is now given by Eqn 5.8

$$r_i = 50 \times 1{,}000 \, \Omega$$

$$= 50 \, k\Omega$$

For open-circuited output terminals the input resistance is equal to r_c, i.e. 1 MΩ.

These numerical examples show that the emitter follower can provide input resistances of 50 kΩ or more for load resistors of the order of 1 kΩ.

Output resistance

We can obtain an expression for the output resistance of a common-collector transistor amplifier by eliminating i_b between Eqns 5.1 and 5.2, so as to obtain a relationship between i_e and v_{in}. From Eqn 5.2 we have

$$i_b = \frac{r_c(1 - \alpha) + r_e + R_l}{r_c} \cdot i_e$$

Substituting for i_b in Eqn 5.1

$$v_{in} = \frac{(r_b + r_c + R_s) \, [r_c(1 - \alpha) + r_e + R_l]}{r_c} \, i_e - i_e r_c(1 - \alpha)$$

$$\therefore i_e = \frac{v_{in}}{\dfrac{(r_b + r_c + R_s) \, [r_c(1 - \alpha) + r_e + R_l]}{r_c} - r_c(1 - \alpha)}$$

Multiplying numerator and denominator by $r_c/(r_b + r_c + R_s)$ we have

$$i_e = \frac{\dfrac{r_c}{r_b + r_c + R_s} v_{in}}{R_l + r_e + r_c(1 - \alpha) - \dfrac{r_c^2(1 - \alpha)}{r_b + r_c + R_s}} \tag{5.9}$$

In a simple circuit containing a signal source of voltage v and internal resistance r_o feeding a load resistance R_l, the current i_e is given by

$$i_e = \frac{v}{R_l + r_o} \tag{5.10}$$

Comparison between Eqns 5.9 and 5.10 shows that the signal voltage effectively acting in the equivalent circuit is given by

$$v = \frac{r_c}{r_b + r_c + R_s} \cdot v_{in}$$

Normally r_c is large compared with $(r_b + R_s)$ and

$$v \approx v_{in}$$

Thus Eqn 5.10 becomes

$$i_e \approx \frac{v_{in}}{R_l + r_o}$$

and, from Eqn 5.13, $r_o \approx 1/g_m$. Thus we have the final important result that

$$i_e \approx \frac{v_{in}}{R_l + 1/g_m}$$

Thus the current in the load resistance R_l is the value which would flow if the signal input to the emitter follower acted directly on the load resistance and had a source resistance of $1/g_m$.

Comparing Eqns 5.9 with 5.10 we obtain the following expression for the output resistance r_o of the emitter follower

$$r_o = r_e + r_c(1 - \alpha) - \frac{r_c{}^2(1 - \alpha)}{r_b + r_c + R_s} \tag{5.11}$$

Combining the second and third terms

$$r_o = r_e + r_c(1 - \alpha) \cdot \frac{r_b + R_s}{r_b + r_c + R_s} \tag{5.12}$$

When the source resistance is very small we have, neglecting r_b in comparison with r_c

$$r_o = r_e + r_b(1 - \alpha)$$

It is often permissible to neglect $r_b(1 - \alpha)$ in comparison with r_e. We then have

$$r_o \approx r_e$$

$$\approx \frac{1}{g_m} \tag{5.13}$$

Normal values of source resistance are large compared with r_b but small compared with r_c

$$\therefore r_o = r_e + R_s(1 - \alpha)$$

$$= r_e + \frac{R_s}{\beta}$$

When R_s is large compared with r_c

$$r_o = r_e + r_c(1 - \alpha)$$

$$= \frac{r_c}{\beta} \qquad\qquad (5.14)$$

Some numerical examples

For a transistor with $r_b = 300\,\Omega$, $r_e = 25\,\Omega$, $r_c = 1\,M\Omega$ and $\beta = 50$ we have, using the above three simple approximations:

For very small source resistances

$$r_o = 25\,\Omega$$

For a source resistance of $1\,k\Omega$

$$r_o = 25 + \frac{1,000}{50}\,\Omega$$

$$= 45\,\Omega$$

For a very high source resistance

$$r_o = \frac{1,000,000}{50}\,\Omega$$

$$= 20\,k\Omega$$

The variation of output resistance with generator resistance is illustrated in Fig. 5.3. For small values of generator resistance, the output resistance is very low, being only slightly greater than the emitter resistance r_e. It is possible, for example, to have an output resistance of less than $50\,\Omega$: a value as low as this is impossible from common-base or common-emitter transistor amplifiers.

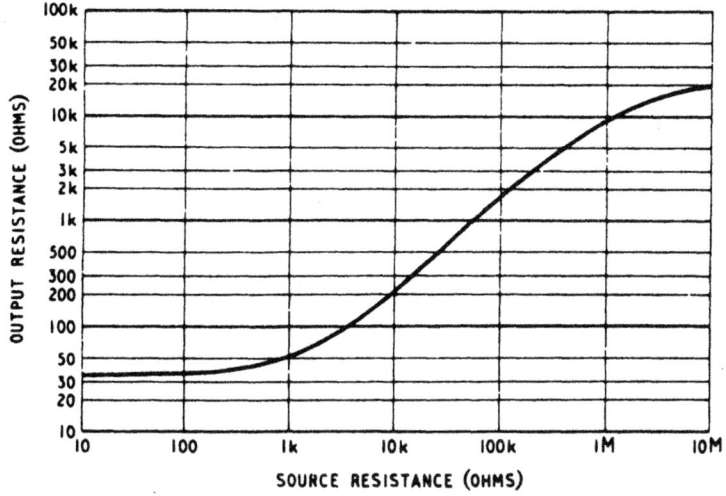

Fig. 5.3. Variation of output resistance with source resistance for a common-collector amplifier

Thus a common-collector amplifier with a low value of generator resistance and a high value of emitter load resistance can have a high value of input resistance and a low value of output resistance, conditions opposite to those normally encountered in transistor amplifiers.

Voltage gain

From Fig. 5.1(b) we can see that the output voltage is given by $r_e R_l$. The voltage gain v_{out}/v_{in} is thus given by $i_e R_l/v_{in}$. From Eqn 5.9 i_e is given by

$$i_e = \frac{\dfrac{r_c}{r_b + r_c + R_s} v_{in}}{R_l + r_e + r_c(1 - \alpha) - \dfrac{r_c^2(1 - \alpha)}{r_b + r_c + R_s}}$$

Hence

$$\frac{v_{out}}{v_{in}} = \frac{\dfrac{r_c R_l}{r_b + r_c + R_s}}{R_l + r_e + r_c(1 - \alpha) - \dfrac{r_c^2(1 - \alpha)}{r_b + r_c + R_s}}$$

$$= \frac{r_c R_l}{[R_l + r_e + r_c (1 - \alpha)] (r_b + r_c + R_s) - r_c^2(1 - \alpha)} \quad (5.15)$$

This can be simplified by ignoring r_e in comparison with the other terms in the first bracket of the denominator and by ignoring r_b in the second bracket. After further evaluation this gives

$$\frac{v_{out}}{v_{in}} = \frac{r_c R_l}{R_l r_c + R_l R_s + r_c(1 - \alpha)R_s}$$

$$= \frac{R_l}{R_l + R_s/\beta + R_l R_s/r_c}$$

The third term in the denominator is normally negligible compared with the other two and thus we have

$$\frac{v_{out}}{v_{in}} = \frac{R_l}{R_l + R_s/\beta} \tag{5.16}$$

Thus the voltage gain is less than unity but R_s/β is small compared with R_l and the gain can usually be taken as unity with very little error.

Applications of emitter followers

The emitter follower has a high input resistance, a low output resistance, high current gain and unity voltage gain. Its main application is as a resistance converter, e.g. an emitter follower is often used as the first stage of a voltage amplifier to give a high input resistance and as the final stage to give a low output resistance. This may alternatively be regarded as a buffering action, e.g. the emitter follower first stage prevents the low input resistance of the second stage shunting the input signal source. Typical practical circuits for emitter followers are given in Chapter 7.

Comparison of bipolar transistor amplifiers

To facilitate comparisons between the three fundamental types of transistor amplifier, the principal properties are summarised in Table 5.1. An approximate expression for each parameter is given in brackets.

Common-drain amplifiers

Circuit properties

The basic circuit for a common-drain amplifier (source follower) is given in Fig. 5.4(a) and its electrical equivalent at (b). Fig. 5.4(a) shows

Table 5.1 Fundamental properties of bipolar transistor amplifiers

	common-base	*common-emitter*	*emitter follower*
Input resistance	very low $(= r_e)$	low $(= \beta r_e)$	high $(= \beta R_l)$
Output resistance	very high $(= r_c)$	high $(= r_c/\beta)$	very low $(= 1/g_m)$
Current gain	unity $(= \alpha)$	high $(= \beta)$	high $(= \beta)$
Voltage gain	high	high	unity
Polarity of output signal relative to input signal }	in phase	antiphase	in phase

that the input resistance is infinite: the output resistance can be calculated in the following way.

As in the common-source circuit the current $g_m v_{gs}$ splits between r_{ds} and R_l the fraction (i_{out}) entering R_l being given by

$$i_{out} = g_m v_{gs} \cdot \frac{r_{ds}}{r_{ds} + R_l}$$

Normally R_l is small compared with r_{ds} and thus we can say

$$i_{out} = g_m v_{gs} \tag{5.17}$$

Now $v_{out} = i_{out} R_l$

$$\therefore \frac{v_{out}}{v_{gs}} = g_m R_l$$

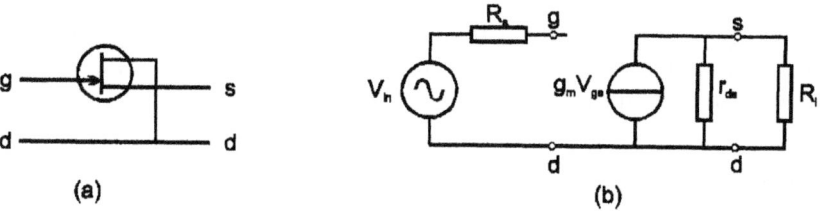

(a) **(b)**

Fig. 5.4. Basic connections for a source follower amplifier (a) and the electrical equivalent with source and load circuit (b)

This is, of course, the voltage gain of the common-source amplifier. Adding unity to both sides we have

$$\frac{v_{out} + v_{gs}}{v_{gs}} = 1 + g_m R_l$$

Now, from Fig. 5.4(a), $(v_{out} + v_{gs}) = v_{in}$

$$\therefore v_{in} = v_{gs} (1 + g_m R_l) \tag{5.18}$$

Substituting for v_{gs} in Eqn 5.17

$$i_{out} = g_m \cdot \frac{v_{in}}{1 + g_m R_l}$$

$$= \frac{v_{in}}{R_l + 1/g_m}$$

which shows that the load is effectively fed from a generator of voltage v_{in} and output resistance $1/g_m$. The source follower thus has unity voltage gain and an output resistance of $1/g_m$. The voltage gain can alternatively be calculated thus. From Fig. 5.4(a)

$$v_{in} = v_{gs} + v_{out}$$

Substituting for v_{gs} from Eqn 5.18

$$v_{in} = \frac{v_{in}}{1 + g_m R_l} + v_{out}$$

from which

$$\frac{v_{out}}{v_{in}} = \frac{g_m R_l}{1 + g_m R_l}$$

Now $g_m R_l$ is the voltage gain of the common-source stage and is normally large compared with unity. Thus the voltage gain of the source follower is very nearly unity.

Applications

The source follower has an infinite input resistance, a low output resistance and a voltage gain of approximately unity. Its only advantage

Table 5.2 Fundamental properties of fet amplifiers

	common-gate	common-source	source follower
Input resistance	very low $(= 1/g_m)$	infinite	infinite
Output resistance	high $(= r_{ds})$	high $(= r_{ds})$	very low $(= 1/g_m)$
Current gain	unity	–	–
Voltage gain	fair $(= g_m R_l)$	fair $(= g_m R_l)$	unity
Polarity of output signal relative to input signal }	in phase	antiphase	in phase

over the common-source amplifier is the low output resistance and it is used wherever such a property is essential: otherwise the common-source amplifier is preferred.

Comparison of field-effect transistor amplifiers

To facilitate comparisons between the three fundamental types of fet amplifier the principal properties are summarised in Table 5.2. An approximate expression for each parameter is given in brackets.

Chapter 6

Bias and d.c. stabilisation

Bipolar transistors

Introduction

It is usual to begin the design of a bipolar transistor amplifying stage by choosing a value of mean collector current which enables the required current swing and/or voltage swing to be delivered with an acceptable degree of linearity. The next step is to devise a biasing circuit to give this particular value of mean collector current.

The most obvious way of biasing a bipolar transistor is by the simple circuit of Fig. 6.1 in which a resistor R_b is connected between the base of the transistor and source of steady voltage V_{bb}. Certainly it is possible by adjustment of R_b and/or V_{bb} to set the mean collector current at the desired value but such adjustments to individual transistors would be tedious if the amplifiers are to be produced in any quantity. The fundamental disadvantage of the circuit of Fig. 6.1 is that it fixes the *base current** not the *collector current*. The collector current is β times the base current and values of β for a given type of transistor can differ by $\pm50\%$ from the nominal value. Thus if the nominal β is 100, values as low as 50 and as high as 150 are possible and, for a given base current, the collector current can have any value within a range of 3:1. Smaller variations in β also occur as a result of temperature changes. Moreover in certain circuits, particularly those using power transistors, the collector current can increase significantly with temperature as a result of variations in the value of V_{be} which, for all bipolar transistors,

---

* Provided V_{bb} is large compared with V_{be}, and R_b is large compared with the transistor input resistance the base current is given approximately by V_{bb}/R_b.

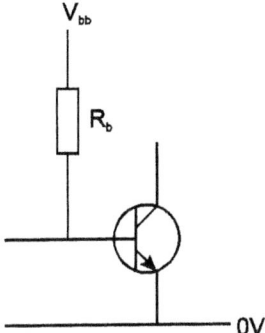

Fig. 6.1. Simple bias circuit which fixes base current

increases at the rate of 2.5 mV per °C. It is these variations in β and V_{be} which make the circuit of Fig. 6.1 unsuitable for use in mass-produced equipment. What is wanted for this purpose is a biasing circuit which, without using preset components:

(a) gives the desired value of mean collector current in spite of manufacturing spreads in transistor parameters and
(b) maintains this mean current in spite of variations in parameters due to temperature changes.

Circuits which fulfil these two purposes are said to provide *d.c. stabilisation* of the operating point and this chapter describes the most commonly-used circuits.

Leakage current

Unfortunately the amplified input current is not the only component of the collector current of a bipolar transistor. There is also an unwanted component, known as the *leakage current*, which is generated by thermal breakdown of covalent bonds in the collector-base junction and is independent of the input current. Leakage current increases rapidly with rise in temperature and, in germanium-transistor common-emitter amplifiers, can be large enough to seriously impair performance even at normal operating temperatures. In fact, unless checked by a stabilising circuit, it can increase regeneratively (an effect known as *thermal runaway*) to a value at which the transistor is destroyed.

Leakage current is negligible in silicon transistors at normal temperatures but d.c. stabilisation is still necessary to minimise the effects of the spreads in transistor parameters and the temperature dependence of β and V_{be}.

Stability factor

The effectiveness of circuits for stabilising mean collector current can be expressed by a stability factor K which may be defined in the following way:

$$K = \frac{\text{collector current in stabilised circuit}}{\text{collector current in unstabilised circuit}}$$

K is thus unity for an unstabilised circuit and is less than unity for stabilised circuits, the smallness of K being a measure of the success of the circuit in limiting changes in mean current.

Basic circuits for d.c. stabilisation

Most stabilising circuits achieve their object by d.c. negative feedback: the collector or emitter current is used to generate a signal which is returned to the base so as to oppose any changes in the mean value of the collector or emitter current. In practice the transistor current is passed through a resistor and the voltage generated across the resistor is used as a source of base bias current. The resistor can be in the emitter circuit as shown in Fig. 6.2(a) or in the collector circuit as in Fig. 6.2(b). The way in which these circuits achieve stabilisation is perhaps more easily understood from Fig. 6.2(b): any increase in collector current causes an increased voltage across R_c, a decrease in the collector-emitter voltage and thus a smaller base current which opposes the initial rise in collector current. Provided R_c

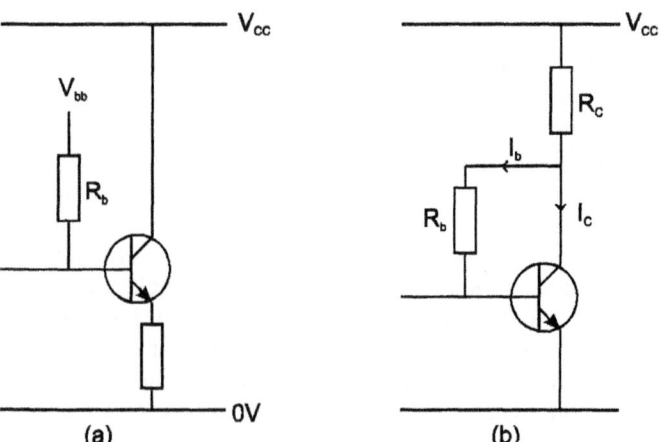

Fig. 6.2. The two basic circuits for d.c. stabilisation

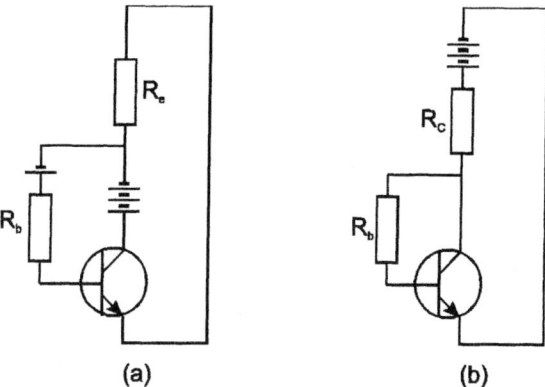

(a)   **(b)**

Fig. 6.3. The two basic circuits redrawn to show their similarity

equals R_e the two circuits of Fig. 6.2 have identical performances: this can be shown by redrawing the two circuits in the form given in Fig. 6.3 from which it is clear that if the d.c. resistance of the batteries is neglected the two circuits are identical.

Calculation of stability factor

Provided R_c equals R_e therefore both circuits have the same stabilising effect and the stability factor can be calculated in the following way in which the equations apply to Fig. 6.2(b). The calculation applies equally to the circuit of Fig. 6.2(a) if $R_c = R_e$ and $V_{bb} = V_{cc}$.

For an unstabilised common-emitter amplifier we have

$$I_c = \beta I_b \tag{6.1}$$

Summing steady potentials in Fig. 6.2(b)

$$R_c(I_c + I_b) + R_b I_b + V_{be} = V_{cc} \tag{6.2}$$

where V_{cc} is the supply voltage and V_{be} is the base-emitter voltage of the transistor.

From Eqn 6.1 we have

$$I_b = \frac{I_c}{\beta}$$

Substituting for I_b in Eqn 6.2

$$I_c[(\beta + 1) R_c + R_b] = \beta(V_{cc} - V_{be})$$

Rearranging

$$I_c = \frac{\beta(V_{cc} - V_{be})/R_b}{1 + (\beta + 1)R_c/R_b}$$ (6.3)

Expression 6.3 shows how the mean value of collector current increases with increase in β.

If R_c is very small or if R_b is very large (to eliminate feedback) we have

$$\text{mean value of } I_c = \frac{\beta(V_{cc} - V_{be})}{R_b}$$ (6.4)

showing that this component is directly proportional to β as would be expected in an unstabilised circuit. On the other hand if R_b is very small or R_c is very large we have

$$\text{mean value of } I_c = \frac{V_{cc} - V_{be}}{R_c}$$

i.e. the current is independent of β and stabilisation is perfect. Expression 6.4 is the numerator of expression 6.3 and thus we have that the stabilisation factor is given by

$$K = \frac{1}{1 + (\beta + 1)R_c/R_b}$$ (6.5)

Estimation of stability factor from circuit diagram

Eqn 6.5 is of the form

$$K = \frac{1}{1 + \beta F}$$ (6.6)

where F is the fraction of the collector current which is fed back to the base. In Fig. 6.2(b) the collector current I_c splits at the junction of R_c and R_b. The current entering R_b is given by

$$I_c \cdot \frac{R_c}{R_b + R_c}$$

and this, of course, is the base current I_b. Thus the fraction of I_c used as feedback is $R_c/(R_b + R_c)$ which agrees with expression 6.3. This is a useful method of deriving stability factors which often enables the factor to be calculated directly from the component values on a circuit diagram. The circuit diagram should be reduced to the form shown in Fig. 6.3 which shows the split of collector current and enables F to be computed from the ratio $R_c/(R_c + R_b)$. The stability factor is then given by Eqn 6.6.

Figure 6.4 shows the circuit diagram of a simple amplifier. There are a few fixed criteria, such as the available supply voltage. For this example, assume that the supply is to be 9 V. The transistor could be any low-power signal transistor, and these typically have a value of β of between 100 and 300 – that is to say, when obtaining one of these, the designer will never know the exact value. The design must accommodate this unpredictability. Assume, then, that $\beta = 200$ (the half-way value). The circuit is to amplify a.c. signals, so the output must be so biased that when no signal is present at the input, the value of the output is $V_{cc}/2$. If this can be achieved, when an a.c. signal is applied, the output can swing positively and negatively by an equal amount as shown in Fig. 6.4. This means that V_{CE} is to be set to 4.5 V (=9.0 V/2).

The next step is to decide on a bias current. All transistors are characterised by the amount of collector current that they can pass. A signal transistor can rarely pass more than 100 mA, and the heating effect with this amount of current would rapidly change the operating condition of the circuit. Signal amplifiers tend to operate with a bias current (or quiescent collector current, as it is often called) in the range of 0.1 mA to 10 mA. If the bias current is less than 0.1 mA, the bias resistors can have large ohmic values. These can be difficult to obtain, and large resistance values in an amplifier increase the amount of electrical (Johnson) noise generated. If the bias current is larger than 10 mA, heating effects occur within the transistor. For this example,

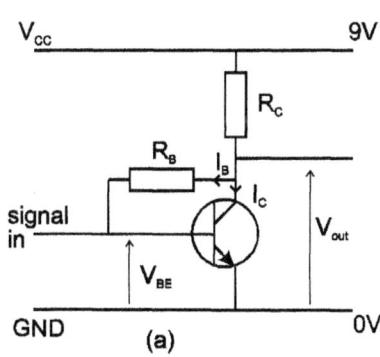 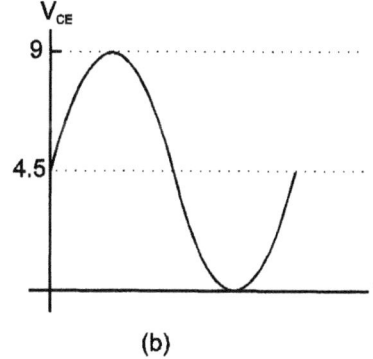

Fig. 6.4. A simple a.c. amplifier

choose a value of bias current I_c of 1 mA. The current flowing through R_c is $I_c + I_B$, where $I_B = I_c/\beta$. Therefore,

$$R_c = \frac{9 - 4.5}{1 + 1/200} = 4.48\,\text{k}\Omega$$

The voltage drop across the base resistor is simply $V_{cc} - V_{BE}$. The bias current through I_B is I_c/β.

The value of R_B is given by

$$R_B = \frac{9 - 0.7}{1/200} = 1.66\,\text{M}\Omega$$

where $V_{BE} = 0.7$ for a transistor.

Minimising signal-frequency feedback

In Fig. 6.2 the feedback circuit cannot distinguish between direct current and signal-frequency current. Thus the gain of wanted signals is reduced in the same ratio as any increase in mean current due to increase in β. If this reduction in signal-frequency gain is not wanted it may be minimised by decoupling. If the signal-frequency output is taken from the collector circuit in Fig. 6.2(a) R_e can be shunted by a low-reactance capacitor to minimise signal-frequency feedback. If the signal-frequency output is taken from the collector circuit in Fig. 6.2(b) it is not practical to decouple R_c and a more usual method is to construct R_b of two resistors R_1 and R_2 in series, the junction being decoupled to emitter as shown in Fig. 6.5. If R_1 is small it reduces the effective collector load resistance of the amplifier and if R_2 is small it lowers the input

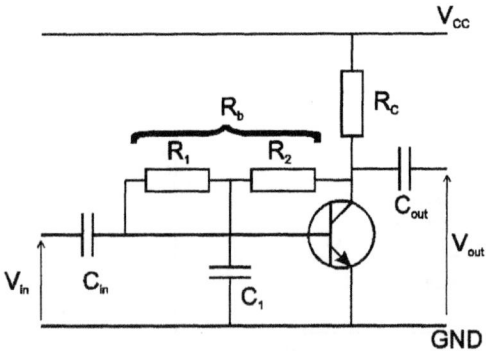

Fig. 6.5. To avoid signal-frequency negative feedback due to R_b, decoupling may be introduced as shown here

resistance. Usually therefore R_1 and R_2 are made approximately equal and the capacitor is chosen to have a reactance small compared with the resistance value at the lowest signal frequency. For example if R_b has a value of $100\,\text{k}\Omega$ R_1 and R_2 can each be $50\,\text{k}\Omega$ and for an a.f. amplifier C_1 can be $2\,\mu\text{F}$ which has a reactance of $1.6\,\text{k}\Omega$ at $50\,\text{Hz}$.

If the transistor is fed from a transformer one end of R_b can be decoupled as shown in Fig. 6.6. Signal-frequency feedback is minimised by the decoupling capacitor C_1 but in this circuit R_b has no shunting effect on the input resistance.

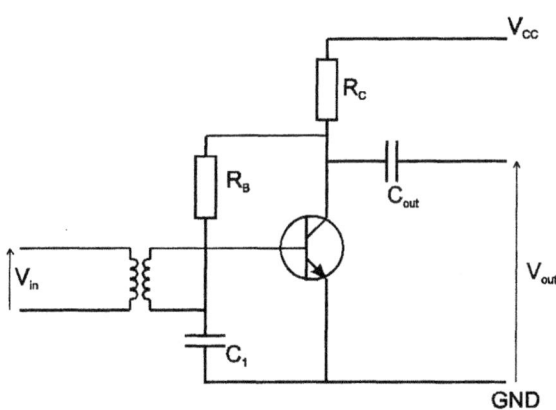

Fig. 6.6. Method of decoupling which can be used with transformer coupling

If a large output voltage swing is required the quiescent collector potential should be made one-half the supply voltage. Thus if in an npn circuit $V_{cc} = 12\,\text{V}$ the quiescent potential should be made 6 V. Suppose the mean collector current is required to be $3\,\text{mA}$. We have immediately that R_c is $6/(3 \times 10^{-3}) = 2\,\text{k}\Omega$. Because a silicon transistor requires a forward bias of 0.7 V, the voltage across R_b is 5.3. Suppose β is 100. I_b is then $3\,\text{mA}/100$, i.e. $30\,\mu\text{A}$ and R_b is given by $5.3/(30 \times 10^{-6}) = 180\,\text{k}\Omega$ approx. The stability factor for this circuit is, from Eqn 6.6, given by

$$K = \frac{1}{1 + 100 \times 2/(180 + 2)}$$

$$= 0.48$$

Without stabilisation $K = 1$ and thus this circuit has succeeded only in halving any unwanted change in mean current – a very poor performance.

The stabilising circuit of Fig. 6.2(a) is more used because signal-frequency decoupling is simple and because the collector is free for signal-frequency components. Most d.c. stabilising circuits are based on Fig. 6.2(a).

Potential divider and emitter resistor circuit

It is not always possible or convenient to tap the supply battery to obtain base bias but an alternative source of steady potential is a potential divider $R_1 R_2$ connected across the collector supply as shown in Fig. 6.7.

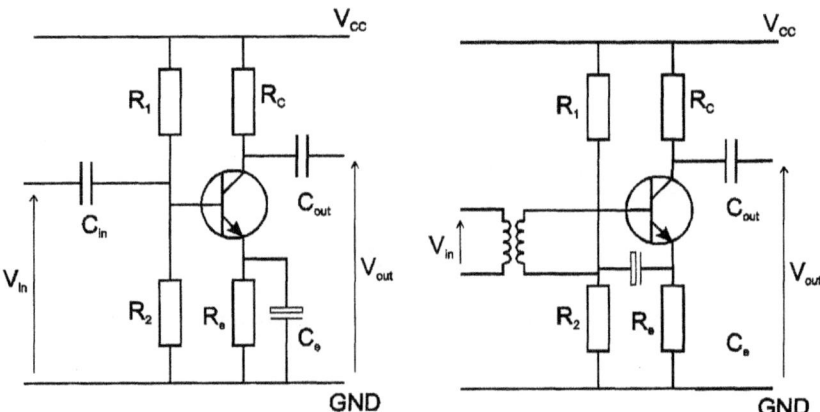

Fig. 6.7. Potential-divider method of stabilising the d.c. conditions in a common-emitter amplifier (a) in an RC-coupled amplifier and (b) in a transformer-coupled amplifier

Two basic types of this circuit exist. In Fig. 6.7(a) the circuit is arranged for RC coupling from the previous stage: a feature of this arrangement is that the resistance of R_1 and R_2 in parallel is effectively shunted across the input circuit of the transistor. This parallel resistance should not, therefore, be too small. In Fig. 6.7(b) the circuit is arranged for transformer coupling from the previous stage: the parallel resistance of R_1 and R_2 does not now enter into input-resistance considerations.

If these circuits are redrawn in the form shown in Fig. 6.8 we can see immediately that they are of the same type as Figs. 6.2 and 6.3 in which R'_b is made up of R_1 and R_2 in parallel. Thus we can at once estimate the

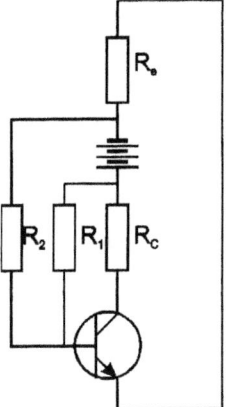

Fig. 6.8. The circuit of Fig 6.7 redrawn to illustrate the feedback due to R_e

stability factor. The output current division ratio F (Eqn 6.7) is $R_e/(R_b' + R_e)$ and hence

$$K = \frac{1}{1 + \beta R_e/(R_b' + R_e)}$$

$$\text{where } R_b' = \frac{R_1 R_2}{R_1 + R_2}$$

Good stability thus requires low values of R_1 and R_2 but this means a heavy drain on the supply. In practice a compromise solution is adopted as indicated in the following numerical example.

Design of a potential-divider circuit

Let us assume R_e to be $1\,k\Omega$: this is a convenient value because, for an emitter current of $1\,mA$, only $1\,V$ of the collector supply voltage is lost. This leaves in a typical circuit with a $6\,V$ supply, $5\,V$ for the transistor and its load resistor. If $\beta = 50$ the base current is $1/50\,mA$, i.e. $20\,\mu A$. This flows through R_1 in addition to the bleed current which flows through R_1 and R_2 from the collector supply. For good d.c. stability the potential at the junction of R_1 and R_2 must be steady in spite of variations in base current and this is achieved by making the parallel resistance of R_1 and R_2 small: this implies that the bleed current must be large compared with the base current. The bleed current can therefore be $200\,\mu A$ which is ten times the base current but is only one-fifth the collector current. Thus the total current in R_1 is $220\,\mu A$.

C_1 should have a reactance small enough to avoid negative feedback and consequent fall in signal-frequency gain even at the lowest operating frequency. To achieve this the reactance must be small compared with r_e, say 25 Ω. In an a.f. amplifier C_1 may be 500 μF which has a reactance of 6.5 Ω at 50 Hz.

The base-emitter voltage is 0.7 (a silicon transistor is assumed) and thus the voltage drop across R_1 is 4.3, giving the value of R_1 as 4.3/(220 × 10⁻⁶) = 20 kΩ approx. The voltage across R_2 is 1.7 and the current 200 μA, making the resistance 8.5 kΩ.

For this circuit R_b' is given by $R_1 R_2/(R_1 + R_2)$, i.e. 20 k × 8.5 k/28.5 k = 6 kΩ approx. The stability factor is thus given by

$$k = \frac{1}{1 + 50 \times 1/(6 + 1)}$$

$$= 0.12$$

implying that variations in mean emitter current are reduced to one eighth of their unstabilised value.

Better stability could be obtained if a greater supply voltage is available and if the value of β is higher. For example, in mains-driven equipment with a 24-V supply and a mean emitter current of 4 mA, a stability factor of 0.034 can be achieved, using a 1.8 kΩ emitter resistor and a potential divider made up of 32 kΩ and 15 kΩ β is assumed to be 150.

D.C. stability in two-stage amplifiers

It is clear from Fig. 6.8 that the performance of the potential-divider circuit could be improved by returning R_1 to the collector: by so doing we ensure that R_c makes a contribution to the d.c. feedback in addition to that provided by R_e. The circuit so produced is shown in Fig. 6.9: in

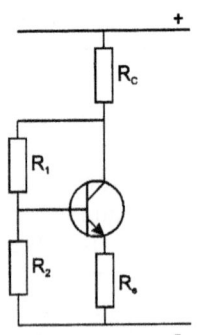

Fig. 6.9. Potential divider returned to collector

Fig. 6.10 it is redrawn to make the feedback paths more obvious. The improvement in stability obtained in this circuit can be estimated in the following way.

The current I_c leaving the collector in Fig. 6.10 splits between R_c and R_1. The current I_1 entering R_1 is given approximately by

$$I_1 = I_c \cdot \frac{R_e + R_c}{R_e + R_c + R_1}$$

in which, for simplicity, the shunting effect of R_2 on R_e is neglected. The current in R_c is now

$$I_c \cdot \frac{R_1}{R_e + R_c + R_1}$$

and this splits at the junction of R_e and R_2. The current I_2 in R_2 is given by

$$I_2 = I_c \cdot \frac{R_1}{R_e + R_c + R_1} \cdot \frac{R_e}{R_e + R_2}$$

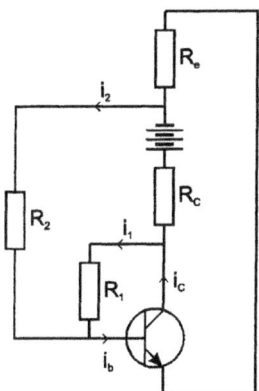

Fig. 6.10. The circuit of Fig. 6.9 redrawn to show feedback paths

I_b is the sum of I_1 and I_2. Thus the feedback fraction F is given by

$$F = \frac{I_b}{I_c} = \frac{R_e + R_c}{R_e + R_c + R_1} + \frac{R_1}{R_e + R_c + R_1} \cdot \frac{R_e}{R_e + R_2}$$

and the stability factor is, from Eqn 6.6:

$$K = \frac{1}{1 + \beta F}$$

Typical component values are $R_e = 1\,\text{k}\Omega$, $R_c = 5\,\text{k}\Omega$, $R_1 = 10\,\text{k}\Omega$, $R_2 = 5\,\text{k}\Omega$ and $\beta = 100$. Substituting these values in the above expressions we have

$$F = \frac{1 + 5}{1 + 5 + 10} + \frac{10}{1 + 5 + 10} \cdot \frac{1}{1 + 5}$$

$$= 0.48$$

$$\therefore K = \frac{1}{1 + 100 \times 0.48}$$

$$= 0.02$$

a considerable improvement in stability over the value obtained when the potential divider is returned to the supply terminal.

Thus this form of potential-divider circuit can give excellent stability. To obtain it, however, the shunting effect of R_1 on R_c must be tolerated. Moreover if signal-frequency feedback via R_1 is to be avoided R_1 must be constructed of two resistors in series with their junction decoupled as shown in Fig. 6.4. Both disadvantages can be overcome by the use of an emitter follower as shown in Fig. 6.11. The high input resistance of TR2 minimises the shunting effect on R_c and the low output resistance makes possible low values of R_1 and R_2. If the emitter circuit of TR2 is decoupled at signal frequencies, this transistor can be used as a common-emitter amplifier and there is no need to include an additional transistor in the circuit purely to give good d.c. stabilisation. If, however, $(R_1 + R_2)$ is decoupled it is necessary to introduce a series

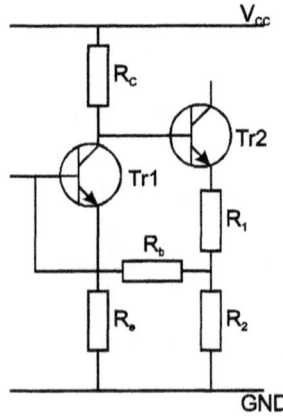

Fig. 6.11. Potential-divider circuit using an emitter follower

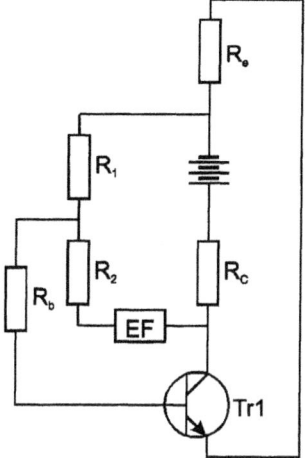

Fig. 6.12. Fig. 6.11 redrawn to facilitate calculation of stability factor

resistor R_b to avoid decoupling TR1 base as shown in Fig. 6.11. Some examples of this circuit are given in the next chapter.

The stability factor obtainable from the circuit of Fig. 6.11 can be calculated from the redrawn version of the circuit shown in Fig. 6.12.

There is effectively no split of collector current at the junction of R_c and the emitter follower because of the high resistance of the emitter follower. It is best to calculate I_b from a knowledge of the voltage applied to R_b. From the emitter follower one component of this voltage is $R_2/(R_1 + R_2)$ of that generated across R_c by I_c: this we can call rI_cR_c. The second component of the voltage is that generated across the parallel resistance of R_e and R_b by I_c: it is assumed that R_1 is negligibly small. Thus we have

$$I_b = \frac{rR_cI_c + \dfrac{R_eR_b}{R_e + R_b} \cdot I_c}{R_b}$$

$$= I_c \left(\frac{rR_c}{R_b} + \frac{R_e \cdot}{R_e + R_b} \right)$$

giving

$$F = \frac{rR_c}{R_b} + \frac{R_e}{R_e + R_b}$$

Thus the stability factor is given by

$$K = \cfrac{1}{1 + \beta \left(\cfrac{rR_c}{R_b} + \cfrac{R_e}{R_e + R_b} \right)}$$

Typical practical values are $r = 0.5$, $R_c = 5\,\text{k}\Omega$, $R_e = 5\,\text{k}\Omega$, $R_b = 3.3\,\text{k}\Omega$ and $\beta = 100$. Substituting in the above expression

$$K = \cfrac{1}{1 + 100 \left(\cfrac{0.5 \times 5}{3.3} + \cfrac{5}{5 + 3.3} \right)}$$

$$= 0.0073$$

This represents exceptionally good stability, variations in mean collector current in TR1 being reduced to less than one-hundredth of their unstabilised value. Transistors are often used in cascade in amplifiers and it is common practice to adopt the circuit of Fig. 6.11 to achieve high d.c. stability.

Use of diodes to compensate for falling battery voltage

The stabilising circuits described above are useful in reducing the effects of spreads in β but they do nothing towards making the mean collector current independent of the supply voltage. Such circuits are useful therefore when a stabilised supply is available but in battery-operated equipment some means is required of compensating for the effects of falling battery voltage. One method which uses the circuit of Fig. 6.7 is to arrange for the resistance of R_2 to increase as the current through it falls, e.g. by using a non-linear device in place of R_2 (or in parallel with it): a suitable device is a forward-biased semiconductor diode. Such diodes are used in battery-operated portable receivers to stabilise the quiescent current of class-B output stages.

Use of diodes for temperature compensation

Diodes connected in the base circuits of transistors are also used to stabilise the collector current against temperature changes. Very good stabilisation can be achieved by using in the compensating circuit a diode of the same material as the transistors to be stabilised. As temperature changes the voltage across the diode changes by the right amount to keep the current constant in the transistors. If the transistors are mounted on a heat sink the diodes should preferably be mounted close to them on the same heat sink.

Diodes are extensively used in integrated circuits for stabilising collector currents and here, of course, they share a silicon chip with the transistors to be stabilised. They are thus in intimate thermal contact with the circuit to be monitored and are capable of maintaining satisfactory performance over a temperature range as wide as −55°C to 125°C.

Use of temperature-dependent resistor for temperature compensation

As an alternative to using a diode in the base circuit it is possible to use a resistor with a positive temperature coefficient in the emitter circuit of the transistor to be stabilised. The technique is often used with power transistors as a means of stabilising collector currents of the order 0.5 A where variations in collector current are due to changes in V_{be} with temperature. The mean value of the emitter current can be kept constant by adjustment of the base-emitter voltage and a change of approximately 2.5 mV per °C is required. Thus the stabilising circuits for power transistors should be designed to apply a correction of this value to the base-emitter voltage.

A simple method of effecting this compensation is to use an external emitter resistor of pure metal. Such resistors have a positive temperature coefficient and a rise in temperature causes the external emitter resistance to increase, thus increasing the voltage across this resistance. This in turn reduces the base-emitter voltage and, if the base potential is constant, tends to maintain the collector current constant.

The temperature coefficient of electrical resistance of copper is approximately 0.004 per °C: if the emitter current is assumed constant, the voltage across a copper emitter resistor therefore increases by 0.004 of its initial value per °C. If the initial voltage is unity the change in emitter voltage is 4 mV per °C. To offset a 2.5 mV change in base-emitter voltage, an initial emitter voltage of 2.5/4, i.e. approximately 0.6 V, is needed. If the mean emitter current is 0.5 A (as is likely in a transistor with 5 W dissipation) the emitter resistance should be 0.6/0.5, i.e. 1.2 Ω, a convenient value to construct of copper wire. The fixed base potential is usually achieved by the use of a resistive potential divider.

Field-effect transistors

Introduction

It is usual to begin the design of an amplifying circuit using a fet by choosing a value of mean drain current which will enable the required output current swing and/or output voltage swing to be achieved with an acceptable degree of linearity. The next step is to devise a biasing circuit to give the required value of mean drain current.

The obvious way to bias a fet is by the simple circuit of Fig. 6.13 which contains a resistor R_g connected between the gate and a source of constant voltage V_{gg}. The gate current in a fet is very small indeed and R_g can be very high, e.g. 100 MΩ, without significantly affecting the gate voltage. As pointed out in Chapter 2, the bias voltage for an enhancement fet lies between the source and drain voltages whereas for a depletion fet it lies outside the range of the drain-source voltage. Typical figures for both types of fet are given in Fig. 6.13.

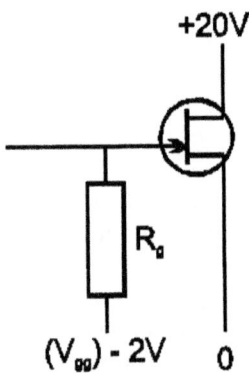

Fig. 6.13. Simple bias circuit for fixing gate-source voltage (a) in depletion and (b) in enhancement fets

Certainly by adjusting V_{gg} it is possible to set the mean drain current at the desired value but such adjustments to individual transistors would be tedious if the circuits are to be produced in any quantity. The fundamental disadvantage of this simple biasing circuit is that it fixes the *gate voltage* not the *drain current*. The drain current for a given type of fet and a given gate-source voltage can have any value within a range of 3:1 due to manufacturing spreads in transistor parameters and to changes in parameters with temperature. In fact drain current may increase or decrease as temperature changes. Ideally what is wanted is a biasing circuit which will enable a desired value of mean drain current to be obtained in spite of manufacturing spreads and which will also maintain this mean current in spite of changes in parameters.

Such d.c. stabilisation can be achieved, as for bipolar transistors, by d.c. negative feedback, e.g. by passing the drain current through a resistor and by using the voltage generated across the resistor as the source of gate bias. The two basic circuits of Fig. 6.2 can be used.

Stabilisation of depletion fets

For depletion fets bias and a measure of d.c. stabilisation can be obtained by including a resistor R_s in the source circuit and by returning R_g to the source supply terminal as shown in Fig. 6.14. Any changes in I_d cause corresponding changes in the voltage across R_s which are applied directly between gate and source so minimising the original change in drain current.

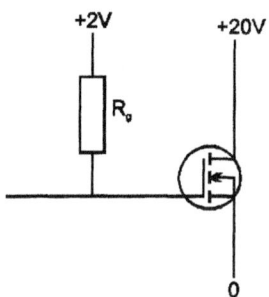

Fig. 6.14. Simple d.c. stabilising circuit which can be used with a depletion fet

The required value of R_s is determined in the following manner. Firstly the mean drain current which will give the desired performance is chosen. Next the gate bias voltage V_g which gives the chosen value of I_d is determined from the I_d–V_g characteristics of the transistor. The required value of R_s is given by V_g/I_d. For example if I_d is 5 mA and V_g is 2 V, R_s is given by $2/(5 \times 10^{-3})$, i.e. 400 Ω.

Calculation of stability factor

For a field-effect transistor without stabilisation the relationship between changes of drain current i_d and changes in gate voltage v_g is simply

$$i_d = g_m v_g \tag{6.7}$$

i.e. the drain current is directly proportional to g_m for a given gate voltage and thus varies with temperature or with change of transistor in the same way and to the same extent as g_m.

For a fet with a source resistance R_s the relationship between i_d and input voltage v_{in} can be deduced in the following way:

$$v_{in} = v_{gs} + v_{fb}$$

Substituting $i_d R_s$ for v_{fb} and i_d/g_m for v_{gs} we have

$$v_{in} = \frac{i_d}{g_m} + i_d R_s$$

from which

$$i_d = \frac{g_m v_{in}}{1 + g_m R_s} \qquad (6.8)$$

Comparison of Eqns 6.7 and 6.8 shows that the effect of the source resistance is to reduce variations in drain current for a given input signal to $1/(1 + g_m R_s)$ of their unstabilised value. The stability factor for this circuit is thus given by

$$K = \frac{1}{1 + g_m R_s} \qquad (6.9)$$

The value of R_s is, of course, fixed by bias considerations and thus the stability factor for the simple circuit of Fig. 6.14 is automatically determined. For example if R_s is 400 Ω as in the last numerical example and if g_m is 2 mA/V, the stability factor is given by $1/(1 + 2 \times 10^{-3} \times 400) = 1/1.8 = 0.55$. This is a very poor performance. However, greater stability can be obtained, without effect on bias, by using two resistors R_{s1} and R_{s2} in series in the source circuit and by returning R_g to their junction as shown in Fig. 6.15. Bias is determined by the value of R_{s1} but stability by $(R_{s1} + R_{s2})$. High stability can thus be obtained by using large values for R_{s2} but such values reduce the voltage available across the transistor and its load resistor.

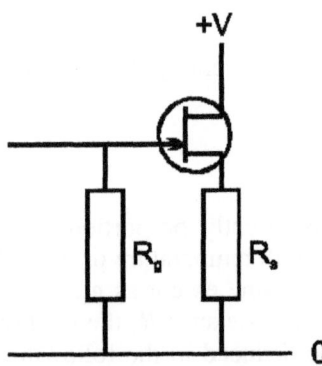

Fig. 6.15. Modification of the previous circuit to give better d.c. stabilisation than is obtainable with the normal value of automatic bias resistor

Minimising signal-frequency feedback

Resistors included in the source circuit give signal-frequency negative feedback (and hence reduced gain) in addition to the d.c. feedback which is responsible for the stabilising effect. Signal-frequency feedback can be minimised by decoupling the source resistors by a low-reactance capacitor as shown in Fig. 6.16.

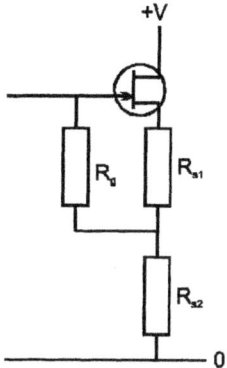

Fig. 6.16. Decoupling to minimise signal-frequency feedback in the circuit of Fig. 6.14

Stabilisation of enhancement fets

The stabilising circuit of Fig. 6.2(b) is suitable for enhancement fets because it gives a gate bias voltage between that of the source and the drain. Fig. 6.17 shows the circuit applied to an n-channel mosfet. The stability factor is given by Eqn 6.9 by substituting R_d for R_s. If the output of the transistor is taken from the drain circuit R_g must be decoupled: this can be done as suggested in Fig. 6.18.

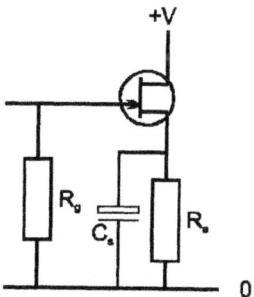

Fig. 6.17. Simple d.c. stabilising circuit for an enhancement fet

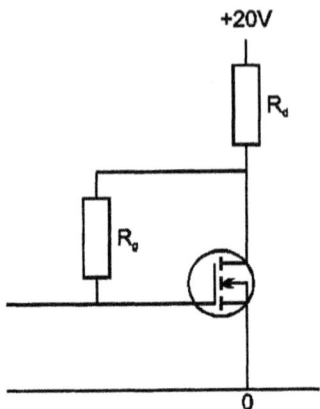

Fig. 6.18. Method of minimising signal-frequency feedback in the circuit of the previous diagram

Potential divider and source resistor circuit

This circuit (illustrated in Fig. 6.19) can be used for enhancement and depletion fets and the stability factor obtained is given by expression 6.9. It is possible to use higher values of R_s than in the simple circuit of Fig. 6.14 and higher values of K are obtainable. This is illustrated in the following numerical example.

Suppose the chosen mean drain current is 1.5 mA and that the gate bias necessary to give this current is −2 V. If the supply voltage is 15, a suitable value for the voltage at the junction of R_1 and R_2 is 3 V. This requires R_1 to be four times R_2. The sum of R_1 and R_2 determines the current taken by the potential divider and this can be kept low in this circuit without sacrificing stability (cf. page 98). For example R_1 could be 4 MΩ and R_2 10 MΩ.

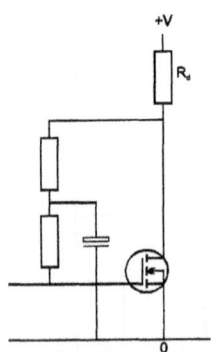

Fig. 6.19. Potential divider and source resistor method of d.c. stabilising a fet

To give the required gate-source voltage the source potential must be 5 V. This gives R_s as $5/(1.5 \times 10^{-3})$, i.e. 3.3 kΩ. If the mutual conductance of the fet is 2 mA/V the stability factor is given by

$$K = \frac{1}{1 + 2 \times 10^{-3} \times 3,300}$$

$$= 0.13$$

Signal-frequency feedback can be minimised by decoupling R_s as in Fig. 6.16.

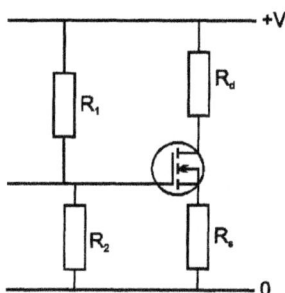

Fig. 6.20.

Chapter 7

Small-signal a.f. amplifiers

Definition of small-signal amplifier

Some amplifiers are required to deliver an output voltage or output current which is small compared with the maximum that the amplifier could deliver. When a voltage output is wanted, the magnitude of the current output is usually of little consequence provided the transistor(s) can supply it without distortion. Similarly if the amplifier is designed to supply a current output, the magnitude of the voltage output is of secondary importance provided it is not sufficient to cause overloading and distortion. Such amplifiers are termed *small-signal amplifiers*: typical examples are microphone head amplifiers and the r.f. and early i.f. stages in a receiver.

Distinction between voltage and current amplifiers

If a stage is to be regarded as a voltage amplifier, the voltage of the signal source must not be affected by the connection of the amplifier across it; thus the input resistance of the amplifier must be high compared with the resistance of the signal source. Moreover the output voltage of the amplifier must be substantially unaffected by the connection of the load: thus the output resistance must be low compared with the load resistance. Provided these resistance requirements are satisfied the voltage gains of the individual stages of an amplifier can be multiplied to give the overall voltage gain of the amplifier or (and this is another way of expressing the same fact) the voltage gains of the individual stages, when expressed in decibels, can simply be added to give the overall gain of the amplifier.

When a current amplifier is connected to a signal source the current from the source must not be affected by the connection of the amplifier: thus the input resistance of the amplifier must be small compared with the resistance of the signal source. The output current from the amplifier should not be affected by the connection of the load resistance: thus the output resistance should be high compared with the load resistance. Provided these resistance requirements are met the current gains of the individual stages of an amplifier can be multiplied to give the overall current gain of the amplifier or (and this is another way of expressing the same fact) the current gains of the individual stages, when expressed in decibels, can simply be added to give the overall gain of the amplifier.

These resistance requirements for voltage and current amplifiers are summarised in Table 7.1. From this it can be deduced that a given amplifier can behave as a voltage or current amplifier. For example if the input and output resistances of the amplifier are both $1\,\mathrm{k\Omega}$, then if $R_s = 100\,\Omega$ and $R_l = 10\,\mathrm{k\Omega}$, the amplifier is best regarded as a voltage amplifier whereas if $R_s = 10\,\mathrm{k\Omega}$ and $R_l = 100\,\Omega$ it is best regarded as a current amplifier.

Table 7.1 Source and load resistance considerations in voltage and current amplifiers

	source resistance R_s	load resistance R_l
voltage amplifier	$\ll r_{in}$	$\gg r_{out}$
current amplifier	$\gg r_{in}$	$\ll r_{out}$

In applying these theoretical considerations to practical transistor circuits, a number of precautions must be taken. For example if a bipolar transistor is fed from a low-resistance signal source (to give a voltage amplifier) severe distortion can occur. To minimise this the source resistance must be made large compared with the input resistance: this is a reminder that the bipolar transistor is inherently a current amplifier. The resistance of a fet is so high that it is impractical to feed it from an even higher source resistance (to give a current amplifier). The input of the fet could, of course, be shunted by a low resistance but this wastes the most important property of the fet: namely its high input resistance: this is a reminder that the fet is essentially a voltage amplifier.

Transistor parameters in small-signal amplifiers

It is characteristic of transistors used as small-signal amplifiers that the signal voltages at the transistor terminals are small compared with the steady potentials at these points. Similarly the signal currents have a magnitude small compared with that of the steady currents on which they are superimposed. When operation is thus confined to small excursions about a mean value, a transistor may be regarded as having substantially constant input resistance, output resistance and current gain.

Single-transistor stages

Bipolar transistor

The circuit diagram for a single-stage bipolar-transistor amplifier is given in Fig. 7.1. It uses the potential divider method of d.c. stabilisation and a method of calculating values for R_1, R_2 and R_e to give a required mean value of collector current is given on page 99. It now remains to calculate the value of R_c.

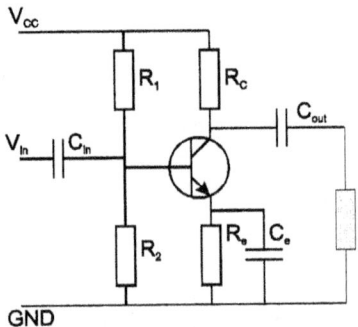

Fig. 7.1. Single-stage bipolar transistor amplifier

The value of R_c depends primarily on whether the amplifier is required to deliver a current or voltage output. If a current output is required, as much as possible of the output current must enter the external load resistor (shown shaded). R_c must hence be large compared with this external load and in practice it is given the highest value possible consistent with maintaining an adequate steady voltage on the collector. For example if the mean collector current is 1 mA, if the base potential is 2 V and if the supply voltage is 9 V, then a suitable value for the collector potential is 3 V (giving a collector-base voltage of 1 V). There are hence 6 V across R_c and its value is $6/(1 \times 10^{-3})$, i.e. 6 kΩ. The value of R_c is not, however, critical provided

(a) it is large compared with the external resistor
(b) it does not greatly exceed 6 kΩ

Provided the parallel resistance of R_1 and R_2 is large compared with the input resistance of the transistor and that R_c is large compared with the external load, the current gain of the amplifier is almost equal to β.

If, however, the transistor is required to give a voltage output then the external resistance will be large compared with R_c: thus R_c is effectively the load into which the transistor is operating. The precise value of R_c is now important and the value of the external resistance is non-critical provided it is large compared with R_c. The value to be given to R_c depends on the magnitude of the output voltage required. If the largest possible voltage swing is required R_c should be given the value which makes the quiescent collector voltage midway between the supply and emitter potentials. Using the values quoted for the current amplifier, the quiescent voltage should be made 6 V which permits upward and downward swings of 3 V amplitude. This fixes the value of R_c at $3/(1 \times 10^{-3}) = 3\,k\Omega$. To minimise distortion it is necessary to include in series with the base a resistor that is large compared with the transistor input resistance. If this resistor is made $30\,k\Omega$ and if β is 100, the voltage gain of the amplifier is, from the first equation on page 75, given by 100×3.30, i.e. 10.

If a smaller voltage output, say 1 V amplitude, is acceptable then R_c can be increased to $6\,k\Omega$. This makes the quiescent collector voltage 3 V so that swings to 2 V and 4 V are possible. The advantage of so increasing R_c is that it doubles the voltage gain.

The operation of the circuit of Fig. 7.1 for various values of R_l can be illustrated by superimposing load lines on the I_c–V_c characteristics as shown in Fig. 7.2. ABC is the load line for a 3-kΩ load. It is drawn

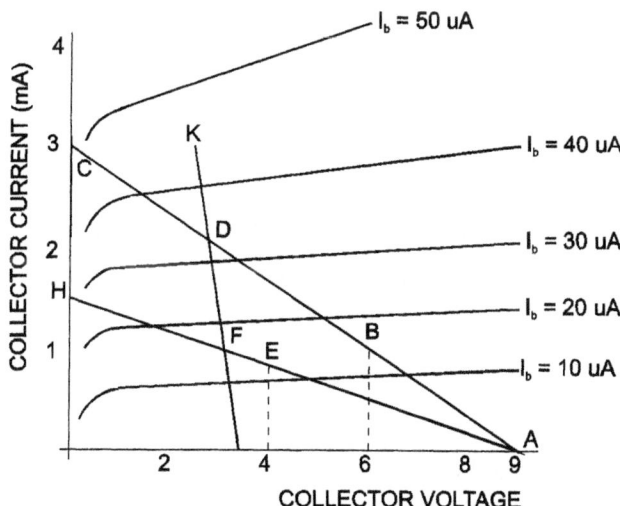

Fig. 7.2. Load lines superimposed on I_c–V_c characteristics

through that point on the V_c axis which corresponds to the supply voltage (9 V) to that point on the I_c axis which corresponds to the current (3 mA) which the supply voltage would drive through the load resistance. The intersections of the load line with the I_b characteristics indicate the current through the series combination of transistor and load, and the voltages across each component. For example point B shows that for a base current of 15 μA, 6 V is developed across the transistor and 3 V across the load resistor. These are, of course, the quiescent conditions required for maximum output voltage. If the input current rises to 32 μA the collector voltage falls to 3 V (point D) and if the input current falls to zero the collector voltage rises to 9 V. During amplification, therefore, the operating point swings between the limits of A and D.

If the load resistance is increased to 6 kΩ the load line becomes AEFGH and the quiescent point is at F corresponding to a collector current of 1 mA and a collector voltage of 3 V. The base current is about 17 μA and when this swings down to 14 μA and up to 19 μA the collector voltage swings up to 4 V and down to 2 V.

The operation of the current amplifier can also be represented on this diagram. The operating point is still at F because the direct-coupled collector resistor is still 6 kΩ for this amplifier. However the external load resistance (assumed capacitance-coupled) is very low and the load line JFK is therefore nearly vertical and passes through F as shown. The intercepts made on this load line by the I_b characteristics show that if the base current changes from 10 μA to 30 μA, the collector current changes from 0.6 mA to 1.9 mA, a current gain of 65 almost equal to β.

When working with load lines, never forget that the graph shown is the characteristic of one sample of one type of transistor. The variation has already been stated as around ±50%. The characteristic also alters according to temperature, collector current and frequency of signal being amplified.

Gain estimation using the mutual conductance g_m

On page 36, it was noted that the mutual conductance of the bipolar transistor was given as approximately

$$g_m \approx \frac{I_e \text{ (mA)}}{25}$$

This permits a method of estimating the a.c. gain of the amplifier. The mutual conductance is a measure of (current out)/(voltage in). To convert it into a voltage gain, the voltage developed across the load

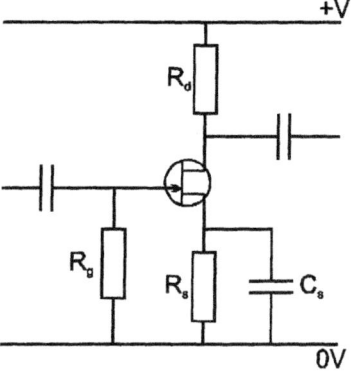

Fig. 7.3. Single-stage amplifier using a jfet

resistance by the current out is required. For a bipolar transistor amplifier with a bias current of 1 mA and load resistor of 1,000 Ω, the voltage gain is approximately

$$Av \approx -g_m \times R_c \approx -\frac{1\,\text{mA}}{25} \times 1{,}000 \approx -40$$

Field-effect transistor

The circuit diagram of a single-stage amplifier using a depletion jfet is given in Fig. 7.3. Because of the extremely-high input resistance such a stage is unlikely to be used for current amplification and we shall assume that voltage amplification is required. The method of choosing the value of R_s is described on page 107 and it remains to calculate the value of R_d. The same technique can be adopted as for the circuit of Fig. 7.1: if the largest possible voltage swing is required R_d is chosen to make the quiescent drain potential midway between the supply and source potentials but if a smaller voltage swing is acceptable R_d can be increased to give higher gain. Suppose R_d is 3 kΩ. The voltage gain is given by $g_m R_d$ and a typical value for g_m is 2 mA/V giving the voltage gain as $3 \times 10^3 \times 2 \times 10^{-3}$, i.e. 6. This is not a high gain and in general it is true that bipolar transistors, although inherently current amplifiers, can give higher voltage gains than fets.

Two-stage amplifiers

If a single-stage amplifier gives insufficient gain two stages can be connected in cascade and Fig. 7.4 gives the circuit diagram of a two-stage amplifier using a transformer to couple the stages, each of which

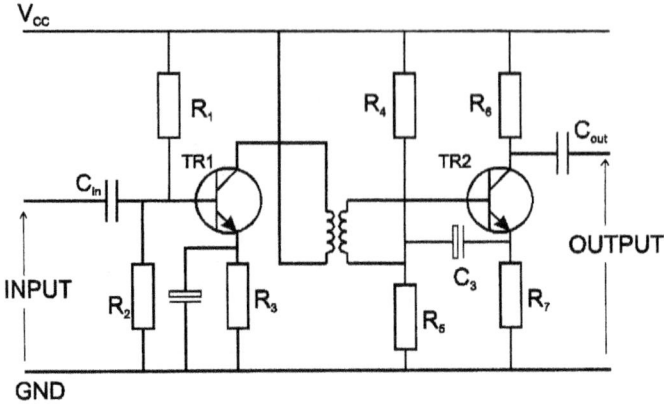

Fig. 7.4. A two-stage small-signal amplifier with transformer coupling

is independently stabilised by the potential divider method. Maximum gain would be obtained from the amplifier if the turns ratio of the transformer were chosen to match the output resistance of TR1 to the input resistance of TR2. This would, however, be a most unsatisfactory design for the following reasons:

(a) TR1 would effectively have a load resistance equal to its output resistance and this large load resistance would seriously limit its output signal amplitude (see page 113).
(b) TR2 would have an effective signal source resistance equal to its input resistance. A larger value of source resistance is necessary to minimise distortion in TR2.
(c) If the amplifier is used at a.f. the transformer would need an inconveniently-large primary inductance to maintain a good bass response. TR1 is likely to have an output resistance of an appreciable fraction of a megohm as for some silicon transistors.

To avoid all these disadvantages the transformer turns ratio is kept small, e.g. 3 to 1. If the input resistance of TR2 is $1 \, k\Omega$ there will be a loss of 3 dB at the frequency for which the reactance of the secondary winding equals this. If the frequency is made 50 Hz we have

$$2\pi fL = 1,000$$

$$\therefore L = \frac{1,000}{6.284 \times 50} \, H = 3.2 \, H \text{ approximately.}$$

If the turns ratio is 3 to 1 the primary inductance is $3^2 \times 3.2 = 29H$.

The circuit of Fig. 7.4 uses a coupling transformer to link the two stages of the amplifier. This is a technique which would only be used at radio frequencies in modern circuits. At audio frequencies, the coupling would be exclusively capacitive because capacitors are lighter, cheaper and more compact. Transformers are only used for the benefits they confer; i.e. impedance matching and the possibility of creating a resonant circuit.

Use of negative feedback

Circuits such as that of Fig. 7.4 suffer from the disadvantage that their properties (gain, input resistance, output resistance, distortion, signal-to-noise ratio, etc.) depend on the characteristics of the transistors. It is thus difficult to make a number of amplifiers with substantially the same performance without very careful choice of transistors and components. Moreover such circuits are wasteful of components: there are simpler circuits giving better d.c. stability.

The difficulties caused by spreads in transistor parameters can be minimised by negative feedback. This technique enables the gain and frequency response to be determined by the constants of a passive network: they are then independent of transistor parameters and hence of changes in temperature. Negative feedback reduces distortion and can be arranged to increase or decrease input and output resistance as desired. The cost of these benefits is reduced gain but the required gain can be restored by the use of additional stages of amplification. The two basic circuits for negative feedback were introduced at the beginning of Chapter 6 and are reproduced in Fig. 7.5. It was shown that these are, in fact, two versions of the same circuit and, for a given value of R_b, give identical performances in respect of d.c. stabilisation provided R_c

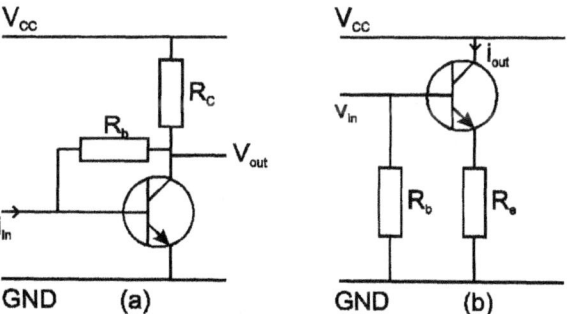

Fig. 7.5. The two basic circuits for applying negative feedback to a common-emitter amplifier

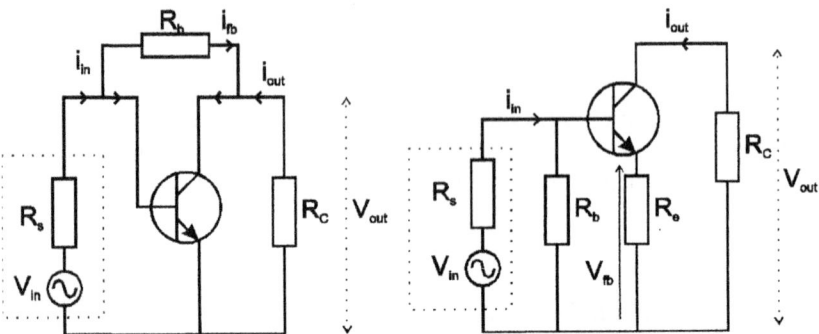

Fig. 7.6. R_b in (a) gives parallel-derived parallel-injected feedback whereas R_e in (b) gives series-derived series-injected feedback

equals R_e. D.C. feedback is, however, independent of external signal-frequency circuits. When the circuits of Fig. 7.5 are used to provide feedback at signal frequencies, account must be taken of the way in which the input signal is fed into the transistor and in which the output signal is taken from it. In fact, as shown in Fig. 7.6 R_b and R_e provide quite different types of signal-frequency feedback.

In Fig. 7.6(a) R_b connects the load resistor R_c and the source resistor R_s *in parallel*: in effect it introduces into the input circuit a current proportional to the output voltage. Such feedback effectively decreases the input resistance and the output resistance of the amplifier, making the circuit suitable as a current-to-voltage converter. In Fig. 7.6(b) the output circuit of the transistor consists of R_c and R_e *in series* and the input circuit consists of v_{in} and v_{fb} *in series*. R_e thus introduces into the input circuit a voltage proportional to the output current. Such feedback effectively increases the input resistance and the output resistance of the amplifier, making this circuit suitable as a voltage-to-current converter. To facilitate reference to the effects of parallel- and series-derived and -injected feedback, these are summarised in Table 7.2.

Table 7.2 Effects of negative feedback on input and output resistance

Type of feedback connection	Effect on input resistance	Effect on output resistance
series-derived		increased
parallel-derived		decreased
series-injected	increased	
parallel-injected	decreased	

Approximate expressions for the essential properties of these two fundamental feedback circuits can be deduced as follows.

Current-to-voltage converter

From Figs. 7.5(a) and 7.6(a) the input current i_{in} is given by

$$i_{in} = i_{fb} + i_b$$

where i_{fb} is given by $v_{out}/(R_b + h_{ie}{}^*)$. R_b is normally very large compared with h_{ie} and thus i_{fb} is approximately equal to v_{out}/R_b. If i_{fb} is large i_b can be neglected in comparison with it and we have

$$i_{in} = i_{fb} = \frac{v_{out}}{R_b}$$

from which

$$\frac{v_{out}}{i_{in}} = R_b \qquad (7.1)$$

Thus the ratio of output voltage to input current for the current-to-voltage converter is given approximately by the value of the feedback resistor R_b.

Now $v_{out} = i_{out}R_c$. Substituting for v_{out} in Eqn 7.1

$$\frac{i_{out}}{i_{in}} = \frac{R_b}{R_c} \qquad (7.2)$$

If R_s is large compared with the input resistance (as it should be to minimise distortion)

$$i_{in} = \frac{v_{in}}{R_s}$$

Substituting for i_{in} in Eqn 7.1

$$\frac{v_{out}}{v_{in}} = \frac{R_b}{R_s} \qquad (7.3)$$

$* h_{ie}$ is the common-emitter input resistance. See Appendix B.

Eqns 7.2 and 7.3 can be derived from inspection of the circuit diagram. In Fig. 7.6(a) the output current from the transistor divides at the junction of R_c and R_b so that a fraction $R_c/(R_c + R_b)$ enters R_b and is returned to the base as feedback. Normally R_b is large compared with R_c and the current division ratio is approximately R_c/R_b. These two resistors determine the current gain of the amplifier which is given by the reciprocal of the current division ratio, i.e. by R_b/R_c.

In Fig. 7.6(a) R_b and R_s form a potential divider across the load resistor R_c and it returns to the base a fraction $R_s/(R_s + R_b)$ of the output voltage. If R_b is large compared with R_s the fraction is approximately R_s/R_b. This potential divider determines the voltage gain of the circuit which is given by the reciprocal of the division ratio, i.e. by R_b/R_s.

Voltage-to-current converter

From Figs. 7.5(b) and 7.6(b) the signal-frequency feedback voltage v_{fb} returned to the input circuit is equal to $i_{out}R_e$ and the input voltage is given by

$$v_{in} = v_{fb} + v_{be}$$

If v_{fb} is large, v_{be} can be neglected in comparison and we have

$$v_{in} = v_{fb} = i_{out}R_e$$

from which

$$\frac{i_{out}}{v_{in}} = \frac{1}{R_e} \tag{7.4}$$

Thus the ratio of output current to input voltage for the voltage-to-current converter is given by the reciprocal of the feedback resistor R_e.

Now $v_{out} = i_{out}R_e$. Substituting for i_{out}

$$\frac{v_{out}}{v_{in}} = \frac{R_c}{R_e} \tag{7.5}$$

If R_s determines the base current $i_{in} = v_{in}/R_s$. Substituting for v_{in} in Eqn 7.4 we have

$$\frac{i_{out}}{i_{in}} = \frac{R_s}{R_e} \tag{7.6}$$

Eqns 7.5 and 7.6 can be confirmed from inspection of the circuit diagram, R_c and R_e constituting the potential divider, R_s and R_e forming the current divider.

From Eqn 7.4 we can obtain a simple expression for the input resistance of the circuit of Fig. 7.6(b). We know that

$$i_{out} = \beta i_{in}$$

and by eliminating i_{out} between this equation and Eqn 7.4 we have

$$r_{in} = \frac{v_{in}}{i_{in}} = \beta R_e \qquad (7.7)$$

Thus if $\beta = 100$ and $R_e = 1\,\text{k}\Omega$ the input resistance is $100\,\text{k}\Omega$. In practice, of course, this may be effectively reduced by other resistors such as those of a potential divider connected to the base.

Two-stage amplifiers

Current amplifier

If a current-to-voltage converter is followed by a voltage-to-current converter, the low output resistance of the first stage feeds into the high input resistance of the second: these are the conditions required to transfer the output voltage of the first stage to the input of the second with little loss. The basic form of the amplifier is shown in Fig. 7.7 in which, for simplicity, direct coupling is employed between the stages. The amplifier so constructed has a low input resistance and a high output resistance and is therefore suitable as a current amplifier.

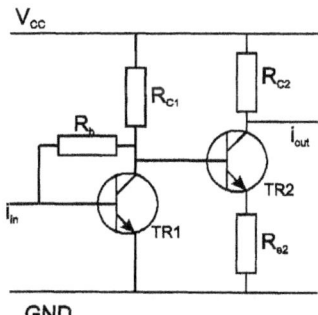

Fig. 7.7. Skeleton form of two-stage current amplifier

If we represent the signal voltage developed across R_{c1} as v, this is the output voltage of TR1 and the input voltage of TR2. For TR1 we have, from Eqn 7.1

$$\frac{v}{i_{in}} = R_b$$

For TR2 we have, from Eqn 7.4

$$\frac{i_{out}}{v} = \frac{1}{R_{e2}}$$

Eliminating v

$$\frac{i_{out}}{i_{in}} = \frac{R_b}{R_{e2}}$$

Because of the undecoupled emitter resistor R_{e2} TR2 behaves as an emitter follower and R_b can be connected to the emitter of TR2 instead of to the collector of TR1 with little effect on the performance of the circuit which now has the form shown in Fig. 7.8.

To make a practical version of this circuit we now need to add measures to ensure d.c. stability of both transistors. One of the most successful methods of stabilisation is that illustrated in Fig. 6.11 and it is easy to add to Fig. 7.8 the few components necessary to give this type of stabilisation. These include a decoupled emitter resistor R_{e1} for TR1 and a decoupled potential divider $R_1 R_2$ in the emitter circuit of TR2. With these additions the circuit has the final form given in Fig. 7.9.

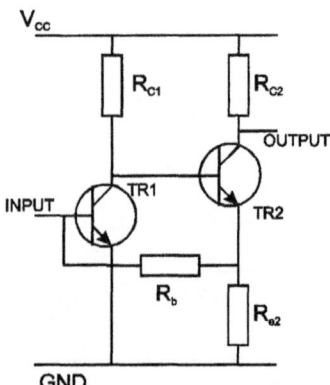

Fig. 7.8. Modification of the circuit diagram of Fig. 7.7

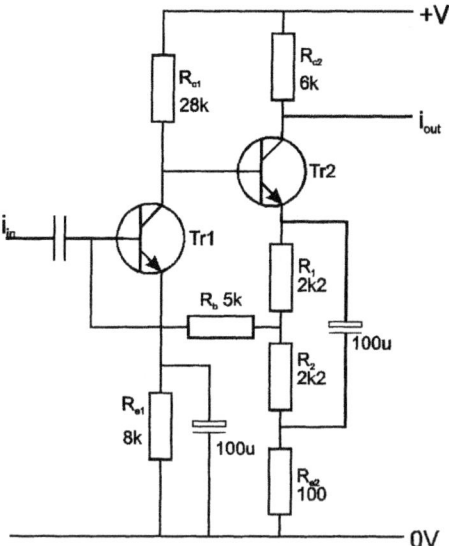

Fig. 7.9. Practical form of the amplifier of Fig. 7.8 designed for a current gain of 50

The current gain of the amplifier is given by R_b/R_{e2} and the resistance values should be chosen to give the required gain. R_b should be small to give good stability and a value of $5\,k\Omega$ is suitable. For a current gain of 50 R_{e2} should then be $100\,\Omega$ as indicated in Fig. 7.9. The remaining component values can be selected in the following way. It is assumed that the transistors are silicon with offset voltages of 0.7 V.

Suppose TR1 is to take 0.5 mA and that the supply voltage is 24 V. Let TR1 collector potential be 10 V. There is then a 14 V drop across R_{c1} and this resistance is thus $28\,k\Omega$. TR2 base voltage is 10 but, because of the offset voltage, the emitter potential is 9.3 V. If TR2 is required to take 2 mA collector current, the total emitter resistance must be $4.65\,k\Omega$. It would probably be sufficiently accurate for R_1 and R_2 to be each $2.2\,k\Omega$ and R_{e2} $100\,\Omega$. We can neglect the very small voltage drop across R_b due to TR1 base current and the base potential is therefore approximately 4.65 V. TR1 emitter voltage is then 3.95 V and the emitter resistance is $8\,k\Omega$. It is probably best however to adjust R_{e1} on test to give the required quiescent currents in TR1 and TR2. As this is a current amplifier R_{c2} should be as large as possible provided distortion does not occur as a result of TR2 collector potential approaching the base potential too closely. A value of $6\,k\Omega$ should be suitable.

Voltage amplifier

If the two stages in Fig. 7.7 are interchanged so that the voltage-to-current converter is followed by the current-to-voltage converter, the high output resistance of the first stage feeds into the low input resistance of the second. These are the conditions required to transfer the output current of the first stage into the input of the second stage with little loss. The basic form of the amplifier so derived is shown in Fig. 7.10 in which, for simplicity, direct coupling is employed between the stages. The amplifier so constructed has a high input resistance and a low output resistance and is therefore suitable as a voltage amplifier.

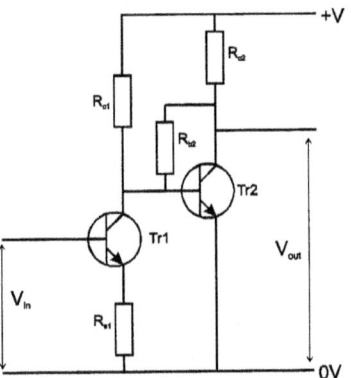

Fig. 7.10. Skeleton form of two-stage voltage amplifier

If we represent the output current of TR1 and the input current of TR2 by i, we have, applying Eqn 7.4 to TR1

$$\frac{i}{v_{in}} = \frac{1}{R_{e1}}$$

For TR2 we have from Eqn 7.1

$$\frac{v_{out}}{i} = R_{b2}$$

Eliminating i between these equations

$$\frac{v_{out}}{v_{in}} = \frac{R_{b2}}{R_{e1}}$$

As shown in Fig. 7.11 it is more usual to connect R_{b2} to TR1 emitter instead of to TR2 base. This makes little difference to the performance

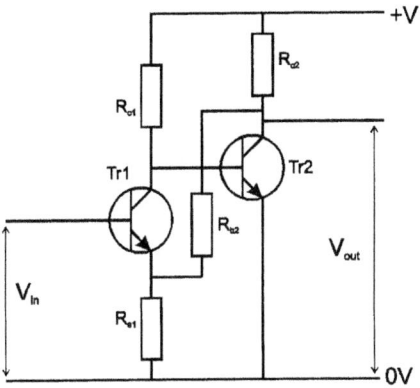

Fig. 7.11. Modification of the circuit diagram of Fig. 7.10

of the circuit because the current injected by R_{b2} for the most part enters TR1 (the resistance R_{e1} being large compared with the internal emitter resistance r_e of TR1). At the collector of TR1 this current mostly enters TR2 base because R_{c1} is, with proper design, large compared with TR2 input resistance. Thus the feedback current, though injected into TR1 emitter, mostly enters TR2 base and, of course, the feedback current suffers no phase inversion or current gain in TR1. To signals injected at the emitter, TR1 behaves as a common-base amplifier, the base being effectively earthed by the low resistance of the signal source.

Finally it is necessary to add means for d.c. stabilisation and once again we can employ the potential divider circuit of Fig. 6.11 as indicated in Fig. 7.12. The gain of the amplifier is given by R_{b2}/R_{e1}: these resistors constitute a potential divider across the output circuit returning a fraction of the output voltage to TR1 emitter.

Suppose that a voltage gain of 100 is required. The amplifier can be designed to have the same mean collector currents and supply voltage as for the current amplifier. As before TR1 quiescent collector potential can be taken as 10 V which gives R_{c1} as 28 kΩ. TR2 base potential is also 10 V, giving the emitter potential as 9.3 V (silicon transistors are assumed) so that R_1 and R_2 can each be 2.2 kΩ as before. The amplifier is required to give a voltage output and TR2 quiescent collector potential can be 17 V permitting upward and downward swings of 7 V amplitude. This gives R_{c2} as 3.5 kΩ. R_{b2} should not unduly shunt R_{c2} (or the supply!) and a suitable value is 50 kΩ. For a voltage gain of 100, R_{e1} must hence be 500 Ω. The total resistance in TR1 emitter circuit should be 8 kΩ to give the required quiescent emitter potential and R_4 should hence be approximately

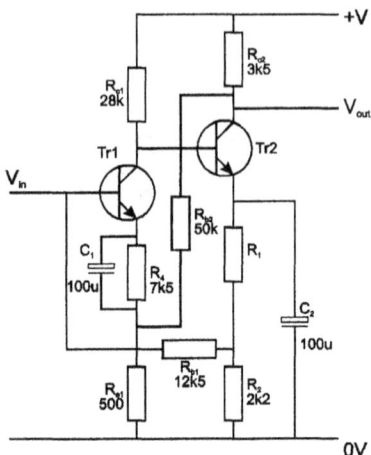

Fig. 7.12. Practical form of the amplifier of Fig. 7.11 designed for a voltage gain of 100

7.5 kΩ. R_1 should preferably be made adjustable (with a range of say 5 kΩ to 10 kΩ) to enable the desired working voltages and currents to be set up. R_{b1} should be small to give good d.c. stability but, as it shunts the input to TR1, it should be large to give a high input resistance to the amplifier. The input resistance of TR1 is given by βR_{e1} (Eqn 7.7), i.e. 50 kΩ if β is 100. To keep the amplifier input resistance above 10 kΩ, R_{b1} should not be less than 12.5 kΩ.

Complementary amplifiers

There are advantages in using complementary transistors in a two-stage amplifier. For example this simplifies the provision of negative feedback as shown in the circuit diagram of Fig. 7.13. The npn common-emitter stage TR1 is direct-coupled to the pnp common-emitter stage TR2. The potential divider $R_{c2}R_{e1}$ is the collector load of TR2 and returns the fraction $R_{e1}/(R_{e1} + R_{c2})$ of the output voltage to TR1 as negative feedback. This parallel-derived series-injected feedback gives the amplifier a high input resistance and a low output resistance so that the amplifier is best suited to voltage amplification. The voltage gain is given by $(R_{c2} + R_{e1})/R_{e1}$ and if R_{c2} is large compared with R_{e1}, as is likely, the gain is given approximately by R_{c2}/R_{e1}. This can be regarded as a complementary form of the amplifier of Fig. 7.11.

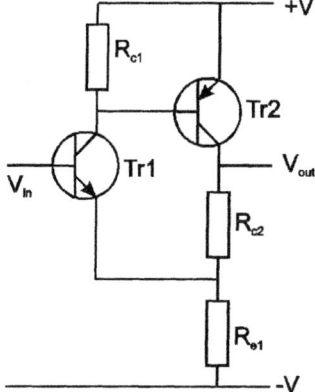

Fig. 7.13. Simple two-stage voltage amplifier using complementary transistors

Emitter follower

An emitter follower is often used as the first stage of a voltage amplifier to provide the required high input resistance. Such a stage is also often employed as the final stage of a voltage amplifier to give a low output resistance.

A typical emitter follower circuit is given in Fig. 7.14 and the resistor values can be estimated in the following way. The aim is to provide a high input resistance and so a good starting point is to make R_1 and R_2 equal and to give them a reasonably high value such as $100\,k\Omega$. This will ensure an input resistance of $50\,k\Omega$. Suppose a 15-V supply is available and that the transistor is a silicon type with $\beta = 100$. If a mean emitter current of $1\,mA$ is assumed then the mean base

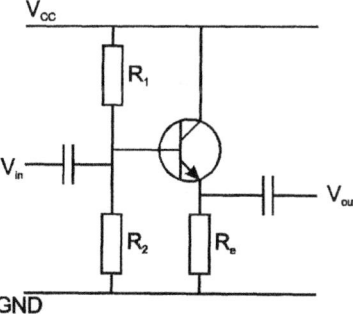

Fig. 7.14. Simple emitter follower circuit

current is $10\,\mu$A and this will flow through R_1 (in addition to the bleed current) giving a voltage drop of $10 \times 10^{-6} \times 10^5$, i.e. 1 volt. The bleed current on its own would give a base voltage of 7.5 V but the base current will reduce this to 6.5 V. Because of the 0.7 V offset voltage in silicon transistors, the emitter voltage is hence 5.8 V and the emitter resistor value is thus $5.8\,\mathrm{k\Omega}$.

With a mean emitter current of 1 mA the transistor has a mutual conductance of approximately 40 mA/V and the output resistance of the circuit is given by $1/g_m$, i.e. $25\,\Omega$.

The stability of the circuit can be calculated from Eqn 6.6 and is approximately 0.08. Though satisfactory this is not as good as can be achieved by the potential divider circuit but is limited in this circuit by the need to keep R_1 and R_2 high to achieve the required high input resistance.

A circuit similar to that of Fig. 7.14 could be used in an emitter follower at the output of an amplifier but it is simpler to use a direct-coupled circuit such as that shown in Fig. 7.15. This is based on the circuit of Fig. 7.12 with an emitter follower stage added. The signal-frequency feedback resistor is taken from TR3 emitter at which the signal voltage is equal to that at TR2 collector.

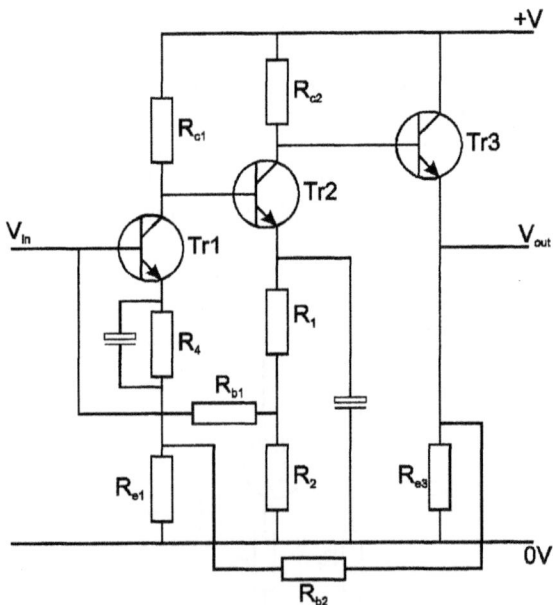

Fig. 7.15. Illustrating the addition of a direct-coupled emitter follower to the output of the circuit of Fig. 7.12

Darlington circuit

A simple and convenient method of connecting two transistors in cascade is the Darlington or super-alpha circuit shown in Fig. 7.16. TR1 is an emitter follower and TR2 a common-emitter amplifier. The collector current of TR1 is in phase with that of TR2 and thus the two collectors can be bonded as shown. The current gain of the emitter follower and the common-emitter amplifier are both approximately equal to β and thus the current gain of the Darlington circuit is given by $\beta_1\beta_2$. The emitter current of TR1 is the base current for TR2: it follows that the collector current of TR2 is β_2 times that of TR1. If TR1 is to have a useful value of β_1 the collector current for TR1 should not be too small. TR2 should preferably therefore have a collector current of at least some tens of mA. This circuit arrangement is well suited for applications where TR2 has to supply appreciable output power.

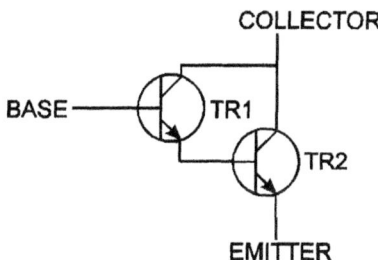

Fig. 7.16. Basic connections for the Darlington or super-alpha circuit

The combination can be regarded as a single transistor with a very high value of β and consequently a high input resistance.

Where more gain is required than is available without undue sacrifice of feedback from the two-stage amplifiers of Figs 7.9 and 7.12 a convenient method of increasing gain is to replace one of the transistors by a Darlington pair. As an example Fig. 7.17 gives the circuit diagram of a current amplifier using a Darlington pair as a second stage. Such an amplifier should easily be capable of a gain of 1,000 whilst still retaining considerable feedback. The values of R_b and R_{e2} should be chosen to give the required value of gain.

Most of the circuits shown so far incorporate npn transistors. A pnp Darlington transistor pair can be implemented as shown in Fig. 7.18. It would be unusual to create a Darlington pair with discrete transistors because it is such a common circuit element that a large range of npn and pnp variants are available in a simple 3-pin package.

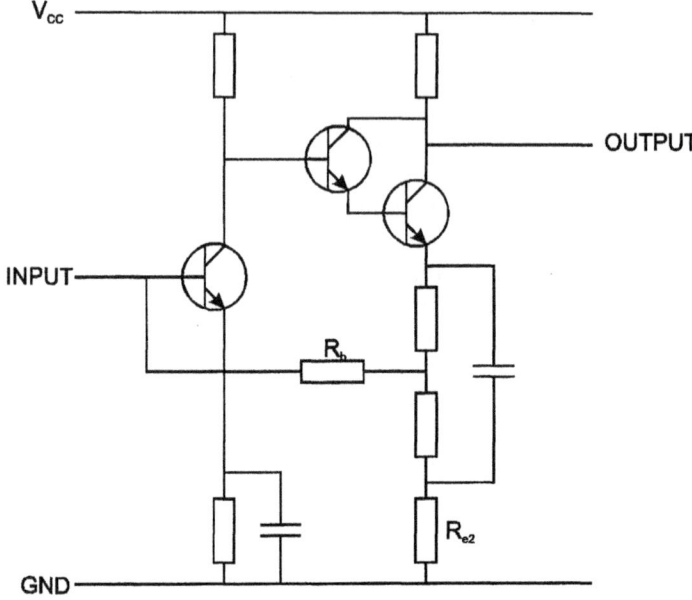

Fig. 7.17. A current amplifier employing a Darlington circuit

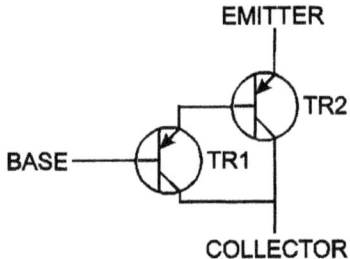

Fig. 7.18. The pnp Darlington transistor pair

Low-noise transistor amplifiers

For certain a.f. amplifiers the input signal is very small and, to obtain a good signal-to-noise ratio at the amplifier output, care must be taken to minimise noise generated in the amplifier itself. An example of such an amplifier is one intended to follow a high-quality microphone. The first stage of an amplifier is, of course, the most likely source of noise and here it is advisable to use a transistor specially selected for use as a low-noise amplifier. It is possible by careful selection of a bipolar transistor and by optimum choice of operating conditions to achieve a noise factor as low as 2 dB. The noise factor is a direct measure of the added noise: thus if the signal-to-noise ratio is 50 dB at the input to an amplifier and 47 dB at the output, the noise factor of the amplifier is 3 dB.

Fets are however preferable for use in low-level stages. Firstly because they use only one type of charge carrier they tend to be quieter than bipolar transistors which employ both types of charge carrier. Jfets are in general quieter than mosfets. Secondly the $I_d - V_g$ characteristics of fets are not linear (they are of square-law shape) so that to minimise distortion the amplitude of the input signal should be kept low. In analogue equipment, therefore, fets should be confined to the early stages where signal levels are lowest.

Circuits employing fet and bipolar transistors

For the reasons just given a fet is often used as the input stage of a.f. equipment and is followed by a bipolar transistor. A typical circuit is illustrated in Fig. 7.19.

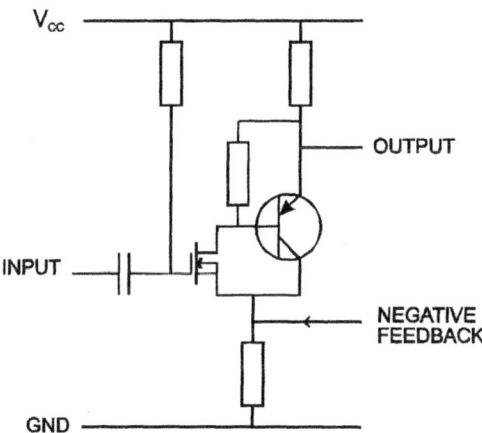

Fig. 7.19. An fet in the input stage of an a.f. amplifier

TR1 is a p-channel enhancement fet used as a common-source amplifier feeding into an npn emitter follower stage TR2. The input resistance is determined by R_1 which can be given a high value such as 1 MΩ to make the amplifier suitable for following crystal pickups or microphones. As shown by the dashed lines the combination of TR1, TR2 and R_2 is available as an integrated circuit.

A similar combination of fet and bipolar techniques is embodied in the igbt described on page 44.

Chapter 8

Large-signal a.f. amplifiers

Definition of a large-signal amplifier

The final stage of an amplifier may be required to drive a loudspeaker, a recording head or some other load requiring appreciable power for its operation. Such stages must give the required power and to obtain it full advantage must be taken of the voltage swing and current swing available from the output transistor(s). Such stages are known as large-signal amplifiers. There is a danger of overloading and one of the problems in designing large-signal amplifiers is how to obtain the maximum power output from the transistors without distortion.

Transistor parameters in large-signal amplifiers

Because of the very large current swings in a large-signal amplifier it is not possible to assume, as in a small-signal amplifier, that the transistor parameters are constant. The input resistance, output resistance and current gain all depend on emitter current and in a large-signal amplifier the variations of these parameters which occur during each cycle of input signal can be very great. Design of such stages must hence be carried out in terms of the mean value of the input resistance, etc., or, better, by evaluation of the extreme values of each parameter and ensuring that the desired performance can be obtained even at the extreme values.

Class-A amplifiers

The circuit of a simple class-A common-emitter stage is given in Fig. 8.1. The load is a.c. coupled to the transistor by a transformer. The

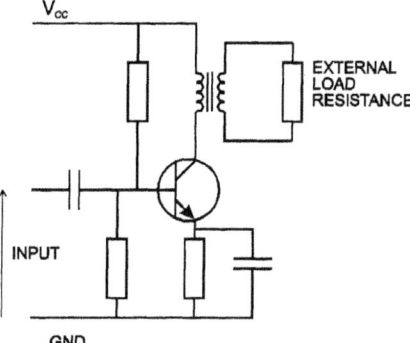

Fig. 8.1. Essential features of a single-ended class-A transistor output stage

most common type of load for a large signal a.f. amplifier are loudspeakers, and these are not tolerant of a constant d.c. current passing through them. The other advantage of the transformer is that it allows matching of the source impedance to the load. The main disadvantage is that the larger the power delivered to the load, the larger and more expensive is the coupling transformer. Capacitive coupling is cheaper, lighter and less bulky. The load line associated with Fig. 8.1 is shown in Fig. 8.2.

The problem is to obtain maximum undistorted power from the transistor without exceeding the maximum safe dissipation prescribed by the manufacturers.

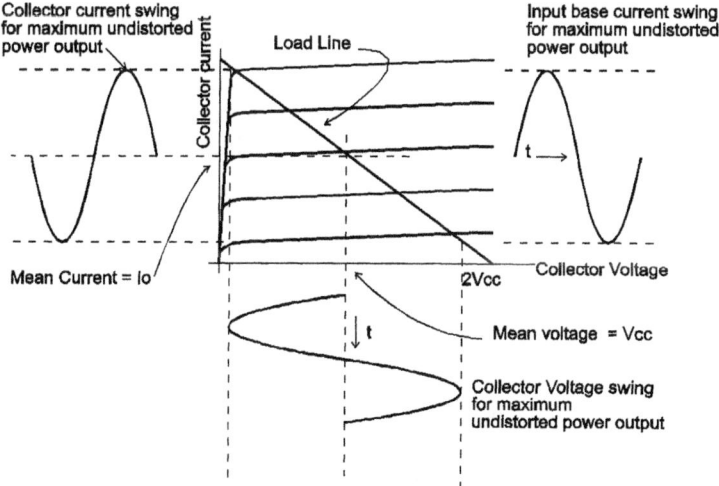

Fig. 8.2. Conditions in a class-A transistor output stage

Efficiency

Suppose the supply voltage is V_{cc} and the steady collector current in the absence of an input signal is I_o. Then the power taken from the supply is $V_{cc}I_o$ and most of this is dissipated as heat in the transistor. This static dissipation must not exceed the maximum value quoted by the manufacturers. If a sinusoidal input signal is applied to the transistor, power is supplied to the load but the power taken from the supply remains constant because no change has taken place in the average or d.c. component of the collector current.

It follows that the power dissipated in the transistor becomes less when the input signal is applied. If, therefore, the transistor does not become too hot in the absence of an input signal, there will be no danger of damaging it by heat when the signal is applied.

In simple generator and load circuits maximum power is delivered to the load when its resistance equals that of the generator. It might be thought, therefore, that maximum power would be obtained from a common-emitter amplifier by making the collector load resistance equal to r_o, the collector a.c. resistance of the transistor (typically several hundred thousand ohms). However, a transistor is not a simple a.c. generator. Its operation depends on a d.c. supply, the voltage of which affects power output. For example, for linear amplification the collector-voltage swing must not exceed the supply voltage and, if the load has a high value, the current swing is limited to a small fraction of what would otherwise be possible. The power output is correspondingly limited. Maximum power output is obtained with a load resistance which makes full use of the voltage and current swings available at the collector: this is known as the optimum load.

When the transistor is delivering its maximum undistorted power, the peak value of the collector voltage is nearly equal to V_{cc} and the peak collector current is I_o. The output power is obtained by multiplying the r.m.s. collector voltage by the r.m.s. collector current: for a sinusoidal output signal these are $V_{cc}/\sqrt{2}$ and $I_o/\sqrt{2}$ and thus

$$\text{maximum power output} = \frac{V_{cc}}{\sqrt{2}} \cdot \frac{I_o}{\sqrt{2}}$$

$$= \frac{V_{cc}I_o}{2}$$

The power taken from the supply is $V_{cc}I_o$ and thus the efficiency is 50 per cent.

This is the theoretical maximum and in practice a transistor class-A stage can approach it very closely. This efficiency applies only for sinusoidal signals and when the transistor is driven to the limit of its output power. For smaller input signals the efficiency is less.

The amplitude of an a.f. signal varies over a range depending on the nature of the signal. For orchestral music the range between the maximum and minimum amplitudes may be as much as 60 dB. For pop music and speech the range is much less.

It can therefore be seen that the efficiency of a class-A amplifier with an a.f. input varies from instant to instant and the average efficiency is in practice much less than the theoretical maximum of 50 per cent.

Optimum load

As can be seen from Fig. 8.2 the slope of the load line for maximum power output is given by V_{cc}/I_o: this is then the value of the optimum load resistance.

As a numerical example, consider a transistor rated for 100 mW maximum collector dissipation and operating as a class-A output stage from a 6-V supply.

The maximum undistorted output power is 50 mW and the mean collector current I_o is given by

$$I_o \times 6 = 100 \times 10^{-3}$$

$$\therefore I_o = \frac{100 \times 10^{-3}}{6} \text{A}$$

$$= 17 \text{ mA}$$

$$R_l = \frac{V_{cc}}{I_o}$$

$$= \frac{6}{17 \times 10^{-3}} \Omega$$

$$= 350 \, \Omega$$

One practical point is that the supply voltage must exceed 6 V to give a 6-V swing of collector potential. This is because the steady voltage across the emitter resistor, that across the transformer primary resistance and the transistor knee voltage must be subtracted from the supply voltage to give the effective collector-emitter voltage.

Output transformer

The output transformer must match the optimum load to the load resistance. Therefore, if the load resistance is 3 Ω, the transformer must have a ratio of

$$\sqrt{(350/3)}{:}1 = \sqrt{117}{:}1$$

$$= 11{:}1, \text{ approximately}$$

The primary inductance determines the low-frequency response which is 3 dB down at the frequency for which the inductive reactance is equal to the optimum load. In an a.f. amplifier the 3 dB loss frequency may be 50 Hz and we have

$$2\pi f L = R$$

$$L = \frac{R}{2\pi f}$$

$$= \frac{350}{6.28 \times 50}\text{henrys}$$

$$= 1.1 \text{ henrys}$$

The transformer should have a primary inductance of this value with 17 mA direct current flowing.

Push-pull operation

In a push-pull amplifier output stage there are two active devices that could be valves; mosfets or bipolar junction transistors. They could be the same type of device or complementary (i.e. 1 npn or 1 pnp). In the example of Fig. 8.3, assume that the audio amplifier is driving a loudspeaker and that it operates from a single power supply. The active devices are bipolar Darlington transistors. The ideal voltage at point X in Fig. 8.3 is 15 V, and C would undoubtedly be an electrolytic capacitor mounted with the polarity as shown. In normal use, TR1 and TR2 would both be conducting, maintaining $V_{CE(Q1)} = 15\,\text{V}$ and $V_{CE(Q2)} = 15\,\text{V}$. This means that the d.c. bias voltages would need to be:

$$V_{B1} = 16.4\,\text{V} \qquad V_E = 15\,\text{V} \qquad V_{BE} = 1.4\,\text{V}$$

$$V_{B2} = 13.6\,\text{V} \qquad V_E = 15\,\text{V} \qquad V_{BE} = 1.4\,\text{V}$$

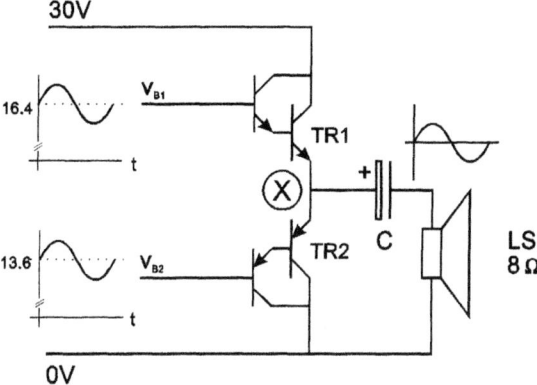

Fig. 8.3. A complementary class-A push-pull output stage

When the signal goes positive, V_{B1} and V_{B2} waveforms alter as shown. To remain as a Class A amplifier, both TR1 and TR2 must conduct continually, i.e. never being cut-off. $V_{CE(Q1)}$ reduces while $V_{CE(Q2)}$ increases for a positive change of output voltage, and $V_{CE(Q1)}$ increases while $V_{CE(Q2)}$ reduces for a negative change of output voltage. The d.c. voltage supply to a class A amplifier remains essentially constant, irrespective of the output power being delivered to the load.

Another configuration uses the same type of transistor (npn in the case of Fig. 8.4). TR1 is a non-inverting emitter follower, whilst TR2 is an inverting common emitter amplifier. These devices are now of the same polarity and a different tactic for driving them is required. Transistor TR3 is called a *phase splitter*. If R1 equals R2, then collector and emitter voltages are equal and opposite. When TR1 is 'driven on' more; TR2 is 'driven off' more. This corresponds to the output at its first positive half

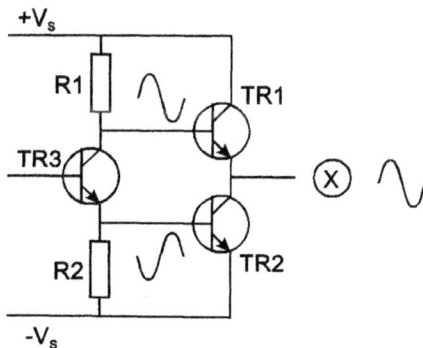

Fig. 8.4. Class-A output with phase splitter

cycle as shown in the waveforms. This is different from the drive circuit needed for complementary output transistors in Fig. 8.3.

In both of these circuits, references to 'driving on more' and 'driving off more' imply a small change of base-emitter voltage bias, which alters the value of V_{CE} to change the output voltage.

Class-B amplifiers

In a class-B amplifier the base is biased almost to the point of collector current cut-off. In the absence of an input signal, therefore, very little collector current flows but the current increases as the amplitude of the input signal increases. This leads to economy in running costs because the current taken from the supply tends to be proportional to the signal amplitude and not independent of it as in a class-A amplifier. Alternate half-cycles of the input signal are, however, not reproduced in a single class-B amplifier and it is essential to use two transistors in push-pull. Such a pair of transistors can give a power output of 2.5 times the maximum permissible collector dissipation of the two transistors: this is 5 times the power output available from the same two transistors operating in class-A push-pull. This may be shown in the following way. Let the peak collector current of each transistor be I_p and the supply voltage V_{cc}. Then the peak collector-voltage swing V_p is equal to V_{cc}. The power output from a push-pull stage is given by the product of the r.m.s. collector current $(I_p/\sqrt{2})$ and the r.m.s. collector voltage $(V_p/\sqrt{2})$ and is thus given by $I_p V_p/2$. If the input signal is sinusoidal, the combined collector current is sinusoidal in waveform as pictured in Fig. 8.5. Such a waveform has a mean value, or d.c. component, equal to $2/\pi$ of the peak value (I_p) and the power taken from the supply is thus equal to $2 V_{cc} I_p/\pi$, giving efficiency as

$$\frac{\text{power output from transistors}}{\text{power taken from supply}} = \frac{I_p V_p}{2} \cdot \frac{\pi}{2 I_p V_p}$$

$$= \frac{\pi}{4}$$

$$= 78.54 \text{ per cent}$$

This is the maximum efficiency of which the stage is capable and to obtain it the optimum load per transistor must be equal to (peak voltage)/(peak current). For class-B push-pull the collector-to-collector load is given by $4 V_p/I_p$.

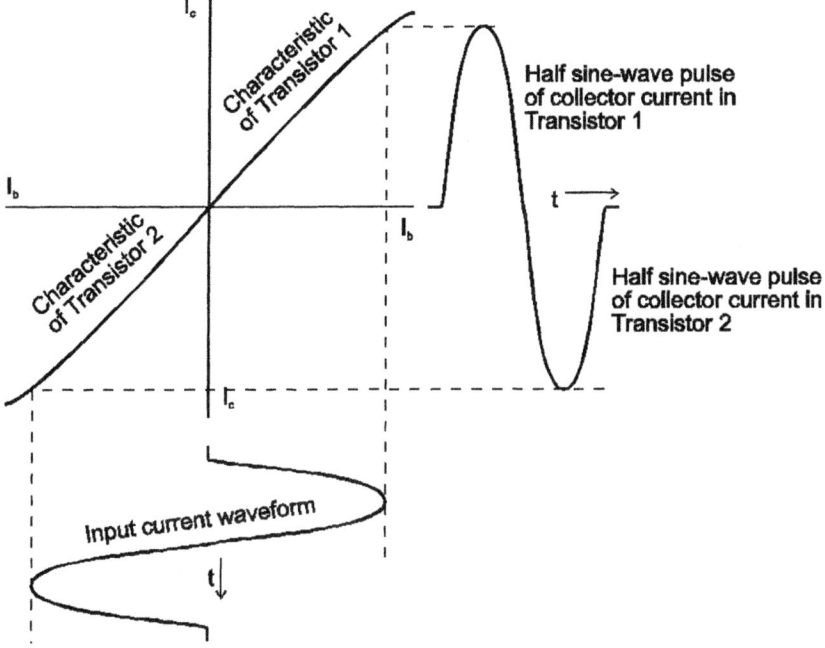

Fig. 8.5. Operation of a push-pull class-B transistor amplifier

If the load and supply voltage are kept constant and the input signal amplitude is reduced the efficiency falls linearly due to the increasing failure to make use of the voltage swing available. This is easily shown. Suppose the input voltage is reduced to a times the value which gives maximum output. Then the collector-current swing falls to aI_p and (since the load is constant) the collector-voltage swing falls to aV_p, giving the power output per pair of transistors as $a^2 I_p V_p/2$. The mean value of the collector current is now $2aI_p/\pi$ and the power taken from the supply is $2aI_p V_p/\pi$. The efficiency for the reduced input signal is given by

$$\frac{\text{power output from transistors}}{\text{power taken from supply}} = \frac{a^2 I_p V_p}{2} \cdot \frac{\pi}{2aI_p V_p}$$

$$= a \cdot \frac{\pi}{4}$$

The efficiency is thus directly proportional to the input signal amplitude.

In transistor amplifiers we are particularly interested in the power P_t dissipated in the transistors themselves, for it is this which causes heating of the junctions and can damage them. Now

$$\begin{matrix} \text{power dissipated} \\ \text{in transistors} \end{matrix} = \begin{matrix} \text{power taken} \\ \text{from supply} \end{matrix} - \text{power output}$$

$$\text{i.e.} \quad P_t \quad = \quad \frac{2aI_pV_p}{\pi} \quad - \quad \frac{a^2I_pV_p}{2}$$

Differentiating this

$$\frac{dP_t}{da} = \frac{2I_pV_p}{\pi} - aI_pV_p$$

Equating this to zero to find the maximum,

$$a = \frac{2}{\pi}$$

The heat in the transistors is thus a maximum when the signal amplitude is $2/\pi$ (approximately 0.63) times that giving maximum output power. Substituting this particular value of a in the general expression given above we have

$$\text{power output from amplifier} = \frac{a^2I_pV_p}{2}$$

$$= \frac{2I_pV_p}{\pi^2}$$

$$\text{power taken from supply} = \frac{2aI_pV_p}{\pi}$$

$$= \frac{4I_pV_p}{\pi^2}$$

By subtraction,

$$\text{power dissipated in transistors} = \frac{2I_pV_p}{\pi^2}$$

These results show that this particular value of a makes the power output one-half that taken from the supply: in other words it coincides with an efficiency of 50 per cent. The maximum power output from the class-B pair is $I_p V_p/2$. The maximum power dissipated in the transistors is $2I_p V_p/\pi^2$. The ratio of these two quantities is $\pi^2/4$, approximately 2.5:1, showing that it is possible to obtain an undistorted output of 2.5 times the rated maximum collector dissipation of the two transistors, i.e. 5 times the maximum collector dissipation of one of them. This applies only for a sinusoidal input.

As a numerical example, consider two transistors each with a maximum collector dissipation of 100 mW. In class-B push-pull it is possible to obtain from these an output power of 500 mW. If the supply voltage is 6 V, this is also the peak value of the collector voltage, and the peak current is given by I_p where

$$P = \frac{1}{2}V_p I_p$$

$$\therefore I_p = \frac{2P}{V_p}$$

$$= \frac{2 \times 500}{6} \text{ mA}$$

$$= 170 \text{ mA approximately}$$

The collector-to-collector load is given by

$$R_l = \frac{4V_p}{I_p}$$

$$= \frac{4 \times 6}{170 \times 10^{-3}} \text{ } \Omega$$

$$= 140 \text{ } \Omega \text{ approximately}$$

Driver stage

The transistor feeding a class-B output stage may be regarded as a large-signal amplifier because it must supply appreciable power to the base circuits of the output stage. The design of the driver stage and the transformer coupling it to the output stage depends on the input resistance of the output stage.

It was mentioned at the beginning of this chapter that the transistor parameters cannot be taken as constant in large-signal amplifiers. This is particularly true of the input resistance of a class-B amplifier. Such an amplifier is biased almost to cut off of collector current in the absence of an input signal and its input resistance is then high. On peaks of input signal, however, the amplifier is driven to large collector currents and at these instants the input resistance is low. The variation may be from, say, $2\,k\Omega$ for small signals to less than $100\,\Omega$ for large signals.

We have seen in earlier chapters that to obtain an undistorted input current in such circumstances the transistors must be driven from a high-resistance source. If too low a source resistance is used, the input current tends to be too low at low values of input. Thus a sine-wave input signal is reproduced with a waveform similar to that shown in Fig. 8.6. This shows symmetrical distortion: such a waveform is typical of odd-order harmonic distortion, i.e. 3rd, 5th, 7th, etc., and unlike even-harmonic distortion is not cancelled out in a push-pull stage. Fig. 8.6 represents the input and output waveforms for an amplifier giving *cross-over distortion*, so called because it occurs when one transistor is being cut off and the other turned on, i.e. when the state of conduction is being transferred from one transistor to the other.

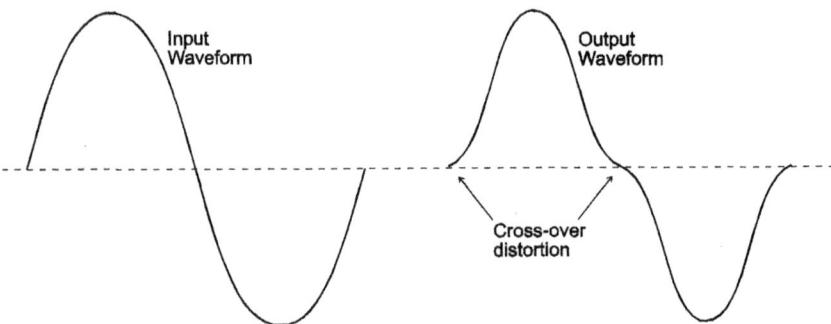

Fig. 8.6. Input and output waveforms for an amplifier giving cross-over distortion

Unfortunately transistors giving appreciable power output often have significant fall off in current gain at high collector currents and this can cause distortion of the output when the signal source is of high resistance. To minimise distortion due to this cause the source resistance should not be too great: on the other hand it should not be too low otherwise cross-over distortion occurs. Between these limits there is a range of source resistance which is satisfactory. Suitable values for a particular class-B amplifier are quoted below.

For transistors rated for $100\,mW$ maximum collector dissipation, the input resistance falls to approximately $100\,\Omega$ at minimum and is probably around $1\,k\Omega$ for a standing collector current of $2\,mA$, a typical value for a class-B output stage for a portable receiver. If the driver stage operates in class-A with 3-mA mean collector current then for a 6-V supply the optimum load is approximately 2 kΩ. If this is accurately matched to 100 Ω we can ensure that the driver delivers maximum power to the output stage just when it is needed. The turns ratio required is given by $\sqrt{(2{,}000/100)}{:}1 = 4.5{:}1$. The effective source resistance (if the driver output resistance is 30 kΩ) is then 1.5 kΩ which is certainly not large compared with the maximum input resistance (1 kΩ) of the output stage for small inputs. Any cross-over distortion which occurs can probably be minimised by reducing the transformer turns ratio to 3:1 or even 2:1. For 2:1 the effective source resistance is 7.5 kΩ.

Bias circuits for class-B operation

The circuit of Fig. 8.7 shows one simple method of biasing the circuit for class-B operation. D_1 provides bias for Q_1; D_2 provides bias for Q_2. A disadvantage occurs when the diode forward voltage drop V_F does not exactly match the transistor V_{BE}. The commonest solution is to replace D_1 and D_2 with the configuration called a V_{BE} multiplier or, more simply, 'a rubber Zener'. This is the circuit in Fig. 8.8.

Setting up is normally achieved with a potentiometer as shown in Fig. 8.8(a) but, for ease of calculation, use Fig. 8.8(b):

$$V_{BE} = 0.6v \quad \text{and} \quad I_{R_2} = \frac{V_{BE}}{R_2}$$

Assuming that I_B is negligible,

$$I_{R_2} = I_{R_1}$$

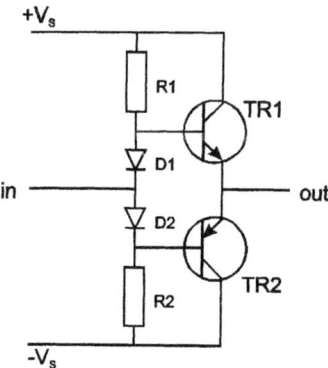

Fig. 8.7. Diode biasing for class-B operation

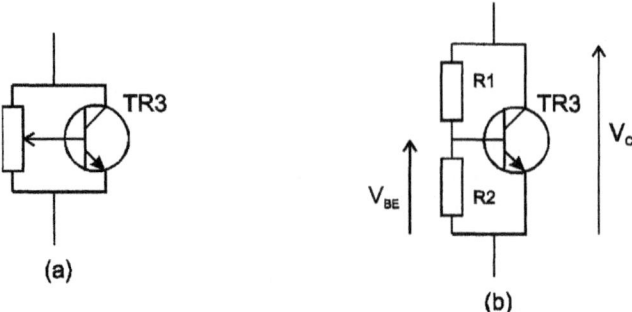

(a)

(b)

Fig. 8.8. Bias circuit for class AB

so $\quad V_{CE} = I_{R_2} \times (R_1 + R_2)$

$$V_{CE} = V_{BE} \times \frac{R_1 + R_2}{R_2}$$

V_{CE} can be adjusted by the potentiometer, its stability being set by the stability of V_{BE}.

Complementary class-B amplifier

The output stage of class-B amplifiers can use the same type of transistor (say, both npn) to give what is called an asymmetrical output stage. The alternative is to use complementary transistors (npn and pnp) to give a symmetrical output stage, as has been suggested in the circuit diagram in Fig. 8.7. A practical circuit is shown in Fig. 8.9.

Although initially daunting to look at, this 1982 complementary mosfet output stage (by Peter Wilson of International Rectifier Ltd) shows all of the details presented so far. The complementary outputs formed by TR1 and TR2 are connected to either side of the V_{BE} multiplier TR5. This, in turn, is in the collector circuit of the class-A driver TR6. Transistors TR3 and TR4 form a differential amplifier, which compares the input with the output. The input is connected to TR3, and TR4 handles the feedback signal. The base of TR4 receives only a proportion of the output as determined by R_6 and R_7. These resistors determine the gain of the amplifier (average approximately 22).

This circuit has the advantage that no phase splitter is necessary. A positive-going signal applied to the base of the pnp transistor reduces its collector current and at the base of the npn transistor increases

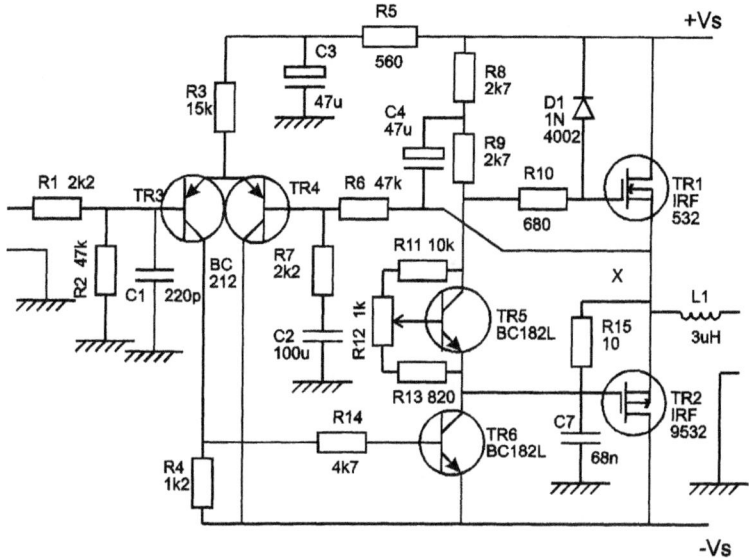

Fig. 8.9. A complete amplifier incorporating a complementary mosfet output stage

collector current. The pnp transistor is in fact turned off and the npn transistor is turned on: thus the potential of their common connection point moves towards that of the positive supply terminal. Similarly a negative-going input signal causes the potential of the common point to move towards the negative supply terminal. Both transistors contribute towards the output voltage and push-pull action is obtained by applying the same input signal to both bases. This basic circuit has something in common with the complementary emitter follower of Fig. 9.13 but in Fig. 8.7 the transistors are used as common-emitter amplifiers by applying the common input between base and emitter of both transistors.

An example of an asymmetric output using two npn transistors and complementary drivers is shown in Fig. 8.10. TR3 is an emitter follower: its output is in phase with its input and is connected directly between base and emitter of TR5. TR4 has the same input as TR3 but is a common-emitter amplifier: its output is in antiphase with the input and is connected directly between base and emitter of TR6. We have seen earlier that the current gain of the emitter follower and the common-emitter amplifier are both approximately equal to β. Thus by choosing complementary transistors with equal βs for TR3 and TR4, and by making R_{10} equal to R_{12} we can arrange for TR5 and TR6 to receive equal-amplitude but antiphase signals so that the output stage operates in push-pull.

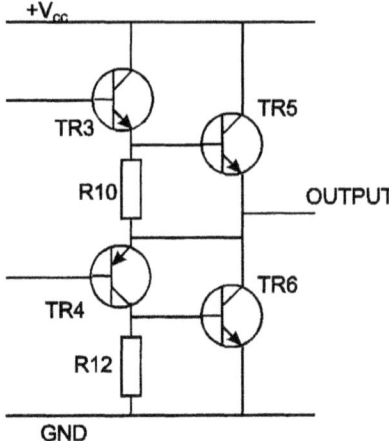

Fig. 8.10. An asymmetric output using two npn transistors and complementary drivers

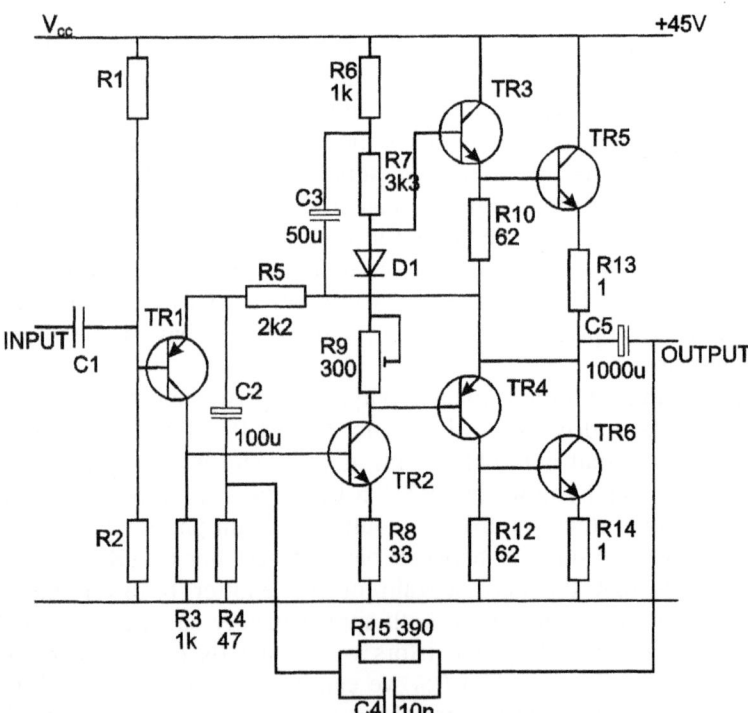

Fig. 8.11. Circuit diagram of an amplifier capable of more than 10 W output

A technique commonly employed to drive TR3 and TR4 is indicated in Fig. 8.11 which gives the circuit diagram with most component values for an amplifier capable of supplying an output of the order of 10 W. TR2 is a common-emitter pnp stage, R_7 is its collector load and R_6 a decoupling resistor. The decoupling capacitor C_3 is, however, returned to the emitters of TR3 and TR4: this ensures that the signals generated by TR2 are applied between base and emitter of TR3 and TR4 as required. R_9 determines the standing voltages on the bases of TR3 and TR4 which in turn control the collector current of the output stage: R_9 is adjusted to give a standing current in TR5 and TR6 sufficient to minimise cross-over distortion. The forward-biased diode D1 in conjunction with the emitter resistors R_{13} and R_{14} stabilise the quiescent current of the output stage as explained in Chapter 6.

TR1 is a common-emitter npn stage used for signal amplification, the collector load R_3 being connected directly between TR2 base and emitter. TR1 is also used as a d.c. amplifier to ensure stability of the quiescent voltage at the amplifier output point P. This voltage must be accurately maintained at $V_{cc}/2$ where V_{cc} is the supply voltage, otherwise the amplifier cannot deliver its maximum output without distortion. The potential at point P is applied to TR1 emitter via R_5, a.f. signals being attenuated by C_2. TR1 base is fed from the potential divider R_1R_2 which provides a constant reference potential. Any variation in the quiescent voltage at point P is amplified by TR1 and the subsequent stages, all of which are direct coupled, so as to minimise the change in potential at P.

Signal-frequency negative feedback is provided by R_{15} and R_4 which return a fraction of approximately 1/9th of the output voltage to TR1 emitter. This means that the overall voltage gain of the amplifier is approximately 9. As the maximum excursion of output voltage is 22.5 V (i.e. $V_{cc}/2$) the input required is approximately 2.5 V peak. C_4 corrects the negative feedback for phase shifts in the amplifier at high frequencies and so avoids a peak in the frequency response curve or possible instability at or above the upper limit of the passband.

In general the maximum peak output voltage is $V_{cc}/2$ and thus the r.m.s. voltage is $V_{cc}/2\sqrt{2}$. If the load resistance is R_l the maximum power output is given by V^2/R_l, i.e.

$$P_{max} = \left(\frac{V_{cc}}{2\sqrt{2}}\right)^2 \cdot \frac{1}{R_l}$$

$$= \frac{V_{cc}^{\ 2}}{8R_l}$$

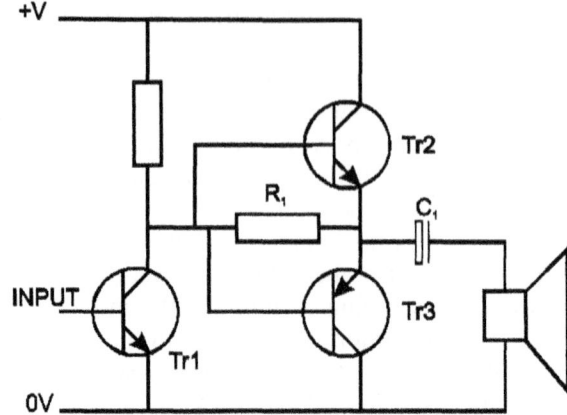

Fig. 8.12. Principle of a current-dumping amplifier

For $V_{cc} = 45$ and $R_l = 15$ we have

$$P_{max} = \frac{45^2}{8 \times 15} \text{ W}$$

$$= 17 \text{ W}$$

The current in the load for maximum output is 1 A r.m.s. and the output transistors should be capable of supplying this without too great a fall in β.

This circuit has been widely used for high-fidelity and public address amplifiers where large power output is required. Direct coupling is used throughout and there are no transformers. As a result, considerable feedback can be used and it is possible to reduce the total harmonic distortion to less than 1%. The circuit was first described by Tobey and Dinsdale.*

Current-dumping amplifiers

In conventional push-pull transistor a.f. output stages the quiescent current in the output pair must be carefully controlled to minimise cross-over distortion. In a high-power stage it can be difficult to stabilise the quiescent current against temperature changes. These difficulties can be overcome by use of the so-called current-dumping technique. A skeleton circuit is shown in Fig. 8.12 and uses a feed-forward system in which a class-A amplifier (or an i.c.) feeds the loudspeaker via resistor R_1 for

* Tobey, R. and Dinsdale, J. 'Transistor Audio Power Amplifiers', *Wireless World*, 67, No. 11, 565 (1961).

small input signals. As the signal level increases, a point is reached on the positive half-cycles where the signal voltage generated across R_1 is sufficient to drive TR2 into conduction: this then feeds current into the loudspeaker via coupling capacitor C_1. At the same time TR3 is driven into conduction on the negative half-cycles. Thus TR2 and TR3 operate in class B at high signal levels. In fact the two transistors form a complementary emitter follower (see page 173) and the 100 per cent negative feedback implicit in the circuit ensures low distortion on high signal levels.

The advantages of this type of circuit, which has also been used in the field output stages of television receivers, are good linearity, no need for closely matched output transistors, good stability and low power consumption.

Bridge output stage

The use of four transistors in a bridge circuit of the type shown in Fig. 8.13 eliminates the need for loudspeaker coupling capacitors, permitting the low-frequency response to extend to d.c. This type of circuit greatly reduces the effect of supply-line hum on the signal and can be used for output powers up to hundreds of watts. Its disadvantage is that any bias or semiconductor failure can cause large direct currents to flow in the loudspeaker, necessitating elaborate protection circuits to

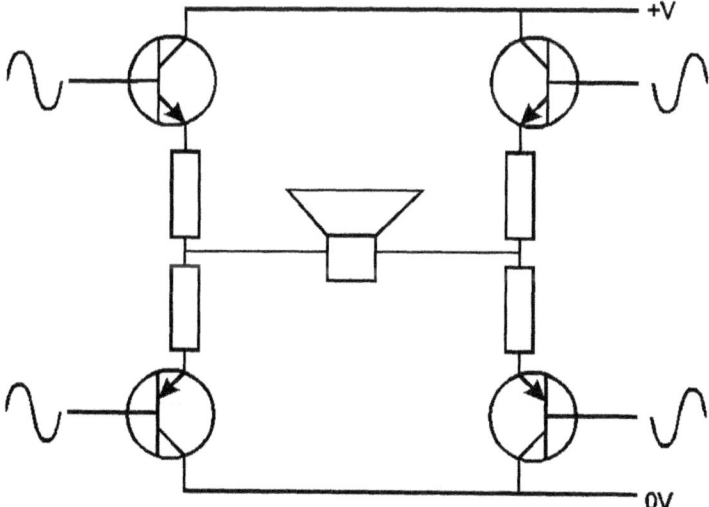

Fig. 8.13. Bridge circuit configuration for a power amplifier

remove the supply voltage whenever a steady out-of-balance current flows through the loudspeaker. The same bridge configuration can be used with i.c.s as described below.

Integrated-circuit a.f. power amplifier

Fig. 8.14 gives the circuit diagram of a monolithic complementary push-pull amplifier capable of 1 W output and designed to operate with a supply balanced about earth potential. TR7 and TR8 form a complementary output pair which, unlike the amplifier just described, share a common-collector connection, the emitters being connected to the terminals of the supply. Because of the small voltage drop across the base-emitter path of transistors, there is a considerable difference in voltage between the bases of TR7 and TR8. This voltage is used as the collector supply for the complementary drivers TR5 and TR6 which are direct coupled to the output stage and have no load resistors. TR5 and TR6 have a common-emitter connection at earth potential and the small difference in voltage between the bases is provided by the voltage drop

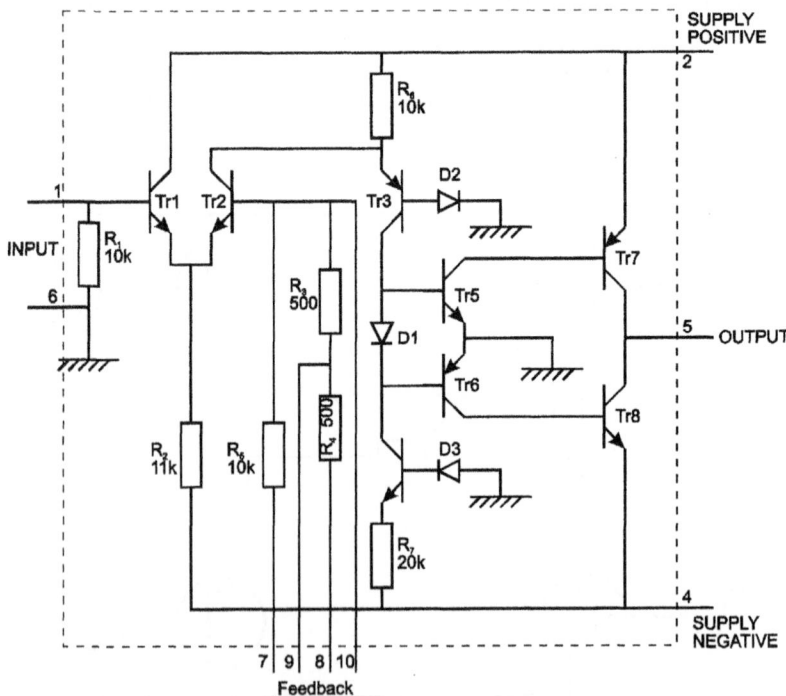

Fig. 8.14. Circuit diagram of a monolithic a.f. power amplifier

across the forward-biased diode D1 which, together with D2, D3 and the transistors TR3 and TR4, ensures d.c. stability in the amplifier.

TR5 and TR6 are driven in phase by the common-base stage TR3 which is fed from the long-tailed pair TR1 and TR2. The advantages of this circuit are discussed in the next chapter. Negative feedback is applied to the base of TR2 by a potential divider across the output terminals of the amplifier. The upper arm of the potential divider is R_5 ($10\,k\Omega$) which is connected into circuit by strapping terminals 5 and 7: the lower arm can be made $1\,k\Omega$ by strapping 8 to earth, $500\,\Omega$ by strapping 9 to earth or $250\,\Omega$ by strapping 8 and 10 and earthing 9. Thus there are three possible values of voltage gain and from inspection of the component values we can see that these are approximately 10, 20 or 40.

The amplifier has an input resistance of approximately $8.5\,k\Omega$ and an output resistance of approximately $0.5\,\Omega$. A supply of $\pm6\,V$ can be used and $1\,W$ can be delivered to a $16\text{-}\Omega$ load. The amplifier is contained in a can about $\frac{1}{3}$-in in diameter which should preferably be cooled by a heat sink.

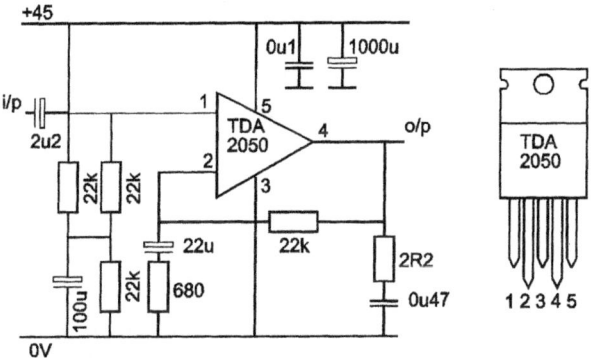

Fig. 8.15. 25-W integrated circuit power amplifier

Figure 8.15 shows the application circuit of an integrated-circuit power amplifier. This is capable of 25 W output over a 20 kHz bandwidth. The distortion is only 0.05% at 15 W.

Heat sinks

As we have seen, transistors used in output stages are often required to dissipate appreciable power in the transistors themselves. In a class-A stage this internal dissipation is a maximum at the instants when there is no input signal: the power given by the product of the standing collector current and the collector-emitter voltage is then entirely dissipated in the transistor. In a class-B stage with a sinusoidal input signal, dissipation in the transistors is a maximum when the input-signal amplitude is $2/\pi$ (approximately 0.63) of that for maximum power output.

With such dissipations there is a real danger of thermal runaway and it is advisable to provide some method of removing heat from the transistors. The usual method is to mount the transistors on a sheet of metal, known as a heat sink, which should have adequate thermal capacity (mass $\times$ specific heat) and the transistors are usually so designed that it is easily possible to bring the collector electrode into good thermal contact with the heat sink.

The primary limitation on the power which can safely be dissipated in a transistor is set by the temperature within the semiconducting material of the transistor. This is known as the collector junction temperature T_j and its maximum value is decided after life tests on the transistor. Operation of the transistor above this maximum value may seriously damage or even destroy the transistor. The circuit designer is not required to measure the junction temperature but it is helpful if he can estimate it. A method of doing this is suggested below.

The junction temperature of a transistor is always higher than that of the air immediately adjacent to the transistor case (the ambient temperature T_{amb}) the difference being proportional to the power dissipated in the transistor. Thus

$$T_j - T_{amb} = P_c \times \text{a constant}$$

The constant has the dimensions of a thermal resistance (the reciprocal of thermal conductivity) and is commonly represented by θ. It may be defined as the temperature rise for unit dissipation and is commonly expressed in °C per mW.

$$T_j = T_{amb} + P_c\theta$$

θ depends on the conductivity between the collector electrode and the external surrounds and is thus a function of the physical construction of the transistor. For a given transistor θ and T_j are constant and the maximum power dissipation decreases linearly with rise in ambient temperature.

For small transistors θ depends almost entirely on the conductivity of the material between the semiconducting material and the case. For such transistors there is little reduction in θ by mounting the case on a heat

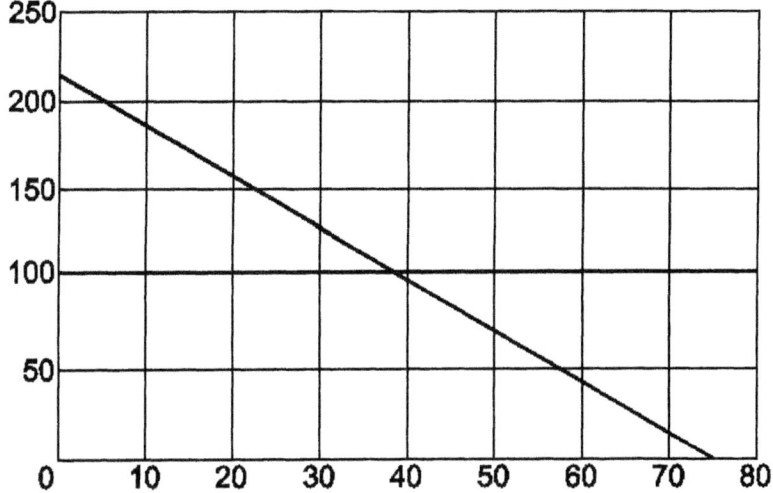

Fig. 8.16. Relationship between P_c and T_{amb} for a transistor for which $T = 75°C$ and $\theta = 0.35°C/mW$

sink and cooling is by convection through the air. A typical value for θ for such transistors is $0.35°C/mW$ and if we take T_j as $75°C$ we can plot P_c against T_{amb} as shown in Fig. 8.16. This is a useful curve for it enables us to calculate the power dissipation permissible at any given value of ambient temperature. At $20°C$, for example, the dissipation is $160\,mW$ but at $40°C$ it has fallen to $100\,mW$.

For transistors intended for larger power outputs the collector electrode is in direct metallic contact with the case which is designed for bolting to a large block of metal such as a chassis to act as a heat sink. This is an example of conduction cooling. The thermal resistance between collector and case of such transistors is very low and the overall thermal resistance depends very largely on the properties of the heat sink.

The total or effective thermal resistance between collector junction and the outside air is made up of the following components:

1 the thermal resistance θ_1 between collector junction and transistor cases: a typical value is $1.2°C/W$;
2 the thermal resistance θ_2 across the mica washer which is generally used between case and heat sink to isolate the case electrically from the heat sink: a typical value is $0.5°C/W$;
3 the thermal resistance θ_3 of the heat sink itself. This depends on the mass and area of the sink.

The total thermal resistance is given by

$$\theta = \theta_1 + \theta_2 + \theta_3$$
$$= 1.7 + \theta_3$$

Suppose, as a numerical example, a dissipation of 6 W is required, that T_j is limited to 80°C and that the ambient temperature is 50°C. These values might be encountered in the design of the output stage of a car radio. From the expression

$$T_j = T_{amb} + \theta \cdot P_c$$

we have

$$\theta = \frac{T_j - T_{amb}}{P_c}$$

But $\theta = 1.7 + \theta_3$

$$\therefore \theta_3 = \frac{T_j - T_{amb}}{P_c} - 1.7 = \frac{80 - 50}{6} - 1.7$$

$$= 3.3°C/W$$

To give a thermal resistance of this order, the heat sink can take the form of a sheet of aluminium about 40 in² in area and at least $\frac{1}{16}$ in thick. If the equipment has a chassis this may satisfy these requirements and the output transistor can be mounted on it. If, however, the equipment has printed wiring and requires no chassis, then a separate heat sink is necessary: the size of the sink can be reduced by providing it with cooling fins to give the required area.

Temperatures at the various junctions in the thermal circuit, e.g. at the junction between chassis and washer or between washer and transistor case, can be calculated in the following manner. As already shown the thermal resistance of the heat sink is 3.3°C/mW. The power transmitted through it is 6 W and thus from the general expression

temp. difference = power × thermal resistance

we have

temp. difference = 6 × 3.3

$$= 20°C \text{ approximately}$$

The temperature of the air is 50°C, that of the chassis at the point of contact with the washer will be 50 + 20 = 70°C. For the mica washer itself the temperature difference across it is given by 6 × 0.5 = 3°C giving the temperature at the junction between washer and transistor case as 73°C.

Darlington power transistors

An output transistor supplying considerable power such as 100 W requires a large heat sink. The transistor driving it is likely to have a dissipation of several watts and it, too, requires a heat sink. The space occupied by the two heat sinks can be reduced by using a single sink carrying an integrated circuit which consists of the output transistor and driver arranged in a Darlington circuit (p. 131) on a single silicon slice and within a single encapsulation. Such an i.c. can be treated as a single power transistor with very high current gain (2,000 is common).

Chapter 9

D.C. and pulse amplifiers

Introduction

The fundamental circuits of Figs. 7.7 and 7.10 for two-stage current and voltage amplifiers are also used for d.c. and pulse amplification. Voltage amplifiers often include an emitter-follower stage to give a high input resistance or a low output resistance, and negative feedback is commonly applied over the three stages.

D.C. amplifiers

These are amplifiers used for magnifying small voltages or currents so that slow variations in them can be more readily seen and measured. To ensure a good response at the very low frequencies of interest the amplifiers are usually direct-coupled.

D.C. amplifiers are used in electro-cardiographs, computers, oscilloscopes, pyrometers and in sensitive measuring instruments such as milli-microammeters.

Direct-coupled cascaded circuit

A simple d.c. amplifier could be constructed as suggested in Fig. 9.1 by using a number of transistors in cascade, each collector being direct-coupled to the base of the following transistor. Such a circuit would, however, suffer from the following serious limitations:

1 Any variations in the collector current of any of the transistors not caused by an input signal will, of course, give rise to a spurious output from the amplifier. Clearly variations in the first transistor are the most important because they are amplified by the two following

transistors. Now the collector current of a transistor is given by βI_b and one possible cause of unwanted variations in I_c is drift in the value of I_b. Such drift occurs and is particularly serious when the external base circuit is of low resistance for there is then no means of defining I_b. Spurious output from this cause can be minimised by arranging for the input signal for each transistor to be supplied from a high-resistance source: the standing base current is then largely determined by the external source voltage and resistance.

2 The gain of a common-emitter stage is determined by the value of β which is itself dependent on mean emitter current and on temperature. The gain of a simple amplifier such as that illustrated in Fig. 9.1 thus depends on the magnitude of the input signal with the result that the output input characteristic of the amplifier is non-linear. By sacrificing sensitivity the characteristic can be made as linear as desired by use of negative feedback.

Provided all the precautions mentioned under (1) and (2) above are taken a circuit based on Fig. 9.1 can be made to give a satisfactory performance. A suitable circuit diagram is given in Fig. 9.2. The emitter bias voltages for TR2 and TR3 must be steady because changes in these would be interpreted by the transistors as input signals. Resistors, if used in each emitter circuit to give the required emitter voltage, would reduce gain by introducing negative feed-back. Individual batteries could be used but this is hardly a satisfactory arrangement because changes in battery voltage and internal resistance with age would upset operation.

A solution to the problem is to use voltage reference diodes with suitable voltage ratings in the emitter circuits. The voltage developed across such a diode is independent of the current in it: thus the diode behaves as a battery with constant e.m.f. and negligible internal

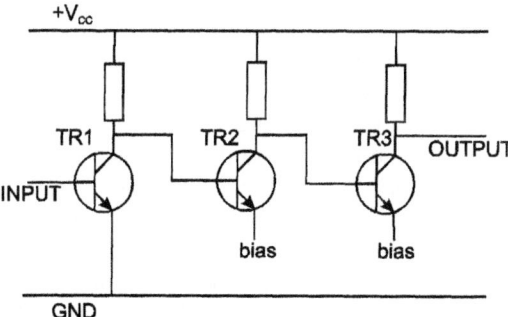

Fig. 9.1. Basic circuit for a three-stage d.c. amplifier. The performance is unsatisfactory for reasons given in the text

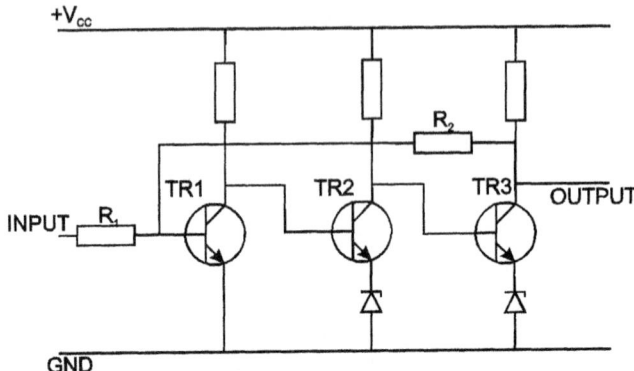

Fig. 9.2. This modification to the circuit of Fig. 9.1 is satisfactory

resistance. There is minimal loss of signal amplification by negative feedback due to the diodes in the emitter circuits.

To give the required gain and linearity, overall negative feedback can be applied to the amplifier by a resistor R_2 connected between TR3 collector and TR1 base. The collector resistors for TR1 and TR2 must be high in value to minimise drift in TR2 and TR3, and this necessitates a fairly high supply voltage. Drift in TR1 is minimised by introducing a high-value series resistor R_1. The feedback circuit is of the type used in the basic current amplifier of Fig. 7.8 and we can therefore say immediately that the current gain from TR1 base to TR3 collector is given by R_2/R_3. Alternatively if the source of input voltage v_{in} is assumed to be of low resistance we can say that R_1 and R_2 constitute a feedback potential divider connected across. The output voltage v_{out} and the voltage gain v_{out}/v_{in} is given by R_2/R_1.

Differential amplifier (long-tailed pair)

An alternative method of minimising drift in direct-coupled amplifiers is to use two similar transistors in a balanced circuit such as that illustrated in Fig. 9.3. This configuration was seen in Chapter 8 where the two inputs were used to compare two signals. The transistors are connected in the common-emitter mode and base bias is provided by similar potential dividers across the supply. The transistors share a common emitter resistor, and a potentiometer enables the emitter currents to be equalised so that equal push-pull signals are received from the output terminals when an input is applied to, say, TR1. The outputs can be applied to a further balanced stage to increase the gain of the amplifier. The merit of this arrangement is that any change in TR1 collector current (caused by an increase in ambient temperature, for

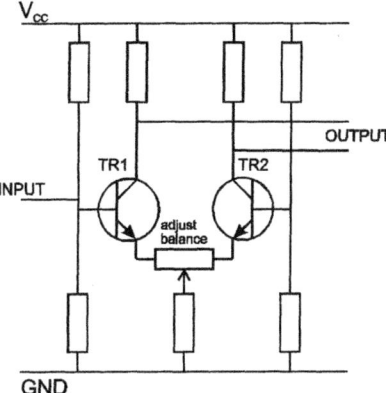

Fig. 9.3. A differential amplifier or 'long-tailed pair'

example) is accompanied by an equal increase in TR2 collector current because this transistor is of the same type as TR1 and shares the same environment. Equal increases in the collector currents cause equal increases in the voltages across the collector resistors. There is therefore no change in voltage between the output terminals as a result of the temperature change. The output of the amplifier is measured by the voltage between the output terminals and such an output can only be produced by an input applied to one of the transistors. A technique sometimes employed in differential amplifiers is to apply the input signal to TR1 base and a linearising negative feedback voltage to TR2 base (as in Fig. 8.9, for example).

Chopper amplifier

When drift must be kept very low or where high d.c. gain is required, a chopper amplifier is sometimes used. In this the signal to be amplified is interrupted, i.e. chopped at regular intervals to produce an alternating signal with an amplitude proportional to the d.c. signal. The alternating signal is then amplified to the required degree in a pulse amplifier, which need not be direct coupled. The output is restored to d.c. form by simple rectification or by commutation which must be synchronous with the interruption of the original signal. Chopping of the input and output signals can be achieved by mechanical means, e.g. by a vibrating reed, or electronically, for example by a gate circuit driven from a multivibrator.

Fig. 9.4 shows the construction of the input stage of such a chopper-stabilised amplifier. Switch **s** is usually a mosfet device, which is driven by the oscillator. The integrated circuit versions of these are now quite

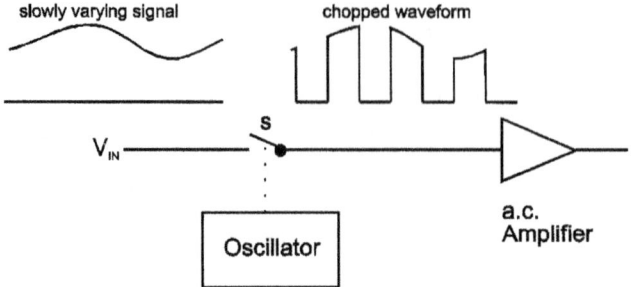

Fig. 9.4. The fundamentals of a chopper-stabilised amplifier

sophisticated, with a signal sample cycle followed by an auto-zero cycle to eliminate any offset effects caused by the signal amplifiers.

Operational amplifiers

General

With modern transistors, it is not difficult to extend the response of a d.c. amplifier to frequencies of the order of 10 MHz. Such amplifiers are available in integrated-circuit form. They are known as operational amplifiers, have high input resistance, low output resistance and very high voltage gain. A typical value for the d.c. gain is 500,000 but the gain falls as frequency rises and may reach unity at, say, 1 MHz. This figure can be treated as the gain-bandwidth product for the amplifier. If a gain of 1,000 is wanted, the amplifier can maintain this up to 500 Hz: if a frequency response up to 10 kHz is wanted, the gain of the amplifier is limited to 50. Operational amplifiers were originally used in analogue computers to carry out mathematical operations such as addition, subtraction and integration but are now widely used for analogue-signal processing.

The circuit diagram of a monolithic operational amplifier is given in Fig. 9.5. To minimise drift the first stage consists of a balanced pair TR2, TR3 direct-coupled to a second pair TR8, TR9. To give a high input resistance the first pair is preceded by emitter followers TR1, TR4. TR9 is followed by two common-emitter stages TR10, TR12 in cascade which drive the complementary output pair TR13, TR14. Direct coupling is used throughout to extend the response to zero frequency: this is essential if the amplifier is to be used for mathematical operations on steady potentials.

It is normally necessary for the output potential to be zero when both inputs are at zero potential. It is difficult to achieve this in mass production and there is normally an output voltage, known as the offset

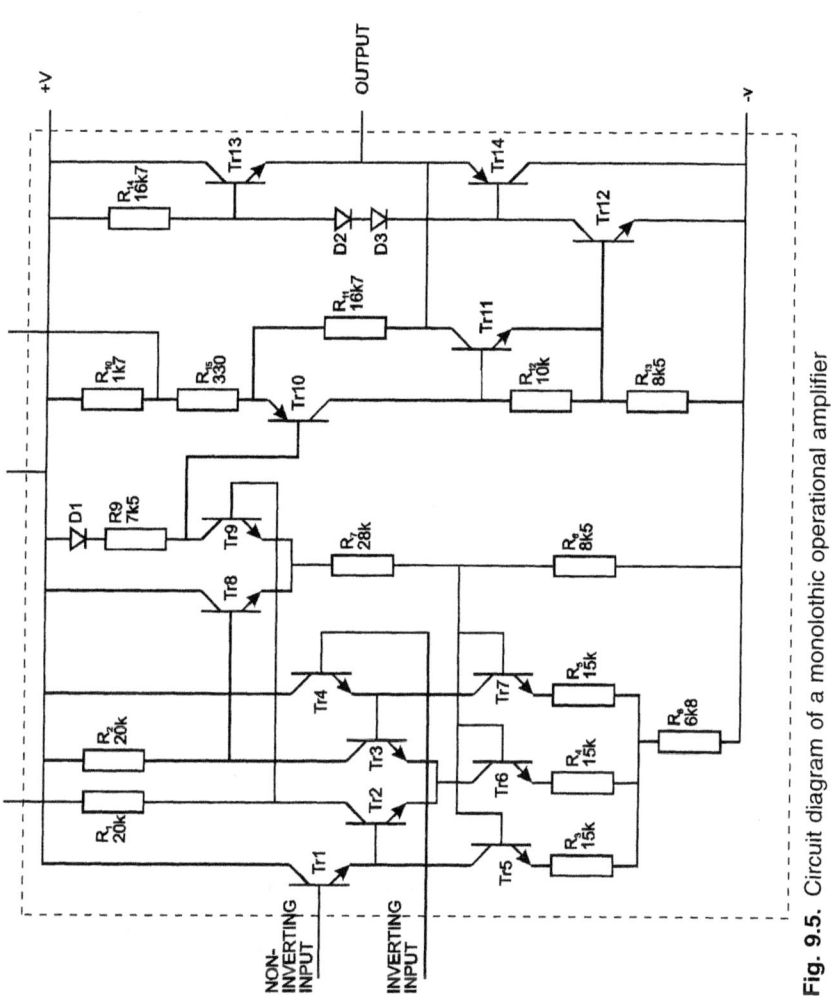

Fig. 9.5. Circuit diagram of a monolithic operational amplifier

voltage, which arises from inherent unbalance in the amplifier. This can be reduced to zero by adjustment of the collector supply voltage for TR2 and the lead from R_1 is brought out to an external connection for this purpose.

The circuit comprising R_3 to R_8 and TR5 to TR7 provides further protection against drift. The stabilisation for the remaining part of the amplifier has something in common with that of the Tobey and Dinsdale amplifier described on page 149. R_{10}, R_{15} and R_{11} constitute a potential divider connected across the output of the amplifier which apply to TR10 emitter a fraction of the output voltage. TR10 compares this fraction with the voltage at TR9 collector and any errors are corrected by TR11 which adjusts its collector current to maintain the standing output voltage at V/2. R_{10}, R_{15} and R_{11} also provide signal-frequency feedback and the voltage gain of the amplifier from TR10 to TR14 is given approximately by $R_{11}/(R_{10} + R_{15})$, i.e. 8.2 for the resistor values shown. The connections to R_{10} are brought out to external terminals to enable this resistor to be short-circuited. By so doing the gain of the latter half of the amplifier can be increased to approximately R_{11}/R_{15}, i.e. 50.

As Fig. 9.5 shows, the amplifier has inputs to both bases of the first long-tailed pair. To obtain a positive signal at the amplifier output requires a negative signal at one input or a positive signal at the other: these are known as the inverting and non-inverting inputs and are marked by minus and plus signs on the block symbol for an operational amplifier. Negative feedback signals are, of course, applied to the inverting input. The same signal applied to both inputs gives no output from the amplifier, a property known as common-mode rejection. The amplifier responds only to the difference between the input signals: it could be described as a differential amplifier.

Applications

It is sometimes necessary to know when a varying signal voltage has a particular value. An operational amplifier can be used to give this information by applying the varying voltage to one input and the fixed voltage to the other. Zero output occurs when the two voltages are equal. This is an example of the use of an operational amplifier as a *comparator*.

Fig. 9.6 illustrates the use of an operational amplifier for non-inverting amplification. Negative feedback is applied via R_1 and R_3, fixing the voltage gain at $(R_1 + R_3)/R_1$. The diagram for a corresponding inverting amplifier is given in Fig. 9.7. An interesting feature of this circuit is that the input voltage applied to R_1 is largely neutralised by the negative-feedback signal from the output via R_3. The signal voltage at

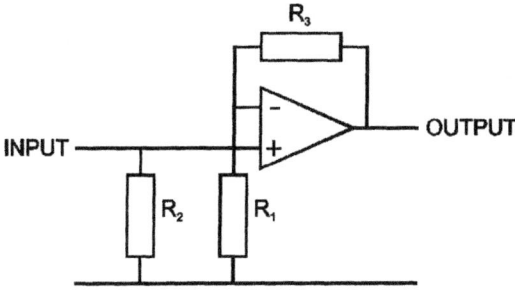

Fig. 9.6. An operational amplifier used for non-inverting amplification

the amplifier input terminal itself is very small. In fact, for an amplifier output of 1 V, the input voltage may be only $2 \mu V$ (corresponding to a voltage gain of 500,000). Thus the voltage at the input terminal hardly moves from zero and can be treated as a source of almost steady potential – a virtual earth, in fact.

This is a useful concept for it shows us immediately that the input resistance in Fig. 9.7 is equal to R_1. It also provides us with a neat solution to Fig. 9.8 in which there are two inputs v_1 and v_2 applied to the

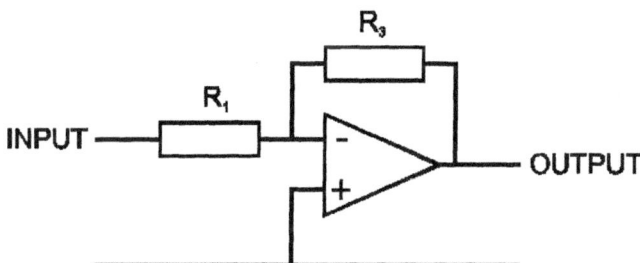

Fig. 9.7. An operational amplifier used for inverting amplification

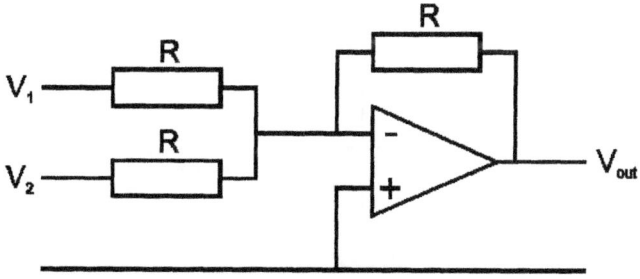

Fig. 9.8. Principle of the summing amplifier

same input terminal via equal resistors of value R. As the inverting input terminal is a virtual earth the current flowing into it from the two inputs must equal the current flowing out via the negative feedback resistor, also equal to R.

$$\frac{v_1}{R} + \frac{v_2}{R} = \frac{v_{out}}{R}$$

giving

$$v_{out} = v_1 + v_2$$

i.e. the output voltage is the sum of the two input voltages. Clearly this circuit arrangement can be extended to embrace any number of inputs and, by reversing the polarity of the inputs, subtraction can be performed.

In the circuit of Fig. 9.9 the feedback current is given by $dQ/dt = Cdv_{out}/dt$ and thus we have, again using the virtual-earth concept

$$\frac{v_{in}}{R} = C\frac{dv_{out}}{dt}$$

from which

$$\frac{dv_{out}}{dt} = \frac{v_{in}}{RC}$$

$$\therefore v_{out} = \frac{1}{RC}\int v_{in} \cdot dt$$

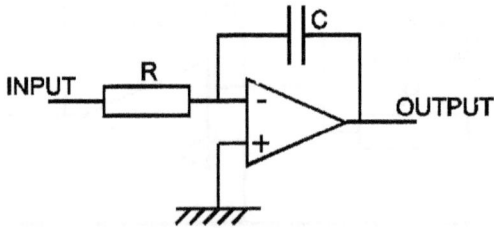

Fig. 9.9. Operational amplifier used for integration

Thus the output of the amplifier is equal to the time integral of the input voltage. This is, of course, the principle of the Miller integrator described on page 283.

As in Fig. 9.5 operational amplifiers may require a positive and a negative supply voltage in order that the input and output terminal voltages shall be near zero. It is possible, however, to use a single supply if the amplifier is to be used for alternating signals thus permitting capacitive coupling at the input and the output terminals. As an example Fig. 9.10 shows a non-inverting amplifier designed for a voltage gain of 100. The non-inverting input is biased to mid-supply voltage by the potential divider $R_1 R_2$ (in which $R_1 = R_2$). Negative feedback is applied to the inverting input by $R_3 R_4$ (R_3 being 100 times R_4 to set the required gain). D.C. isolation at the input and output is provided by C_1 and C_2.

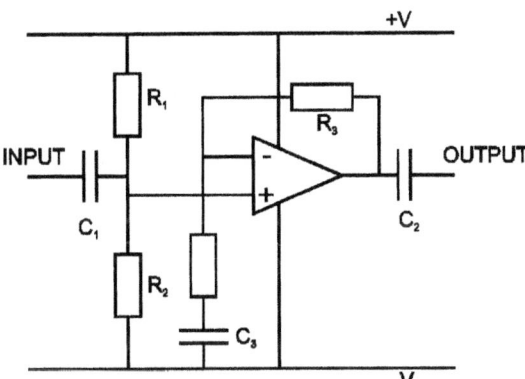

Fig. 9.10. Operational amplifier with a single supply voltage used for alternating-signal amplification

Pulse amplifiers

General

Amplifiers such as oscilloscope Y-amplifiers and video amplifiers are required to handle signals which may have steep, almost vertical edges and also long, almost horizontal sections. A signal which has both features is a rectangular pulse and is commonly used in tests of such amplifiers which are usually known as pulse amplifiers.

The ability of an amplifier to reproduce rapid changes such as a steep edge in a signal waveform is determined by the high-frequency response of the amplifier: in fact such amplifiers must have a response good up to the frequency given by

$$f = \frac{1}{2t}$$

where t is the rise time* of the steepest edge. If the rise time is 0.1 μs the upper frequency limit is given by

$$f = \frac{1}{2 \times 0.1 \times 10^{-6}} \text{ Hz}$$

$$= 5 \text{ MHz}$$

The ability of an amplifier to reproduce very slow changes such as an almost horizontal section in a signal waveform is determined by the low-frequency response of the amplifier: the longer the section of the waveform, the better must be the low-frequency response of the amplifier to reproduce it without distortion. As a numerical example, an amplifier required to reproduce a 50-Hz square wave with less than 2 per cent sag in the horizontal sections must have a low-frequency response which is good down to at least 1 Hz. Sometimes pulse amplifiers are direct-coupled to extend the low-frequency response to zero frequency.

To summarise the above we may say that pulse amplifiers are characterised by an extremely wide frequency response: for a video amplifier, for example, the useful frequency response may extend from very low frequencies to 5.5 MHz. A statement such as this of the steady-state amplitude response of the amplifier does not, however, give complete information about its performance as a pulse amplifier. In general a pulse signal contains many components which must not only undergo similar magnification but also must be maintained in their original phase relationship when distortionless amplification is required. The shunt capacitance which is inevitable in any amplifier causes the phase of high-frequency components of a complex signal such as a pulse to lag behind that of low-frequency components. Such a lag would be comparatively unimportant in an a.f. amplifier but can seriously degrade the performance of a pulse amplifier by increasing the rise time. Phase response is thus important in pulse amplifiers and for a good performance both the amplitude and the phase response must satisfy

---

* The rise time of a step is defined as the time taken for the signal to rise from 10 to 90 per cent of the final steady value.

certain stringent requirements over the spectrum occupied by the signals to be amplified.

For amplifiers using simple inter-transistor coupling networks, such as are likely to be used for pulse amplification, there is a mathematical relationship between the amplitude response and the phase response so that, given one, it is possible to deduce the other. In general the flatter the frequency response the better is the phase response also and it is thus possible to ensure an adequate phase response by making the amplitude response of the amplifier sufficiently good. When a pulse amplifier is designed in this way the amplitude response must satisfy standards far more exacting than if phase response were not under consideration. For example to design a video amplifier with a passband extending to 5.5 MHz it may be necessary to make the amplitude response of the amplifier flat to say 35 MHz.

Use of transistors

There is no difficulty in achieving bandwidths of this order using planar transistors as conventional RC-coupled amplifiers: even the transistors intended for a.f. applications have transition frequencies of about 200 MHz and transistors with f_T greater than 1 GHz, i.e. 1,000 MHz, are available.

The transition frequency f_T specifies the product of the current gain and bandwidth of the transistor. As an example if a transistor with $f_T = 500$ MHz is required to give a current gain of 100, it can do this up to 5 MHz. If, on the other hand, the response must be maintained up to 25 MHz the current gain is limited to 20.* Clearly in designing pulse amplifiers some means is required of dividing the product f_T between gain and bandwidth as required. This is possible by choice of load-resistor value but a better method is by use of negative feedback.

Use of negative feedback

Feedback is extensively used in pulse amplifiers principally to make the performance independent of the inevitable differences in characteristics between transistors of the same type and hence to make performance also independent of temperature. If a large degree of feedback is used the gain and bandwidth are largely determined by the constants of the feedback loop. Thus the performance of the amplifier can be

---

* In tuned amplifiers it is possible to achieve some current or voltage gain in the inter-transistor coupling circuits and this enables a transistor stage to give worthwhile gain above f_T: in pulse amplifiers, however, extra gain cannot be obtained from the resistance coupling normally used and f_T may be taken as the highest usable frequency.

predetermined and it is possible to manufacture a number of amplifiers all having the same performance within very close limits. This is important, for example, in colour television where the amplifiers handling the red, green and blue components of the colour signal must have identical performances.

It is not easy to apply a large degree of feedback to an amplifier without producing oscillation or a sharp peak in the frequency-response curve at frequencies near the extremes of the passband. One method of avoiding such effects in a multi-stage amplifier is to employ a number of independent feedback loops each embracing only two stages (three if one is an emitter follower).

Negative feedback reduces the gain and extends the bandwidth of an amplifier in approximately the same ratio: thus a 6 dB reduction in gain is accompanied by a doubling of bandwidth. This is illustrated in Fig. 9.11. The upper frequency-response curve is that of an amplifier without feedback and is 3 dB down at the frequency f_1 which can be taken as the limit of the passband. When feedback is applied the gain falls at all frequencies but the fall is greater at middle than at high and low frequencies and the new 3-dB loss point is now at f_2 representing an upwards extension of the passband. The low-frequency response is similarly extended but this is not illustrated in Fig. 9.11.

If the negative feedback circuit is so designed that the feedback voltage or current becomes less as frequency approaches the limits of the passband it is possible to achieve a greater extension of frequency response as suggested by the dotted line in Fig. 9.11. This, however, is usually accompanied by a deterioration in phase response, manifested by overshoot in reproduced pulses. Where a very good pulse response is

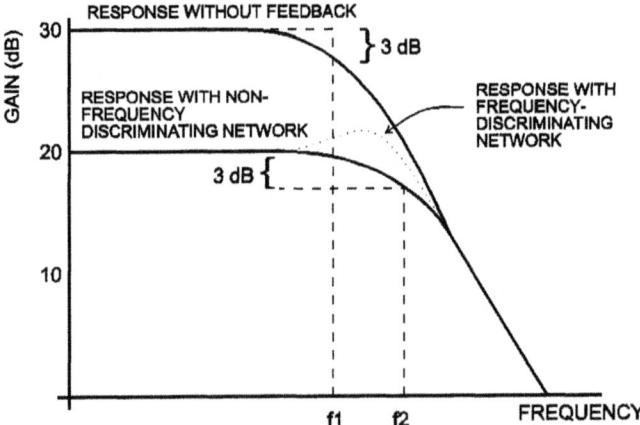

Fig. 9.11. Effect of negative feedback on the frequency response of an amplifier

essential it is advisable to adhere to the aperiodic, i.e. non-frequency-discriminating feedback represented by the solid curve. In a television receiver slight overshoot in the video amplifier may be tolerable (even advantageous by slightly exaggerating the transitions), and frequency-discriminating feedback is often used.

A number of examples of typical pulse amplifier circuits will now be considered.

Voltage pulse amplifiers

Fig. 9.12 shows the circuit diagram of a voltage pulse amplifier designed to give a gain of 20 dB from a very low frequency up to approximately 6.6 MHz. Two common-emitter stages in the basic arrangement of Fig. 7.15 provide the required voltage gain and an emitter follower gives the required low output resistance. Direct coupling is used throughout and signal-frequency feedback is injected into the emitter circuit of TR1 to give the required high input resistance (approximately $10 \, k\Omega$).

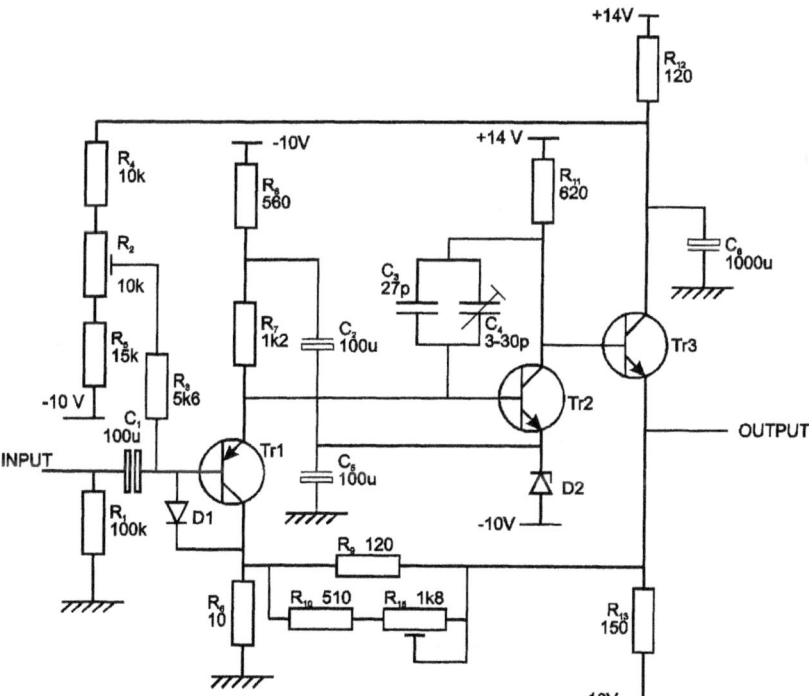

Fig. 9.12. A voltage pulse amplifier

Signal-frequency feedback is provided by two loops:

(a) R_9, R_{10}, R_{15} between TR3 emitter and TR1 emitter
(b) C_3, C_4 between TR2 collector and base.

Resistors R_9, R_{10} and R_{15} together with R_6 determine the gain of the amplifier over the lower middle of the passband and R_{15} is adjusted to give the required gain at 10 kHz. The same value of gain at 5.5 MHz is obtained by adjusting C_4 which, with C_3, determines the gain near the top of the passband. This rather complex feedback circuit is used to control the phase response of the amplifier at and above the upper extreme of the passband so as to eliminate any possibility of instability at these frequencies.

An essential requirement of this amplifier is that the mean output voltage should not depart significantly from zero. It is designed to operate from a supply unit giving stabilised voltages of −10 and +14, and a very high degree of d.c. stability in the amplifier is clearly necessary. In a direct-coupled amplifier any variations in collector current of the first stage (due to changes in ambient temperature, for example) are interpreted as signals and are amplified by the following stages to give large unwanted changes in voltage at the output. To achieve high stability therefore a very low value of zero-frequency gain is wanted. For this purpose the zero-frequency gain is reduced to a very low value by the direct-coupled feedback loop R_2, R_4, R_5, R_{12} between TR3 collector and TR1 base. This loop must not, however, affect the gain of the amplifier in the passband (which extends to a very low frequency) and accordingly R_{12} is bypassed by the 1,000-μF decoupling capacitor C_6. The degree of direct-coupled feedback can be adjusted within limits by the preset potentiometer R_2 which is set to give zero voltage output from TR3 emitter. Although this feedback reduces amplification of unwanted changes in TR1 collector current, it is still possible for zero-frequency voltages in the amplifier to drift and there is need to stabilise the potential at some point in the signal path. Applying the technique mentioned at the beginning of this chapter, such stabilisation is achieved by including a voltage reference diode in TR2 emitter circuit. With the assistance of feedback this method restricts the variations of mean voltage output to less than 0.03 V in the circuit shown.

The low-frequency response of the amplifier is not required to extend to zero frequency but must nevertheless be very good. As just explained C_6 causes some fall-off in low-frequency response and the only other source of low-frequency loss is C_1 which is made very large (100 μF). The frequency for 3 dB loss is 0.4 Hz and the amplifier can reproduce a 50-Hz square wave with less than 1 per cent sag on the horizontal sections.

Diode D1 is included to prevent the base of TR1 being driven appreciably positive with respect to its emitter potential. Such input voltages can occur and if they exceed a particular value can damage or even destroy TR1 by taking the base-emitter potential past the breakdown voltage for this junction.

Complementary emitter follower

Simple emitter followers such as that illustrated in Fig. 7.14 are satisfactory with sinusoidal signals but may not be suitable for use with pulse signals. Consider, for example, a rectangular-wave voltage input: the output waveform is produced by the sequential charge and discharge of the shunt capacitance in the output by the emitter current of the transistor. Provided the transistor remains conductive throughout the cycle, the charge and discharge time constants are equal. The output waveform is then symmetrical and a reasonable copy of the input. If, however, the output time constant is comparable with the rise or fall times of the input, the output cannot change as rapidly as the input. It is then possible for positive-going input steps to cut off the transistor (assumed pnp). This considerably increases the output time constant by making r_o effectively infinite, leaving only R_e as the discharge path for the output capacitance. Thus for steep-sided input signals the output can have markedly dissimilar rise and fall times.

This effect can be eliminated by using a symmetrical circuit, e.g. a complementary pair consisting of one pnp and one npn transistor feeding the same output terminal as suggested in the basic circuit of Fig. 9.13. Positive-going input steps, even if they cut the pnp transistor off, drive the npn transistor into conduction: similarly negative-going input steps turn the pnp transistor on. In a practical version of this circuit it is necessary to provide the transistors with base bias: a suitable circuit is applied to TR3 and TR4 in Fig. 8.11.

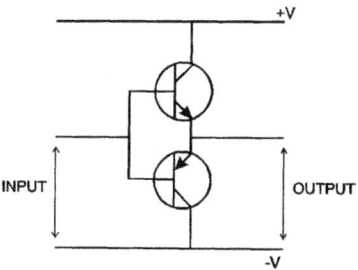

Fig. 9.13. Basic form of complementary emitter follower

Video amplifier for television receiver

As an example of another type of pulse amplifier we will consider the design of a video amplifier for a 625-line monochrome television receiver. This has to supply a maximum voltage swing of about 70 to the cathode of the picture tube and an output of 1.5 V (from black to white level) can be expected from the vision detector. The input resistance of the amplifier must be at least 30 kΩ to minimise damping of the detector circuit. We therefore need a voltage amplifier with a gain of approximately 45, high input resistance and a response up to 5.5 MHz.

A single common-emitter stage could provide the voltage gain and frequency response but a second stage is needed to give the high input resistance and this is usually in the form of an emitter follower preceding the common-emitter stage. Thus the amplifier has the basic form shown in Fig. 9.14. To keep dissipation in the output stage at a minimum it is usual to arrange for TR2 to be almost cut off for black-level signals and to be driven into conduction by the picture signal. As the collector is direct-coupled to the tube cathode it follows that TR2 must have a positive h.t. supply and must be of npn type. TR1 gives no phase inversion and thus the diode must give an output in which increasing whiteness is portrayed by a positive excursion of voltage. This is achieved by arranging for the diode to give a negative output voltage: with negative modulation this voltage approaches zero for peak white signals and this is the required form of operation.

The component values required in the circuit which couples TR2 to the picture tube are determined by the total shunt capacitance C_s at this point. An approximate value for C_s is 10 pF made up of contributions from tube input, TR2 output and strays. At 5.5 MHz the reactance of 10 pF is nearly 3 kΩ. If R_6 is made 3 kΩ the frequency response of the

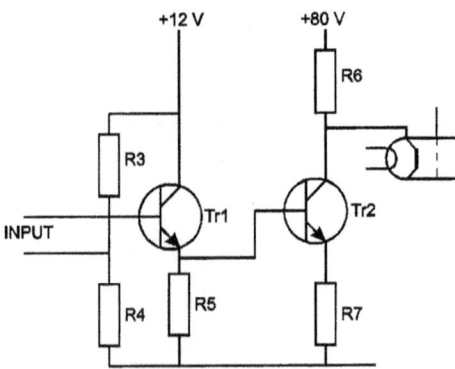

Fig. 9.14. Basic form of video amplifier

output circuit is 3 dB down at 5.5 MHz. It is preferable, however, to use a larger value for R_6 to reduce the current swing required from TR2 and to improve the gain at low and medium frequencies. A value of 3.9 kΩ is suitable. Knowing this we can determine the value of the undecoupled emitter resistor R_7. TR1 gives unity voltage gain and thus TR2 must provide the gain of 45 required. From Eqn 7.5 the gain is given approximately by R_6/R_7. Thus R_7 is equal to R_6/gain, i.e. 3,900/45 about 87 Ω.

With R_6 equal to 3.9 kΩ the 3-dB loss frequency is 4 MHz but the response can be extended to the required 5.5 MHz in two ways:

(a) by use of an inductor L_2 in series with R_6. This offsets the effect of the shunt capacitance as frequency rises. The value of L_2 can be calculated from the expression

$$L_2 = aR_6^2C_s$$

By putting $a = 0.5$ the response can be levelled at 4 MHz at the cost of approximately 7 per cent overshoot. If $R_6 = 3.9$ kΩ and $C_s = 10$ pF L_2 is 80 μH.

(b) by shunting R_7 by a capacitor C_4. This reduces the feedback due to R_7 as frequency rises. A suitable value for C_4 can be found by equating the time constants of the collector and emitter circuits of TR2. We thus have

$$R_7C_4 = R_6C_s$$

$$\therefore C_4 = \frac{R_6C_s}{R_7}$$

$$= \text{voltage gain} \times C_s$$

$$= 45 \times 10\,\text{pF}$$

$$= 450\,\text{pF}$$

The output from the detector varies between a minimum value for peak white signals and −1.5 V at the tips of synchronising pulses. When the detector output is almost zero, the voltage across R_6 must be 70 V and across R_7 1.5 V. If, as likely, TR2 is a silicon planar transistor, the base-emitter voltage is, say, 0.7 so that the voltage across R_5 is 2.2 V. A suitable value for R_5 is 1.5 kΩ and the current in this resistor is then 1.5 mA. If TR1 is also silicon planar the base-emitter voltage is again 0.7 and the base-earth input voltage for the amplifier is 2.9 V. As the detector provides almost zero voltage approximately 2.9 V must be obtained from R_4. The potential divider R_3R_4 must be proportioned to give this voltage. It is usual to make one of the resistors preset so that correct operation of the amplifier can be obtained in spite of differences in the parameters of the transistors used.

The input resistance of TR2 is given approximately by βR_7 and if β is taken as 50, is equal to 4.35 kΩ. This is in parallel with R_5 (1.5 kΩ) giving a net external emitter resistance for TR1 of 1.1 kΩ. If β for TR1 is also taken as 50 the input resistance of the amplifier is given by 50 $\times$ 1.1 = 55 kΩ, nearly double the minimum value required. This is, of course, the input resistance to alternating signals but the resistance for zero-frequency signals is of the same order. We have assumed that β for both transistors is maintained at 50 up to 5.5 MHz, equivalent to f_t = 275 MHz. It would probably be satisfactory, however, to use transistors with f_T = 200 MHz because the only effect of this reduction would be to lower the input resistance of the amplifier at the upper end of the video-frequency range: a little reduction is acceptable.

To deliver the required output of 70 V TR2 requires an h.t. supply of, say, 80 V. The collector-current swing required is 18 mA and a dissipation of approximately 0.5 W is likely. TR2 must hence be capable of working under these conditions. If the receiver is intended for operation from a 12-V battery the 80-V supply can be obtained by rectifying the output of a winding on the line output transformer. TR1 can operate satisfactorily from a 12-V supply.

As pointed out above, the limiting factor in the design of video amplifiers is the capacitance shunting the output. This capacitance is rapidly discharged by the collector current of the video stage when this is driven into the increasing conduction by a positive-going input signal (corresponding to increasing brightness). But on negative-going input signals (decreasing brightness) the rise in collector voltage is controlled by the shunt capacitance as this is charged from the supply via the collector load resistor. The video transistor can usually discharge the shunt capacitance more rapidly than the load resistor can charge it and as a result reproduced images tend to be blurred.

To overcome this in many video amplifiers provision is made to provide increased charging current on negative-going input signals. For example, single-ended push-pull can be used with a class-B output of (say) two npn transistors. In this case, one output transistor conducts on positive input signals and the other on negative input signals. In an alternative circuit shown in basic form in Fig. 9.15 the collector load R1 is paralleled by a transistor TR2 arranged to conduct on negative input signals and thus to connect a low-value resistor R2 in parallel with TR1 load resistor. A diode D1 is connected in series with the amplifier output and TR2 base-emitter junction is connected across D1. When TR1 base is driven positive its collector voltage falls

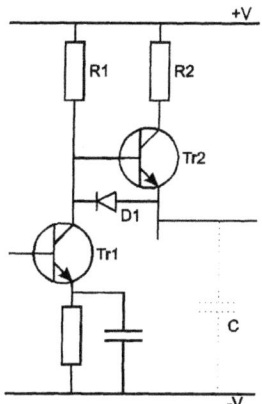

Fig. 9.15. Basic form of video amplifier with provision for increasing the charging current for the stray capacitance

and D1 conducts: when TR1 base is driven negative its collector voltage rises and D1 is cut off, causing TR2 to be turned on and R2 to be connected in parallel with R1 to provide the increased charging current required for C. This circuit enables values of R1 approaching $10\,k\Omega$ to be used whilst still permitting a frequency response extending to 5 MHz.

In a black-and-white receiver the only input to the picture-tube cathode is the luminance signal as just described. In a colour receiver, however, the picture tube requires inputs representing the three television primary colours red, green and blue. These inputs are applied to the three cathodes of the electron gun in the tube, the three corresponding grids being commoned and held at a fixed voltage which determines the brightness of the picture. Thus three video amplifiers are required and these must be closely matched in gain and output level to ensure that the colour components are correctly balanced throughout the amplitude range of the amplifiers. Preset controls are included to permit these adjustments. The circuits used are often based on that in Fig. 9.16 which can give the required frequency response without need for peaking inductors provided that frequency-discriminating negative feedback is used in the emitter circuit of the video amplifier.

A desirable precaution against possible damage to the video output transistors by internal flash-over in the picture tube is to include resistors in the tube driving circuits. Most picture tubes have spark gaps around the electrode pins which are connected via low-inductance paths to the earthed conductive coating of the tube. This arrangement ensures that transient currents due to flash-overs are confined to the low-inductance paths.

Integrated-circuit pulse amplifier

Fig. 9.16 gives the circuit diagram of a voltage pulse amplifier manufactured in monolithic form. It is contained in a can $\frac{1}{3}$-in in diameter.

The circuit consists of three common-emitter stages and an emitter follower all direct-coupled. Signal-frequency negative feedback is applied by the network R_2, R_3, R_4 and the resistor R_7: the connections to R_4 are brought out to pins 3 and 4 to enable gain to be adjusted by an external feedback resistor.

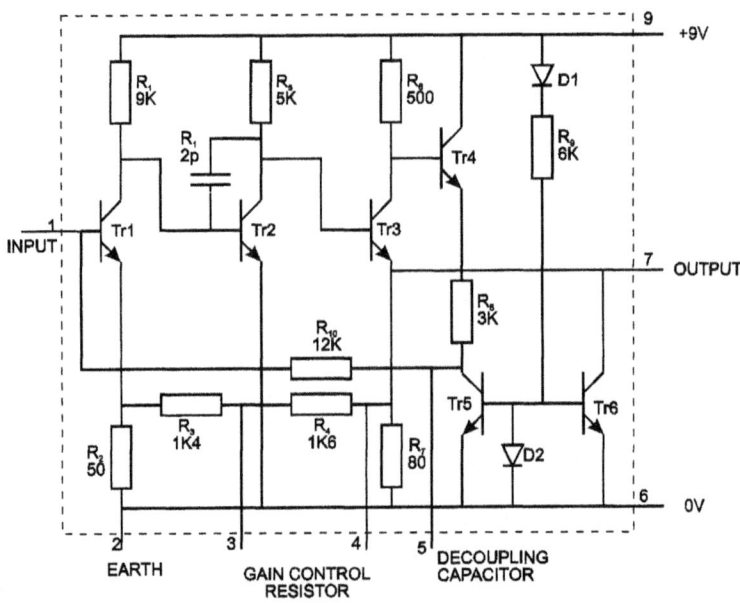

Fig. 9.16. Circuit diagram of a monolithic pulse amplifier

The method of ensuring stability of d.c. operating conditions has something in common with that used in Fig. 9.12 on page 178. The large degree of direct-coupled feedback provided by R_{10} reduces the zero-frequency gain of the amplifier to a very small value. (An external decoupling capacitor connected between pins 5 and 6 eliminates this feedback loop at signal frequencies.) The diodes D1 and D2 in conjunction with R_9, determine the base bias on TR5 and TR6. By suitable choice of diode characteristics the current in TR5 and TR6 can be kept constant in spite of variations in temperature. In this way the current in the emitter follower is stabilised as well as the d.c. conditions throughout the amplifier. So effective is this circuit that the amplifier operates satisfactorily at any temperature between −55°C and +125°C, the variation in standing output voltage being less than 0.05%.

The upper frequency limit of the amplifier depends on the values of f_T for the transistors TR1 to TR4: it extends to 35 MHz at a gain of 200 and to 15 MHz at a gain of 400. The low-frequency response depends on the value of the external decoupling capacitor and on the values of any capacitors connected in series with the input or output. The input resistance of the amplifier is 10 kΩ and the output resistance 16 Ω. The maximum output voltage swing is 4.2 into the recommended load resistance of 1 kΩ.

Chapter 10

R.F. and I.F. amplifiers

Introduction

The amplifiers discussed so far have had a passband beginning at zero or a very low frequency and extending over several octaves. Such a passband is termed the *baseband* and baseband amplifiers are characterised by their use of direct or RC inter-stage couplings (aperiodic couplings, in fact). We shall now consider amplifiers in which the passband is small compared with the mid-band frequency. Here LC circuits can be used as coupling elements. If the r.f. signal to be amplified is sinusoidal and of fixed frequency, the passband can be made as narrow as desired: this is possible in the carrier-wave amplifiers in transmitters. If, however, the r.f. signal is modulated, the passband must be wide enough to accept the significant side frequencies of the signal. This is necessary in the r.f. and i.f. amplifiers of radio receivers. The r.f. amplifiers are required to have a little gain so as to improve the signal-to-noise ratio by raising the signal level at the input to the frequency changer. They also provide some protection against second-channel interference and contribute to the selectivity of the receiver although most of the adjacent-channel selectivity is achieved in the i.f. amplifier.

Class-C operation

The r.f. and i.f. amplifiers used in receivers generally operate in class A. The low efficiency of such stages is no disadvantage because only a small output power is required. By contrast, the r.f. amplifiers in transmitters may be required to give considerable output power and the use of class A would be uneconomic. Higher efficiency is possible by using class-B push-pull and such stages were at one time used to amplify amplitude-modulated r.f. signals in transmitters where modulation was carried out at a low power level. Linearity is essential in such amplifiers to minimise distortion of the modulation envelope but it limits efficiency to, say, 35%. Much higher efficiency is possible where constant-amplitude r.f. signals are to be amplified and a class-C amplifier can be used. In this mode of operation the active device is biased beyond the point of output-current cut-off and is driven into conduction by the peaks of the sinusoidal input signals as suggested in Fig. 10.1. The output of the device thus consists of a series of rounded pulses with a waveform bearing little resemblance to that of the input signals. The output has, however, a large fundamental component at the frequency of the input signal accompanied by a host of harmonics. The fundamental component can be selected at the device output by use of a tuned circuit resonant at the fundamental frequency. The signal generated across this tuned circuit is, of course, sinusoidal in form so that, in effect, distortion-free amplification is obtained. Efficiencies approaching 80 per cent can be obtained from a class-C amplifier and medium-frequency transmitters often use one as the final stage, amplitude modulation being achieved at very high power level by varying the supply voltage to the amplifier.

Still higher efficiency is possible if the peaks of the input signal to a class-C amplifier are flattened (by the addition of, say, 15 per cent of third harmonic) so that the active device is switched on and off very

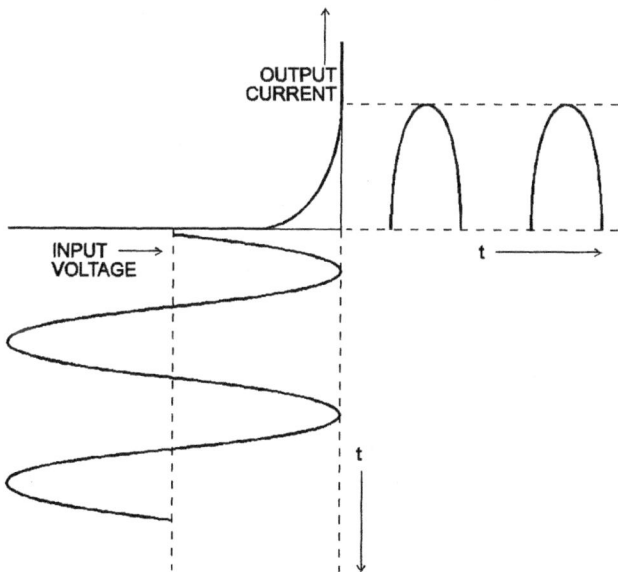

Fig. 10.1. Illustrating the principle of class-C amplification

rapidly, its output current approximating to a train of rectangular pulses. The device is thus used as a switch and is at all times either off (with zero output current) or on (with zero voltage across it). Dissipation is zero in both states and the only losses occur during the brief instants of transition, making possible efficiencies approaching 90%.

Common-emitter r.f. amplifiers

The common-emitter amplifier with its ability to give current gain and voltage gain simultaneously is the obvious first choice for r.f. amplification and the narrow-band frequency response can be obtained by tuning one or both windings of the inter-transistor coupling transformers. The resonance frequencies, coupling coefficients and Q values can be chosen to give an approximation to the wanted shape and extent of the frequency response. The effective Q value of an inductor depends on its physical construction and also on any damping caused by the components connected to it. Transistor damping is sometimes heavy and, to obtain a desired Q value, the designer must know the input and output resistances of the transistors. Transistor manufacturers usually quote these parameters at frequencies of interest. Over the restricted passband of an r.f. or i.f. amplifier the input and output resistances of a transistor may usually be taken as constant. Thus a bipolar transistor used for such amplification may be represented by the simple equivalent circuit of Fig. 10.2. The resistances r_i and r_o can be used to provide the required damping whilst the capacitances c_i and c_o can be used to provide some of the capacitance required in the input and output tuned circuits. The components r and c represent the internal collector-base resistance and capacitance. These provide feedback between the output and input of a common-emitter amplifier, feedback which clearly increases as frequency rises. Its effect on an amplifier with a tuned collector circuit can be assessed in the following way.

At resonance the external collector circuit is effectively resistive and the signal voltage at the collector is in antiphase with the collector

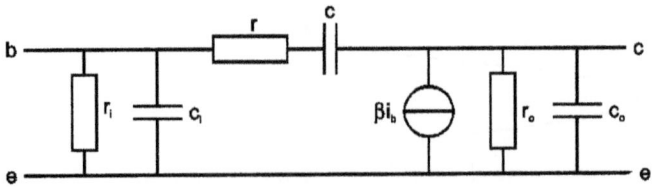

Fig. 10.2. Equivalent circuit of a bipolar transistor used as an r.f. or i.f. amplifier

current. The current fed back to the base via the reactive path provided by c is therefore in quadrature with the external base input current.

Above the resonance frequency the external collector circuit is effectively capacitive and the phase of the collector voltage, with respect to the collector current, lags the antiphase condition by 90°. The current in c is in antiphase with the external base current, giving negative feedback and reduced gain from the transistor.

Below the resonance frequency the external collector circuit is effectively inductive. The phase of the collector voltage relative to the collector current leads the antiphase condition by 90°. The current in c is in phase with the external base current giving positive feedback and increased gain from the transistor. If the feedback current exceeds the external input the circuit will oscillate. Even when the feedback is insufficient to cause oscillation, the change from positive to negative feedback at the resonance frequency of the collector circuit can cause asymmetry in the frequency-response curve.

The precise condition for instability in a common-emitter amplifier is evaluated in Appendix C.

Use of a field-effect transistor in place of a bipolar does not solve the instability problem because the drain-gate capacitance is of the same order as the collector-base capacitance, causing the common-source r.f. amplifier also to be inherently unstable.

The problems caused by this instability can be solved in several ways (described later in this chapter) but at v.h.f. and u.h.f. they can be dodged by using the transistor as a common-base (or common-gate) amplifier.

Common-base r.f. amplifier

In this arrangement (Fig. 10.3) any positive feedback occurs via the collector-emitter capacitance of the transistor and this is effectively eliminated by the base region which screens the emitter from the collector, making stable amplification possible up to thousands of MHz. Conventional inductors and capacitors can be used for tuning up to v.h.f., the inductance and capacitance becoming smaller as frequency is raised. Above about 600 MHz, however, lumped components can no longer be used for tuning and are replaced by distributed inductance and capacitance in the form of lengths of u.h.f. transmission line. These can be used with silicon planar transistors for which the dimensions and physical construction are designed to fit the line exactly so avoiding the need for connections which would add undesirable reactance. Even transmission lines have to be abandoned above about 2,000 MHz and cavity resonators are used in their place. Again the transistors are

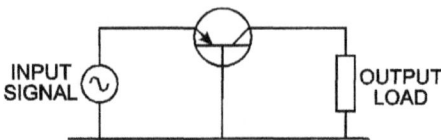

Fig. 10.3. In the common-base circuit the base acts as a screen between input and output circuits

designed for direct connection to the resonator. At these frequencies gallium arsenide transistors are preferred to silicon to take advantage of the better electron mobility in GaAs. These transistors have a reversed Schottky diode as the gate and are known as mesfets (metal-semiconductor field-effect transistors).

The basic form of circuit for a common-base r.f. amplifier is given in Fig. 10.4. It normally delivers its output signal into a resistive load, often the input resistance of the following stage. Thus the inductor $L1$ of the collector tuned circuit is used as the primary winding of a matching transformer coupling the external load to the transistor. The turns ratio of the transformer should be chosen to ensure that the collector circuit is presented with a load value into which the transistor can deliver the required output-signal amplitude.

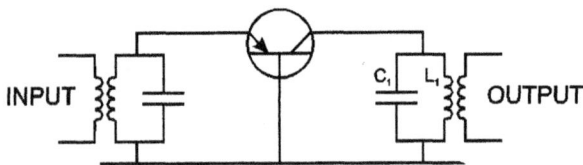

Fig. 10.4. Basic form of tuned common-base amplifier

Insertion loss

The amplifier has a low-frequency loss caused by the shunting effect of the low reactance of L_1: there is also a loss at high frequencies because of the low reactance of C_1. Ideally there should be no loss around the resonance frequency of L_1C_1 because these frequencies constitute the required passband. However, the dynamic resistance of L_1C_1 would need to be infinite to give zero loss in the passband. In practice the finite value of the dynamic resistance inevitably introduces a loss, known as the insertion loss.

If the dynamic resistance of the tuned circuit is equal to the load resistance, its effective value is halved by the connection of the load. The effective resistance into which the transistor feeds is also halved, giving a loss of 6 dB in the level of the output signal. This is therefore the insertion loss. If the undamped dynamic resistance is much greater than the load resistance, the effect of connecting the tuned circuit to the transistor is to reduce Q considerably (equivalent to very heavy damping) but the effective resistance into which the transistor works is only slightly reduced, implying a small insertion loss. For very heavy damping the insertion loss is often negligible.

In general the inclusion of the tuned circuit reduces the transistor load from R_c to $R_c R_d/(R_c + R_d)$ where R_d is the dynamic resistance of the tuned circuit. The factor by which the load is reduced is therefore $R_d/(R_c + R_d)$. If constant-current operation is assumed this is also the factor by which the gain is reduced, i.e. the insertion loss. We thus have

$$\text{insertion loss} = 20\log_{10} R_d/(R_c + R_d)$$
$$= 20\log_{10} [1 - R_c/(R_c + R_d)]$$
$$= 20\log_{10} (Q_u - Q_d)/Q_u$$

where Q_u is the undamped value of Q and Q_d is the damped value.

V.H.F. amplifier

As an example of a small-signal r.f. amplifier Fig. 10.5 gives the circuit diagram of the first stage of a v.h.f. f.m. receiver. The common-base stage TR1 is stabilised by the potential-divider method, the lower arm of

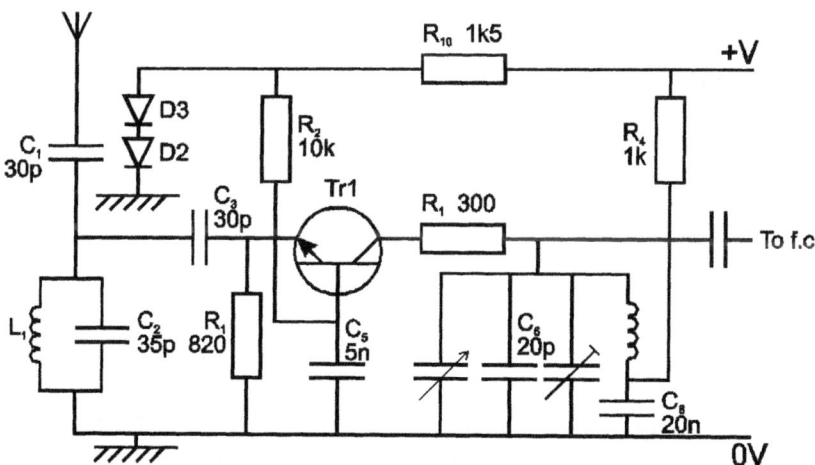

Fig. 10.5. V.H.F. r.f. stage using a common-base amplifier

the divider consisting of the two series-connected diodes D2 and D3. R_{10} is the upper arm. As mentioned on page 104, this technique is used to compensate for fall in battery voltage. R_1 is the emitter resistor. The base is decoupled to chassis by C_5 and biased by R_2.

TR1 input resistance is, of course, very low, say 25 ohms, which would give very heavy damping of the input tuned circuit $L_1 C_2$ if connected directly across it. Damping is therefore reduced by using a low-value (30 pF) capacitor C_3 to couple $L_1 C_2$ to the emitter. C_3 and the input capacitance of TR1 form a capacitive tapping across the tuned circuit so that the resistance effectively damping $L_1 C_2$ is greater than the emitter a.c. resistance and may be, say, 150 ohms. This still constitutes fairly heavy damping and advantage is taken of it by adopting fixed tuning for the damping, broadening the response to cover the whole of Band II (87.5 to 108 MHz). An estimate of the value of C_2 required can be obtained as follows.

Because of the heavy damping the effective dynamic resistance of $L_1 C_2$ is 150 ohms. The required damped value of Q is given by

$$Q = \frac{\text{centre frequency}}{\text{bandwidth}}$$

$$= \frac{97.75}{20.5}$$

$$= 4.77.$$

Now

$$R_d = \frac{Q}{\omega C}$$

which gives

$$R_d = \frac{Q}{\omega R_d}$$

Substituting numerical values

$$C = \frac{4.77}{6.284 \times 97.75 \times 10^6 \times 150}\, \text{F}$$

$$= 52\,\text{pF}.$$

Towards this capacitance the coupling to the aerial and to TR1 together with the inevitable stray capacitance probably contribute about one third and C_2 (at 35 pF) provides the rest.

There is unity current gain in a common-base amplifier and the voltage gain, between emitter and collector terminals, is equal to the ratio of the collector load to emitter a.c. input resistance r_i. If the load is 2000 ohms the voltage gain is approximately 80 from the emitter terminal. The capacitive attenuator at the transistor input reduces the signal input by a factor of, say, 3 so that the net voltage gain is 27.

Cascode

One way of avoiding the instability of the common-emitter amplifier at r.f. is to use it to drive a common-base stage and the two stages can conveniently be connected in series across the supply as shown in Fig. 10.6. The collector load for the common-emitter stage is the input resistance of the common-base stage which is so low that the gain of the first stage is insufficient to sustain oscillation. The base of the common-base stage is earthed at r.f. so this stage, too, is stable. The combination is thus a stable r.f. amplifier with the input resistance, the current gain and the voltage gain of a common-emitter amplifier. It is known as a cascode.

Fets can also be used in a cascode. A common-source stage feeding into a common-gate stage forms a stable r.f. amplifier with an input resistance much higher than that of a bipolar cascode. In the

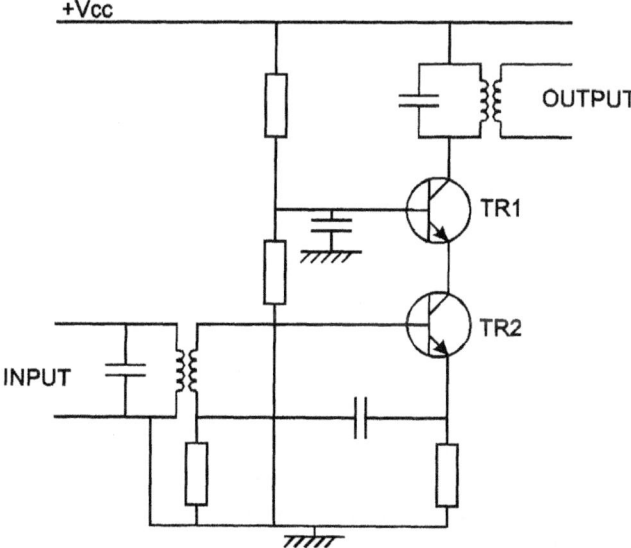

Fig. 10.6. A cascode r.f. or i.f. amplifier using two bipolar transistors

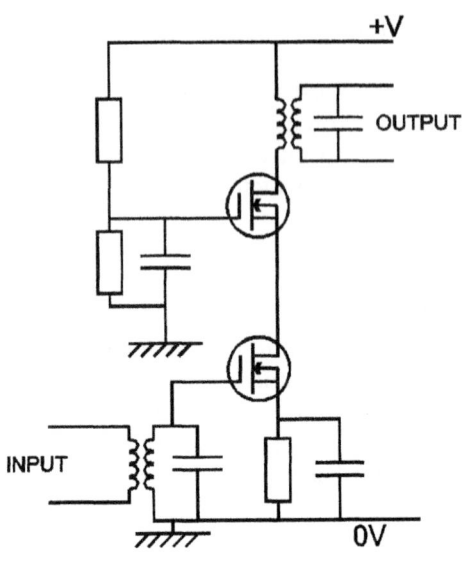

Fig. 10.7. A cascode formed of two discrete mosfets

arrangement shown in Fig. 10.7 using mosfets, it is significant that there is no external connection to the inter-transistor link. This suggests that there is no need for two separate transistors. The same effect, i.e. stable r.f. amplification, can be achieved using a single transistor with two gates controlling the channel conductivity as in Fig. 10.8. Such dual-gate fets are frequently used for r.f. amplification in Band-II receivers. The circuit bears a striking resemblance to that of a tetrode or pentode valve r.f. amplifier: in fact dual-gate fets are often termed tetrode transistors. An advantage of the circuit is that its gain can be controlled by adjustment of the upper-gate voltage – useful for a.g.c. purposes.

I.F. amplifiers

The i.f. amplifier in a superheterodyne receiver uses principles similar to those of an r.f. amplifier but the tuning is fixed and it has the task of providing most of the gain and the adjacent-channel selectivity of the receiver. The shape of the passband is thus very important. Moreover the final i.f. stage has to deliver appreciable power into the detector if

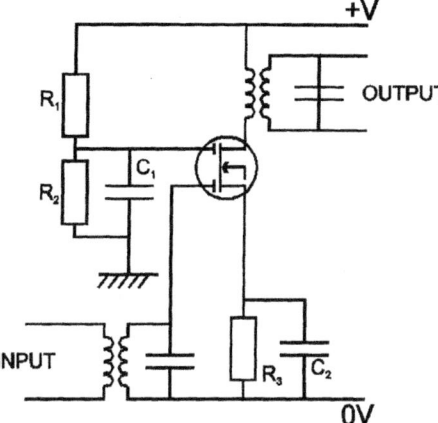

Fig. 10.8. An r.f. amplifier using a dual-gate mosfet

this is a diode. The coupling to the diode must therefore be designed to present the final i.f. stage with its optimum load (i.e. the quotient of available voltage swing and available current swing).

Neutralisation and unilateralisation

The i.f. amplifiers in early transistor receivers used bipolar common-emitter stages and stable operation was achieved by neutralisation, a technique in which the internal positive feedback in the transistors was offset by an equal amount of external negative feedback. This is the second method of avoiding the instability of common-emitter r.f. amplifiers.

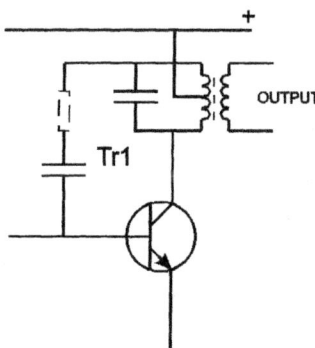

Fig. 10.9. Neutralisation of common-emitter i.f. amplifier using a centre-tapped primary winding in the i.f. transformer. The resistor in dashed lines may be used to improve neutralisation

One popular circuit which was used for neutralisation is shown in Fig. 10.9. The collector supply voltage is introduced via a centre tap on the inductor. At the top end of the inductor the signal voltage is in antiphase with that at the collector and a signal taken from this point gives negative feedback when applied to the base. To make this signal equal to the positive feedback, the negative-feedback path must include a neutralising capacitor C_n equal to the collector-base capacitance. In fact, as shown in Fig. 10.2, the internal feedback path includes resistance in series with the capacitance and to make neutralisation more perfect an equal resistor R_n was sometimes included in series with C_n as shown in dashed lines: this process was termed unilaterisation since it made the transistor truly a one-way device with no feedback from output to input.

456 kHz i.f. amplifier

As a numerical example let us consider the design of a 465 kHz i.f. transformer required to couple one common-emitter stage to another in an a.m. medium-wave and long-wave receiver as shown in Fig. 10.10.

A suitable value for the passband is 7 kHz and the required damped value of Q is given by

$$Q_d = \frac{\text{centre frequency}}{\text{passband}} = \frac{465}{7} = 66.5$$

The turns ratio of the transformer should be chosen to present the collector with a suitable value of load, say $10\,\text{k}\Omega$. The input resistance of the following stage is taken as $3.3\,\text{k}\Omega$. This gives us immediately that the turns ratio of the transformer from TR1 collector to TR2 base is given by $\sqrt{(10/3.3)} = 1.74$.

For a practical miniature inductor of the type likely to be used in an i.f. transformer a typical value of undamped Q is 100. We can now calculate the required value of inductance and capacitance in the following way. The dynamic resistance R_d of an undamped tuned circuit is given by

$$R_d = L\omega Q_u$$

where Q_u is the undamped Q value. When a resistance R is connected across the tuned circuit the dynamic resistance falls to

$$\frac{R_d R}{R_d + R} = \frac{RL\omega Q_u}{R + L\omega Q_u}$$

This is therefore the damped value of dynamic resistance $L\omega Q_d$ where Q_d is the damped Q value

$$\therefore L\omega Q_d = \frac{RL\omega Q_u}{R + L\omega Q_u}$$

which gives

$$L = \frac{R(Q_u - Q_d)}{\omega Q_u Q_d}$$

Now

$$C = 1/\omega^2 L$$

$$\therefore C = \frac{Q_u Q_d}{\omega R(Q_u - Q_d)}$$

In this derivation we have assumed that R (effectively the input resistance of TR2) is the only resistance damping the tuned circuit. In practice the output resistance of TR1 is also in parallel with LC but practical values of output resistance are likely to be very large compared with R and damping from this source has therefore been neglected.

Substituting the values appropriate to our numerical example we have

$$C = \frac{66.5 \times 100}{6.284 \times 465 \times 10^3 \times 10^4 \times 35.5} \text{ F}$$

$$= 6790 \, \text{pF}.$$

This is the value of capacitance needed in the circuit of Fig. 10.10. There is, however, no need to use such an awkward value. A more convenient value such as 250 pF can be used if the collector is connected to a tapping point on the inductor as shown in Fig. 10.11. L is, of course, adjusted empirically to give resonance at 465 kHz but we can calculate its value by substitution in the expression $1/\omega^2 C$ which gives L as approximately 500 µH. The position of the tapping point must be such that the collector is presented with an effective capacitance of 6,790 pF. The position is given by $\sqrt{(250/6,790)}$, i.e. approximately 0.2, implying that the collector tap should be at 1/5th the length of the coil measured from the earthy end.

To calculate the voltage stage gain we must know the effective collector load resistance. TR2 input resistance appears at TR1 collector

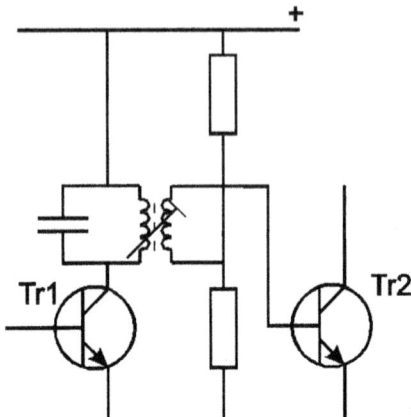

Fig. 10.10. Simplified circuit diagram of two successive stages of i.f. amplification

as a resistance of $10\,k\Omega$ but we must also take into account the dynamic resistance of the tuned circuit. This is given by $Q/\omega C$ and for a Q value of 100 and a capacitance of 6,790 pF gives a dynamic resistance of about $5\,k\Omega$. The effective load resistance is thus $10\,k\Omega$ in parallel with $5\,k\Omega$, i.e. $3.3\,k\Omega$. If we assume TR1 mean emitter current to be 1 mA its mutual conductance is approximately 40 mA/V, giving the voltage gain from base to collector as 40×3.3, i.e. 132. The transformer reduces this to 132/1.74 approximately 75 at TR2 base. If the theoretical value of collector load resistance of $10\,k\Omega$ could be realised the voltage gain would be 40×10, i.e. 400 to TR1 collector and 230 at TR2 base. The

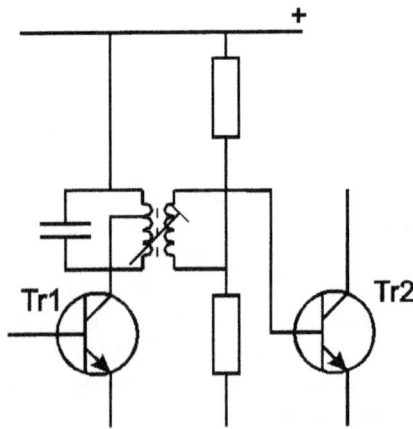

Fig. 10.11. Modification of Fig. 10.10 to permit use of a standard value of tuning capacitor

insertion loss is hence considerable, reducing the gain to one third. This agrees with the factor deduced earlier $(Q_u - Q_d)Q_u$ which, if we substitute the appropriate values, gives 0.33.

Stability by gain limitation

The third method of avoiding the instability of tuned common-emitter amplifiers is to limit the gain, by choice of source and load resistance, to values at which instability cannot occur. Although gain is limited it is still possible using this design technique to obtain adequate i.f. gain for a television receiver from three cascaded stages.

Stability factor

Appendix C shows that a transistor with similar tuned circuits tuned to the same frequency in collector and base circuits will oscillate if

$$\omega c g_m R_b R_c = 2 \tag{10.1}$$

where R_b and R_c are the resistances effectively in parallel with the base and collector circuits and c is the internal base-collector capacitance.

To be certain of stability the left-hand side of Eqn 10.1 must be less than two and the factor by which the expression must be multiplied to give two is known as the *stability factor*. For a stability factor of 4

$$\omega c g_m R_b R_c = 0.5$$

Stability factors between 2 and 8 are used in designing amplifiers without neutralising, the chosen value depending on likely spreads in g_m and c and on such considerations as whether single or double-tuned circuits are used. Such amplifiers are said to be designed to have *stability-limited gain*. We shall now consider the design of an i.f. amplifier of this type for an f.m. sound receiver.

10.7 MHz i.f. stage

For f.m. receivers the standard intermediate frequency is 10.7 MHz and for transmissions with a maximum rated deviation of ±75 kHz a bandwidth (to the −3 dB points) of 200 kHz is needed to accept the significant sidebands of the transmission. For adequate selectivity the response is required to be −40 dB at 600 kHz bandwidth. To give such a response it is common to use ceramic filters but three double-tuned transformers can be used instead. Each transformer is thus required to be −1 dB at 200 kHz and −13.3 dB at 600 kHz bandwidth. The flattest response is obtained from a pair of identical coils when these are critically coupled and for this degree of coupling, universal selectivity curves show that the response is −1 dB when

$$\frac{Q \times \text{bandwidth}}{\text{centre frequency}} = 1$$

This gives the required value of damped Q as

$$Q_d = \frac{\text{centre frequency}}{\text{bandwidth}} = \frac{10.7}{0.2}$$

$$= 53$$

Consider now one of the i.f. transistors situated between two of the critically coupled bandpass filters.

For a stability factor of 4 we have

$$R_b R_c = \frac{0.5}{\omega c g_m}$$

and for the type of transistor likely to be used for such an i.f. amplifier a value of c of possibly 0.15 pF can be assumed. We can also assume that g_m is 35 mA/V. Substituting these values in the above expression we have

$$R_b R_c = 1.42 \times 10^6$$

We can let R_c be 8 kΩ for this enables the transistor to deliver a good output-signal amplitude. This gives R_b as approximately 180 ohms.

The problem now is to design the bandpass filters to present the transistor with these values of source and load resistance. An important point is that the dynamic resistance of a tuned circuit is halved when a similar circuit is critically coupled to it. If therefore we design the tuned circuits to give the required source and load resistances in the absence of coupling to their companion tuned circuits, the resistances decrease and stability increases when the couplings are present. This means that instability is not possible even if the tuned circuits are grossly mistuned (as during alignment), mistuning being equivalent to reduction of coupling coefficient. Decrease of load resistance means that the transistor cannot deliver its maximum power but this is unlikely to cause difficulty in early i.f. stages. For the final stage the design should aim at maintaining the optimum load.

The calculated source resistance is small compared with the input resistance of the transistor and the chosen load resistance is small compared with the output resistance. Thus the transistor does not significantly damp the preceding or following tuned circuit and to give the required passband the coils can be designed to have undamped Q values of 53.

If the whole of the tuned circuit is connected directly in the collector circuit R_d must be 8 kΩ. As Q is 53 we can calculate the required tuning capacitance as follows:

$$R_d = \frac{Q_d}{\omega C}$$

$$\therefore C = \frac{Q_d}{\omega R_d} = \frac{53}{6.284 \times 10.7 \times 10^6 \times 8 \times 10^3} \, \text{F}$$

$$= 100 \, \text{pF approximately.}$$

The inductance can similarly be calculated

$$R_d = L\omega Q_d$$

$$\therefore L = \frac{R_d}{\omega Q_d} = \frac{8 \times 10^3}{6.284 \times 10.7 \times 10^6 \times 53} \, \text{H}$$

$$= 2.245 \, \mu\text{H}$$

A tuned circuit with these values of L and C is also used to feed the base of the transistor. The circuits have an (uncoupled) dynamic resistance of 8 kΩ and to obtain the 180 Ω effective source resistance the secondary winding must be tapped, the position of the tapping point being given by

$$\sqrt{\left(\frac{180}{8,000}\right)} = 0.15$$

from the earthy end of the tuned circuit.

Calculation of gain

To calculate the gain of the 10.7-MHz stage suppose that the following stage has a similar transistor fed from a similar tapping point on the secondary winding of the double-tuned transformer. The effective resistance of the primary circuit is 4 kΩ so that the voltage gain between base and collector is given by

$$g_m R_c = 35 \times 10^{-3} \times 4 \times 10^3 = 140$$

The voltage across the secondary winding is equal to that across the primary and the gain to the tapping point is thus $140 \times 0.0337 = 9.13$ (18 dB).

Use of capacitance tapping

Inductors of the order of $2\mu H$ such as are required in 10.7 MHz i.f. amplifiers are unlikely to have more than approximately 10 turns and it is therefore difficult to obtain desired positions of tapping points with precision. An alternative solution is to tap the capacitive branch of the tuned circuit instead. For example to obtain a tapping point 0.15 along an inductor as required in the last numerical example we can proceed thus. If the capacitor of 100 pF is replaced by a series combination of two capacitors C_a and C_b then to keep the tuning capacitance at 100 pF we have

$$\frac{C_a C_b}{C_a + C_b} = 100 \text{ pF} \tag{10.2}$$

The 'tapped down' output is of course taken from the larger of the two capacitors. If this is C_b we have

$$\frac{100}{C_b} = 0.15$$

This gives

$$C_b = 650 \text{ pF}$$

Substituting in Eqn 10.2 above

$$\frac{650 C_a}{650 + C_a} = 100$$

from which

$$C_a = 120 \text{ pF}$$

A typical 10.7 MHz i.f. stage using capacitance tapping is shown in Fig. 10.12.

Television i.f. amplifiers

Television i.f. amplifiers require bandwidths wider than any so far described. For example, for 625-line television a bandwidth of

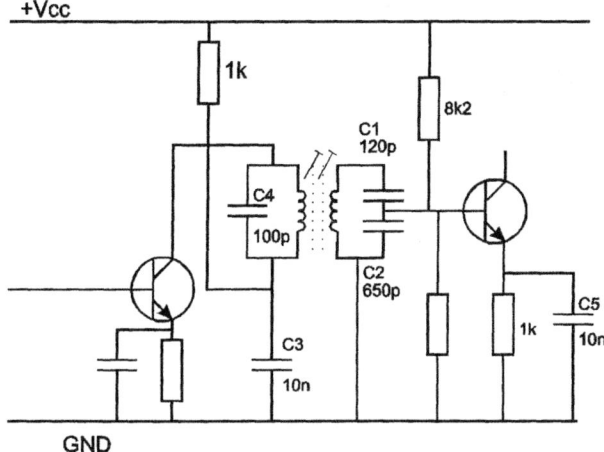

Fig. 10.12. A stage of a 10.7 MHz i.f. amplifier using double-tuned transformers and capacitive tapping

approximately 5.5 MHz is required to accommodate the vision sidebands but the sound signal is also normally included and this extends the bandwidth to 6 MHz. The intermediate frequency (corresponding to the vision carrier frequency) is 39.5 MHz and the required passband is from 33.5 to 39.5 MHz, the centre frequency being 36.5 MHz.

The i.f. amplifiers so far considered have required a symmetrical frequency response: in fact their ideal shape would be a rectangle. However, the frequency response of a television i.f. amplifier must satisfy particular requirements. For example, the vestigial sideband transmission used for television broadcasting requires the low-frequency response to fall off according to a particular law and to be 6 dB down at the frequency corresponding to the vision carrier. Steep notches are also required to reject the vision and sound transmissions on adjacent channels. These requirements dictate the shape of the i.f. response which is achieved in modern receivers by the use of a s.a.w. (surface-acoustic-wave) filter, the frequency characteristics of which can be controlled during manufacture.

In earlier receivers an approximation to the ideal i.f. response was achieved by use of conventional LC circuits in a three-stage common-emitter i.f. amplifier with some rejectors to give the necessary notches. Some idea of the L and C values required can be obtained in the following manner.

A single tuned circuit with a centre frequency of 36.5 MHz and a bandwidth of 6 MHz needs a damped Q value given by 36.5/6, i.e. 6.1, much lower than any so far considered. Undamped Q values are very large compared with this and thus insertion losses are low.

In a mains-driven receiver the final i.f. stage might be designed to take a mean collector current of 10 mA from a 20-V supply. The optimum load is then approximately 2 kΩ and this is also the approximate input resistance of the vision detector (a load resistance of 4 kΩ is assumed). The coupling circuit can thus take the form of a 1:1 transformer with the detector circuit providing the damping. The output resistance of the final i.f. stage, even at 10-mA collector current, is likely to be very large compared with 2 kΩ. The damped dynamic resistance of the collector load is thus 2 kΩ: from this and the damped Q value of 6.1 we can calculate the tuning capacitance and inductance required in the following way:

$$R_d = \frac{Q_d}{\omega C}$$

$$\therefore C = \frac{Q_d}{\omega R_d}$$

$$= \frac{6.1}{6.284 \times 36.5 \times 10^6 \times 2 \times 10^3} \text{ F}$$

$$= 13.3 \text{ pF}$$

a small capacitance but a feasible value to use in view of the extremely low values of output capacitance of silicon planar transistors. The inductance value can be calculated thus

$$L = \frac{R_d}{\omega Q_d}$$

$$= \frac{2 \times 10^3}{6.284 \times 36.5 \times 10^6 \times 6.1} \text{ H}$$

$$= 1.43 \mu\text{H}$$

As the transformer has unity turns ratio the current gain of the final i.f. stage is equal to β, e.g. 50. To avoid instability in an unneutralised stage (as this is assumed to be) the effective value of the source resistance should be calculated as indicated earlier in this chapter and the input circuit designed to provide this.

The earlier stages of the i.f. amplifier could be designed in a similar manner to that just described, using the input resistance of the following stage to provide the required damping. It would be difficult however

Table 10.1

Position	Type of tuned circuit	Resonance frequency (MHz)	Q
Between tuner and 1st i.f. stage	Bandpass pair	36.5	6
Between 1st and 2nd i.f. stages	Single	36.0	20
Between 2nd and 3rd i.f. stages	Single	38.0	20
Between 3rd i.f. stage and det.	Single	36.5	6

with a succession of synchronously tuned circuits each with a Q of 6, even with additional rejector tuned circuits, to provide the required shape of response curve. This difficulty can be resolved by using higher Q values such as 20 or 30 in the earlier i.f. stages and by tuning them to frequencies displaced from the centre frequency. By suitable choice of Q values and resonance frequencies, this method of stagger tuning can give an approach to the required shape of response curve although it is still necessary to use additional rejector circuits. The design of these earlier stages can follow the procedure indicated above for 'stability limited gain' stages, the external collector resistance being so chosen that the transistor can supply the required amplitude of signal, the external base resistance being made low enough to provide an adequate margin of protection against instability.

Typical values of Q and resonance frequencies are given in Table 10.1.

Automatic gain control (A.G.C.)

Most receivers use some form of automatic gain control to minimise the effects of ionospheric or man-made* fading and to ensure that all signals, irrespective of their input amplitude, are reproduced at substantially constant amplitude. A.g.c. is achieved by controlling the gain of i.f. and r.f. stages by a voltage (or current) derived from the signal at the detector output or at a post-detector point. Some means is therefore required of adjusting the gain of a transistor amplifier by a control signal.

The gain of a transistor falls at low collector currents and at low collector-emitter voltages and there are two corresponding ways of achieving gain control. The first is by applying the control signal to the base as a reverse bias: this is known as *reverse control* and an example, using an npn transistor and negative-going control bias, is given in

_____
* e.g. aircraft flutter in television reception.

Fig. 10.13. A pnp transistor would, of course, require a positive-going bias for reverse control. For both types of transistor the control bias increases when a strong signal is received thus biasing back the transistor and reducing the gain. An unfortunate feature of this type of control is that the signal-handling capacity of a transistor is reduced by reverse bias but the circuit has the advantage that collector current is reduced when strong signals are received: this is important in battery-operated receivers where current economy is desirable.

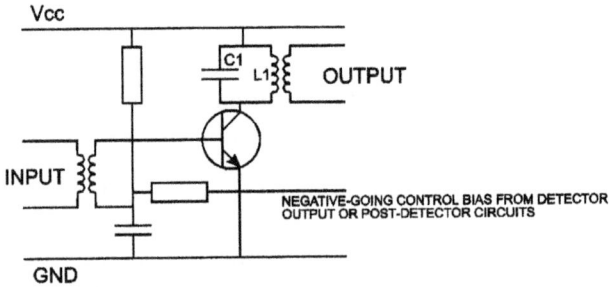

Fig. 10.13. An example of reverse a.g.c.

In the second method, known as *forward control*, the gain of the transistor is reduced by increasing the forward bias, thus increasing collector current. An essential feature of the circuit, illustrated in Fig. 10.14, is the decoupling circuit R_2C_2: as the collector current increases, the voltage drop across R_2 is increased thus reducing the collector-emitter voltage and forcing the operating point to move into the knee of the I_c-V_c characteristics where the characteristics are more

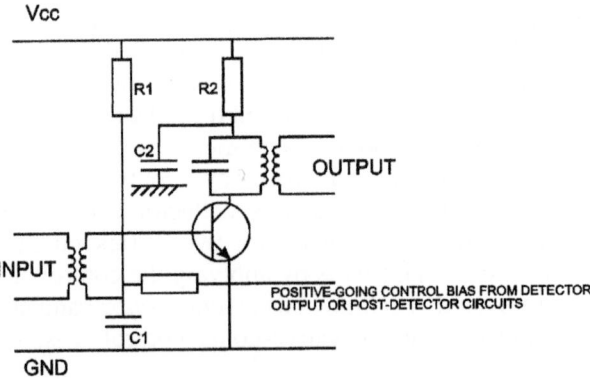

Fig. 10.14. An example of forward a.g.c.

crowded and the g_m therefore lower. For this type of control a pnp transistor requires a negative-going bias and an npn a positive-going bias. Forward control has the advantage of increasing the signal-handling capacity of the transistor when this is needed for large-amplitude signals. Not all transistors are suitable for forward control and it is important to select a type which has been specifically designed for use in this type of circuit. By increasing the collector current of a suitable transistor from 4 to 13 mA, it is possible to reduce the gain by more than 40 dB.

Both methods of a.g.c. are in common use, sometimes in the same receiver. Both forms of control reduce the power output of the controlled stage. This is of little significance in early i.f. stages but it is not usual to apply a.g.c. to the final i.f. stage because this is required to supply appreciable power to the detector.

Decoupling

In Fig. 10.12, the source of base input signal (i.e. the secondary winding of the i.f. transformer) and the emitter decoupling capacitor are both returned to the negative terminal of the power supply. This minimises the impedance of the external base-emitter circuit which is essential for maximum performance. The primary winding of the i.f. transformer must be returned to the positive terminal of the supply to provide the necessary collector bias but should be returned to the negative terminal of the supply in order to minimise the impedance of the external collector-emitter circuit which is also essential for best results. This difficulty is usually overcome by connecting a low-reactance capacitor C_3 (the collector decoupling capacitor) between the primary winding and the emitter (or the negative terminal of the emitter decoupling capacitor). If the collector decoupling capacitor is omitted, the signal output current of the transistor must flow through the impedance of the supply in addition to the primary winding of the transformer. In this way, signal voltages are developed across the battery and these, if impressed upon other stages of the amplifier or receiver of which this stage is part, can distort the shape of the frequency-response curve or even cause instability. To avoid this, the collector decoupling capacitor C_3 is included to short-circuit the supply at signal frequencies.

However, by a simple alteration to the circuit such a capacitor becomes unnecessary. If the source of base input signal (i.e. the secondary winding) and the emitter decoupling capacitor are both returned to the positive terminal of the supply, as shown in Fig. 10.15,

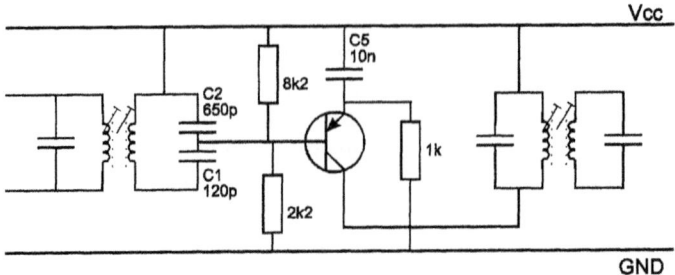

Fig. 10.15. A modified version of the circuit of Fig. 10.12 which eliminates the need for collector decoupling components

then the impedance of the external base-emitter circuit and of the external collector-emitter circuit are minimised simultaneously and no additional decoupling capacitor is necessary.

Use of integrated circuits

Modern radio and television receivers and video recorders use integrated circuits to provide the i.f. gain. These i.c.s embody several stages of amplification and often contain circuitry for other functions such as detection, a.g.c., a.f.c., limiting and noise reduction. External components provide the i.f. selectivity. *LC* circuits are still used at

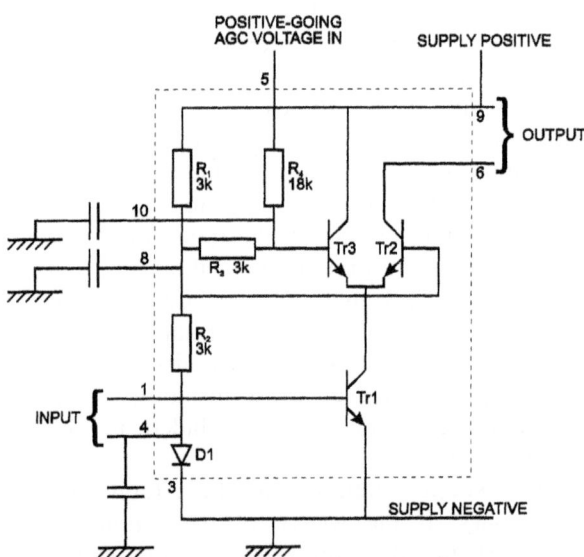

Fig. 10.16. Circuit diagram of a monolithic i.f. or r.f. amplifier

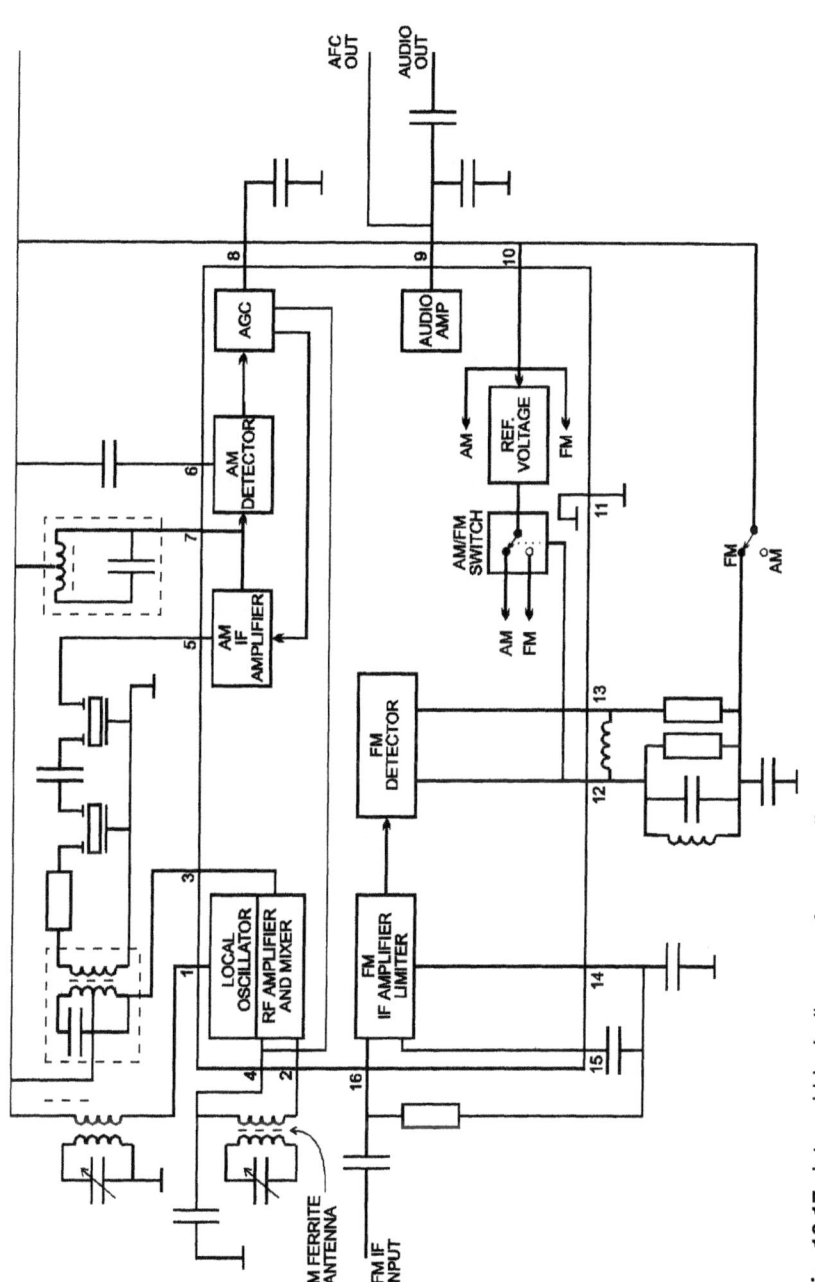

Fig. 10.17. Internal block diagram of an a.m./f.m. radio receiver i.c.

465 kHz in spite of the need for alignment. They are also used at 10.7 MHz but are often supplemented by ceramic filters which have sharper cut-offs at the edges of the passband and do not require alignment. In television receivers and video recorders the i.f. selectivity is commonly provided by a surface-acoustic-wave filter which is manufactured to have the particular shape of passband required and needs no subsequent adjustment. A ceramic filter is used in the 6 MHz inter-carrier sound i.f. amplifier.

The circuit diagram of an i.f. integrated circuit is far too complex to reproduce here but an illustration of the type of circuitry employed can be gained from the diagram of an early and very simple i.c. given in Fig. 10.16. This has only a single stage of amplification.

TR1 and TR2 form a cascode amplifier and TR3, the a.g.c. transistor, is in parallel with TR2 and its output load. TR2 and TR3 are biased from the same point on the potential divider $R_1 R_2$ and TR1 is biased from the diode D_1 which ensures constancy of mean current in TR1.

When TR3 is forward biased it shunts the output load and reduces the output-signal level. The shunting effect becomes more significant as the positive a.g.c. voltage on TR3 base is increased. This is an effective method of controlling gain and has the merit of leaving the input resistance and input capacitance of the amplifier substantially constant.

A block diagram of a typical a.m./f.m. receiver i.c. together with some of the associated external circuitry is given in Fig. 10.17. The i.c. incorporates all the a.m. stages from aerial input to demodulated audio output, together with a four-stage f.m. amplifier/limiter and quadrature synchronous demodulator (see page 236) for f.m. reception. The complete circuit diagram of a receiver using a similar i.c. is given in Fig. 12.21.

Chapter 11

Sinusoidal oscillators

Introduction

Sinusoidal oscillators have widespread applications in electronics. To name only a few obvious examples, they provide the carrier source in transmitters and form part of the frequency changer in superhet receivers. They are used for erasing and biasing in magnetic recording and time the clock pulses in computers. Many electronic measuring instruments incorporate oscillators and a variable-frequency oscillator forms part of a phase-locked-loop detector.

There are many different types of sinusoidal oscillator but all consist essentially of two parts:

1 A frequency-determining section which may be a resonant circuit or an RC network. The resonant circuit can be a combination of lumped inductance and capacitance, a length of transmission line or a cavity resonator depending on the frequency required.

   RC networks have, of course, no natural resonance but the phase shift of the network can be used to determine the frequency of oscillation.

2 A maintaining section which supplies energy to the resonant circuit to keep it in oscillation. The maintaining section itself requires a d.c. source of power. In many oscillators this section is an active device, e.g. a transistor which delivers regular pulses to the resonant circuit.

   An alternative form of maintaining section for a resonant oscillator is a source of negative resistance, i.e. a device or an electronic circuit for which an increase in applied voltage brings about a decrease in current through it. There are a number of semiconductor devices or circuits with such a characteristic.

The maintaining section of an RC oscillator must provide the energy needed to sustain the oscillation controlled by the phase-shift network and sufficient gain to offset the attenuation in the network.

Positive-feedback oscillators

A parallel tuned *LC* circuit resonates at a frequency of approximately

$$f = \frac{1}{2\pi\sqrt{LC}}$$

Energy oscillates between the inductor and capacitor. The inductor stores the energy as a magnetic field, and the capacitor stores it as an electric field. At resonance, the impedance is at a maximum, and little current is drawn from the supply (Fig. 11.1).

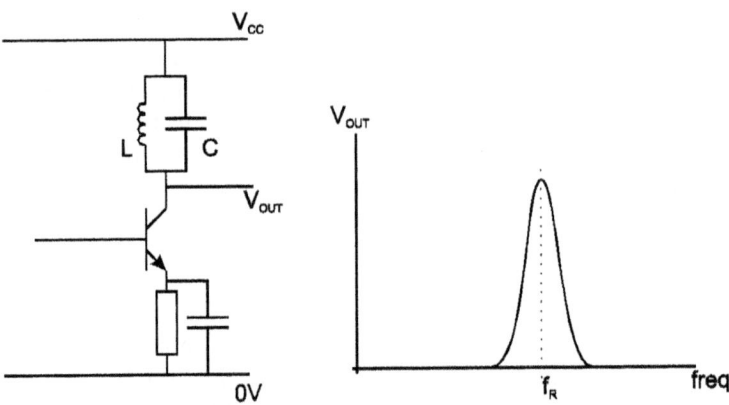

Fig. 11.1. Resonance effects with a parallel *LC* circuit collector load

Now consider the effect of tapping the inductor part-way along its length. The circuit still resonates at a frequency of

$$f = \frac{1}{2\pi\sqrt{LC}}$$

In this circumstance, the voltage across the inductor 'see-saws' about the earthed tap. (The tap can also be taken to the supply voltage, or any point with a low a.c. impedance.) This means that the ends of the inductor present voltages which are in antiphase with each other.

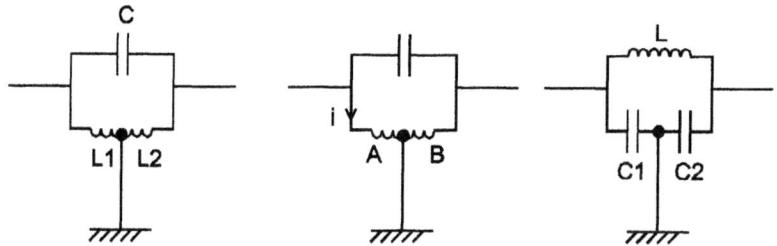

Fig. 11.2. Tapped inductors and capacitors

When the current i passes through the inductor in the direction shown (Fig. 11.2), end A is positive while end B is negative with respect to the tap. This forms a basic Hartley circuit. A Colpitts circuit is obtained by constructing the capacitor of two parts, C_1 and C_2. Since they are in series, the total capacitance is

$$\frac{1}{C_r} = \frac{1}{C_1} + \frac{1}{C_2}$$

In both of the circuits of Fig. 11.3, the voltage BE is in antiphase with the voltage CE. This is one of the requirements for an inverting amplifier to oscillate. The other is that the amplifier should amplify the signal by at least the amount given by the ratio of the collector-emitter voltage to the base-emitter voltage.

	Colpitts	*Hartley*
Gain	$= \dfrac{V_{CE}}{V_{BE}}$	$= \dfrac{V_{CE}}{V_{BE}}$
	$= \dfrac{i \cdot X_{C_1}}{i \cdot X_{C_2}}$	$= \dfrac{i \cdot X_{L_1}}{i \cdot X_{L_2}}$
	$= \dfrac{1}{\omega C_1} \times \dfrac{\omega C_2}{1}$	$= \dfrac{\omega L_1}{\omega L_2}$
	$= \dfrac{C_2}{C_1}$	$= \dfrac{L_1}{L_2}$
Frequency	$\dfrac{1}{2\pi \sqrt{L \dfrac{C_1 C_2}{C_1 + C_2}}}$	$\dfrac{1}{2\pi \sqrt{(L_1 + L_2)C}}$

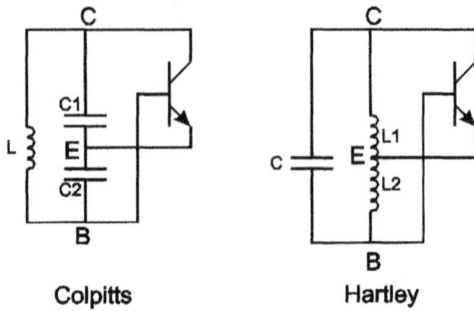

Fig. 11.3. The basic structure of Colpitts and Hartley oscillators

Hartley oscillator

This is an example of a low-frequency oscillator using an *LC* circuit for frequency determination and a transistor supplying maintaining pulses. The circuit diagram (Fig. 11.4) shows a common-emitter amplifier with the *LC* circuit connected between collector and base, the centre tap of the inductor being effectively connected to the emitter (the power supply being regarded as having zero resistance). A common-emitter amplifier inverts its input signal and its output signal is inverted by the centre-point-earthed inductor before it is applied to the base. The circuit can thus be regarded as that of an amplifier which supplies its own input: in other words there is considerable positive feedback and this causes oscillation, the amplitude of the signal (at the resonance frequency of L_1C_1) building up rapidly. The resulting pulses of base current charge up C_b, the polarity of the voltage thus generated biasing the base negatively. As the signal amplitude grows, so does the voltage across C_b until equilibrium is reached when the losses from the *LC* circuit due to output loading, ohmic resistance and base current equal the energy supplied to it from the collector. In this final condition the transistor may well be cut off for most of each cycle of oscillation, taking a burst of base current (and collector

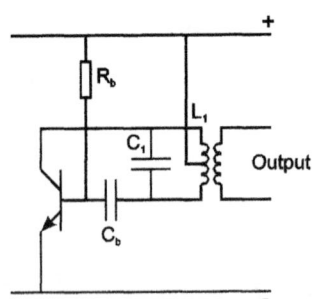

Fig. 11.4. One form of Hartley oscillator using a bipolar transistor

current) on each positive peak at the base. In the interval between successive peaks C_b begins to discharge through R_b but, if the time constant R_bC_b is large compared with the period of oscillation, little of the voltage across C_b is lost in the interval and C_b can be regarded as a source of steady negative bias. Such a biasing system is used in many oscillators. It has the advantage of compensating to some extent for any reduction in oscillation amplitude caused, for example, by increased output loading or a fall in supply voltage. Reduction in oscillation amplitude causes reduced bias so that the transistor takes larger current pulses which tend to maintain the amplitude. This is, in fact, an example of class C operation (see page 180).

Colpitts oscillator

It is significant that three connections were needed between tuned circuit and transistor in Fig. 11.4 to provide positive feedback. The emitter connection is effectively connected to the centre point of the inductor but it could equally well be connected to the capacitive branch of the LC circuit by using two equal capacitors in series as shown in Fig. 11.5(a). This oscillator uses a jfet with a resistor R_d in the drain circuit, the LC circuit being coupled to the drain by a capacitor C_d. Thus the LC circuit is shunt-fed by contrast with the direct feed of Fig. 11.4.

The frequency-determining capacitors C_1 and C_2 are in parallel with the input and output capacitances of the transistor which therefore have an effect on oscillation frequency. The effect can be minimised by making C_1 and C_2 as large as possible. On the other hand if oscillation is required at a high frequency necessitating a very small tuning capacitance, the input and output capacitances of a transistor can be used as C_1 and C_2, a small variable capacitor being connected across L for tuning as shown in Fig. 11.5(b). This technique is sometimes used in

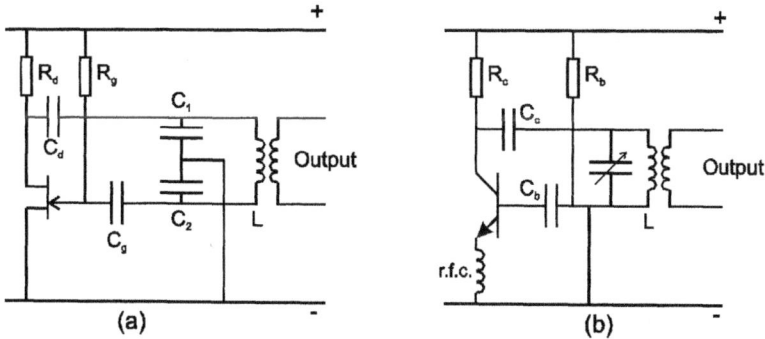

Fig. 11.5. A Colpitts oscillator using (a) a jfet and (b) a bipolar transistor

v.h.f. and u.h.f. oscillators. Biasing is again automatically provided by C_b which is charged by pulses of gate current and discharges through R_b. To permit the moving vanes of the tuning capacitor (and hence the transistor base) to be earthed, an r.f. choke with high impedance at the operating frequency is included in the emitter circuit.

All three oscillators described above operate in class C for large oscillation amplitudes. To obtain a sinusoidal waveform the output must be taken from the *LC* circuit, e.g. by a coil inductively coupled to the *LC* circuit as shown in Figs. 11.4 and 11.5. If the output is taken from the transistor itself, e.g. from a resistor in the emitter or source circuit, it would consist of a pulse train with a pulse repetition frequency equal to the resonance frequency of *LC*.

The automatic bias system used in the above oscillators, relying as it does on the flow of base or gate current, clearly cannot be used with mosfets which do not take gate current. Instead bias can be provided by the potential-divider circuit of Fig. 6.7 and, for small oscillation amplitudes, the transistor operates in class A.

Reinartz oscillator

The Reinartz oscillator (Fig. 11.6) is of interest because it, or a derivative of it, is frequently used in transistor receivers. Positive feedback is achieved here by coupling the collector circuit to the emitter circuit via the mutual inductance between L_1 and L_2. Both inductors are also coupled to the frequency-determining circuit $L_3 C_3$. This oscillator is stabilised by the potential-divider method but in this circuit the lower arm of the potential divider must be decoupled by a low-resistance capacitor as shown to ensure that the signals developed across L_2 are

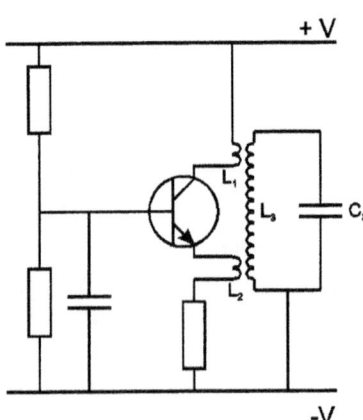

Fig. 11.6. Reinartz oscillator

applied directly between base and emitter. The frequency-determining section of the Reinartz oscillator appears at first sight to have four connections (two from L_1 and two from L_2) but the connections to the positive and negative terminals of the supply are effectively common because the supply has, or should have, negligible impedance at the frequency of oscillation.

This circuit can be simplified by omitting the components L_3 and C_3 and connecting a tuning capacitor across L_1 or L_2 but the form illustrated here is often preferred in receivers because it permits the moving vanes of the tuning capacitor to be earthed: in addition the loose coupling between $L_3 C_3$ and the transistor helps to improve frequency stability.

Crystal-controlled oscillator

There are certain applications in which oscillators require great frequency stability. Examples are carrier sources in transmitters, clock source generators in computers, and in frequency-synthesis tuning systems in receivers. A common method of achieving the required stability is by using a piezoelectric crystal to control the frequency of oscillation. Such crystals have a natural frequency of resonance (depending on the size and shape) and at that frequency behave as a tuned circuit with a very high Q. The very high Q is associated with a very narrow bandwidth, so the frequency of oscillation can be very precise. The frequency can be 'pulled' by a small fraction of a per cent of the resonant frequency with a small tuning capacitor or varactor diode. In use a crystal is mounted between two metal plates (sometimes in the form of a metallic coating on the faces) which provide electrical contact with the crystal. There are many ways in which a crystal can be connected into an oscillator circuit: one

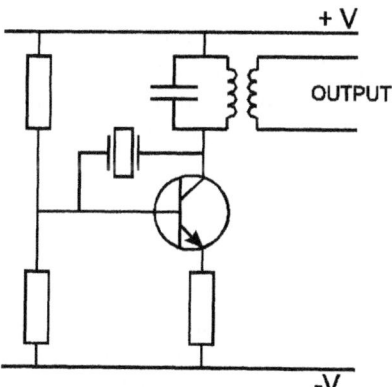

Fig. 11.7. Example of a crystal-controlled oscillator circuit

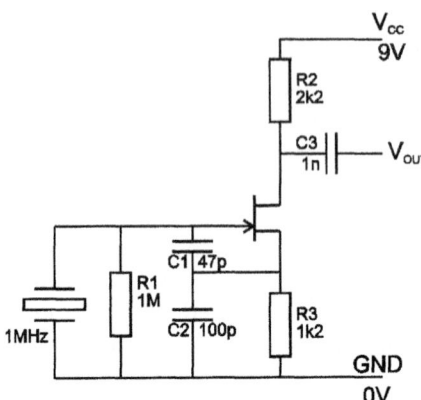

Fig. 11.8. A simple Jfet-based crystal oscillator

example is given in Fig. 11.7. Here the crystal is connected between collector and base of a bipolar transistor to form a Colpitts oscillator, the collector-base and base-emitter internal capacitances providing positive feedback. The collector circuit need not be tuned but the secondary winding of the transformer provides a convenient output point.

Fig. 11.8 shows an application of a crystal-stabilised oscillator which uses a jfet as the active device.

Dielectric resonator oscillator

Allied to the crystal oscillator is the type used as local oscillator in first-stage converters sited at the head-end of satellite broadcast receivers, a typical circuit for which is given in Fig. 11.9. TR1 is a GaAs fet working on the principle of reflection or feedback oscillation. Its operating frequency is typically 10 GHz, stabilised by the dielectric resonator located in the region of the microstrip line in its gate circuit; it has no electrical connection to the circuit at all, mutual coupling being achieved by radiation from the stripline. A characteristic of the dielectric resonator is its excellent frequency stability, in spite of wide temperature changes: drift is held below 1 MHz, representing a stability of 100 parts per million.

The dielectric resonator is formed as a small circular pill of high-Q ceramic material with a high dielectric constant; it simulates a cavity resonator and is usually inductively coupled into the circuit so that the adjacent stripline takes on the characteristics of a circuit tuned to the frequency of the resonator.

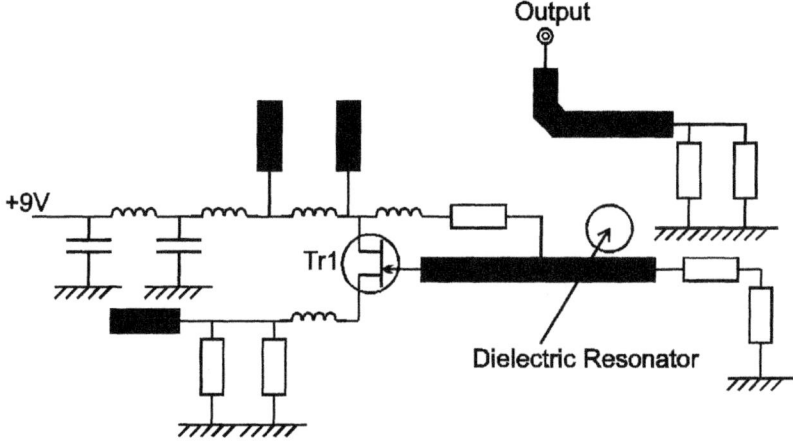

Fig. 11.9. Microwave oscillator for low-power use at 10 GHz

Phase-shift oscillators

Introduction

In all the oscillators so far described, apart from the dielectric resonator type, three connections are necessary to the frequency-determining section to achieve the signal inversion which, with the inversion in the maintaining section, gives the positive feedback essential for oscillation. For a sinusoidal signal the effect of 180° phase shift is identical to that of waveform inversion. This suggests that an oscillator could be constructed by using a 180°-phase-shift network as a frequency-determining section provided that the maintaining section can offset the attenuation of the phase-shift network at the operating frequency.

An alternative approach is to use a non-inverting amplifier as maintaining section and a phase-shift network which gives zero phase shift at the wanted oscillation frequency.

Both principles can be used successfully in oscillators as the following descriptions show. And again three connections are needed between phase-shift network and maintaining system.

RC oscillators

A single RC circuit such as that illustrated in Fig. 11.10(a) introduces a phase shift which increases as frequency falls and approaches a limiting value of 90° as frequency approaches zero. At such low frequencies, however, the attenuation introduced by the circuit is very great. Two such sections in cascade could produce 180° phase shift but the attenuation would be enormous. To give 180° phase shift with

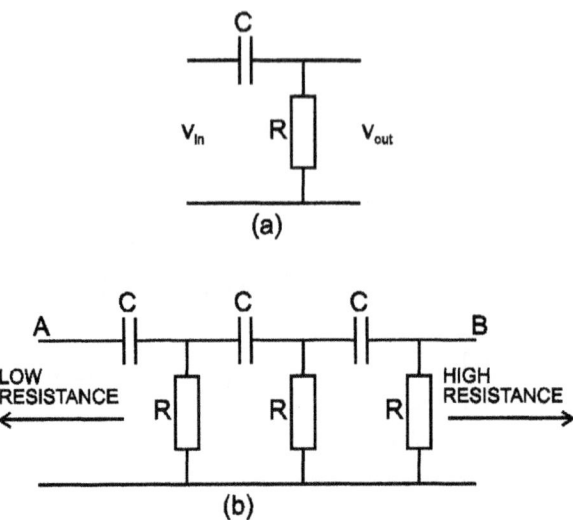

Fig. 11.10. A simple section of an RC network (a), and a three-section network (b)

reasonably small attenuation, a network of at least three sections is necessary as shown in Fig. 11.10(b). For successful results end A of this network should be terminated in a low resistance and end B in a high resistance. Thus if the input resistance of the amplifier is large compared with its output resistance, as in a voltage amplifier such as a field-effect transistor, A should go to the output and B to the input.

If, however, the amplifier input resistance is small compared with its output resistance, as in a current amplifier such as a bipolar transistor, A should go to the input and B to the output as shown in the circuit diagram of a complete RC oscillator shown in Fig. 11.11.

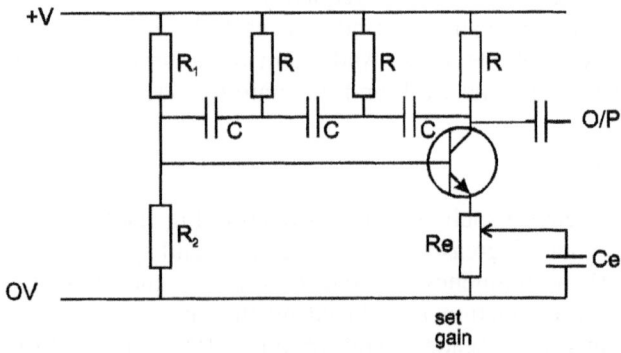

Fig. 11.11. Simple phase-shift oscillator suitable for single frequency operation

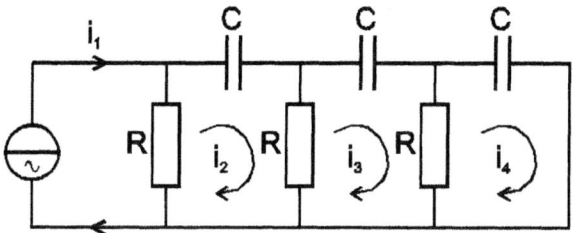

Fig. 11.12. A three-section RC network

The frequency of oscillation and the attenuation of the network can be determined as follows. To simplify the analysis we shall assume the network to be supplied from a constant-current source and that the network is terminated in a short circuit. Thus the circuit has the form shown in Fig. 11.12 in which i_1 is the signal current from the collector circuit and i_4 is the signal current into the base circuit. From Kirchhoff's law we have

$$i_1 R = i_2(2R + jX) - i_3 R \tag{11.1}$$

$$i_2 R = i_3(2R + jX) - i_4 R \tag{11.2}$$

$$i_3 R = i_4(R + jX) \tag{11.3}$$

where $X = -1/\omega C$.

Eliminating i_2 between Eqns 11.1 and 11.2

$$i_1 R = \frac{(2R + jX)^2}{R} \cdot i_3 - (2R + jX)i_4 - i_3 R \tag{11.4}$$

Eliminating i_3 between Eqns 11.3 and 11.4

$$i_1 R = \frac{(2R + jX)^2(R + jX)}{R^2} \cdot i_4 - (2R + jX)i_4 - (R + jX)i_4$$

from which

$$\frac{i_4}{i_1} = \frac{R^3}{R^3 + 6jXR^2 - 5RX^2 - jX^3} \tag{11.5}$$

For 180° phase shift between i_4 and i_1, the terms in j vanish and we have

$$6jXR^2 = jX^3$$

giving

$$X = \sqrt{6}R \tag{11.6}$$

$$\therefore -\frac{1}{\omega C} = \pm \sqrt{6}R$$

and

$$\omega = \frac{1}{\sqrt{6}RC}$$

The frequency of oscillation is thus given by

$$f = \frac{1}{2\pi\sqrt{6}RC} \tag{11.7}$$

At this frequency we have, from Eqn 11.5

$$\frac{i_4}{i_1} = \frac{R^3}{R^3 - 5RX^2}$$

But, from Eqn 11.6

$$X^2 = 6R^2$$

$$\therefore \frac{i_4}{i_1} = \frac{R^3}{R^3 - 30R^3}$$

$$= -\frac{1}{29} \tag{11.8}$$

Thus the transistor must give a current gain of at least 29 to achieve oscillation. This is the ratio of base current to current in the collector load resistor and some of the output current is shunted through the parallel paths presented by the network. A ratio of i_4 to i_1 of 29 thus represents a high order of gain and it is desirable to select a transistor with a high value of β to give oscillation. On the other hand, to obtain a sinusoidal output the transistor must not oscillate too strongly and the gain should be adjusted to give only a small amplitude of oscillation. One way of controlling the gain is by adjustment of the value of emitter resistor R_e as shown in Fig. 11.11. Reduction in R_e increases I_e, hence increases g_m and gain. If excessive gain is used the oscillation amplitude

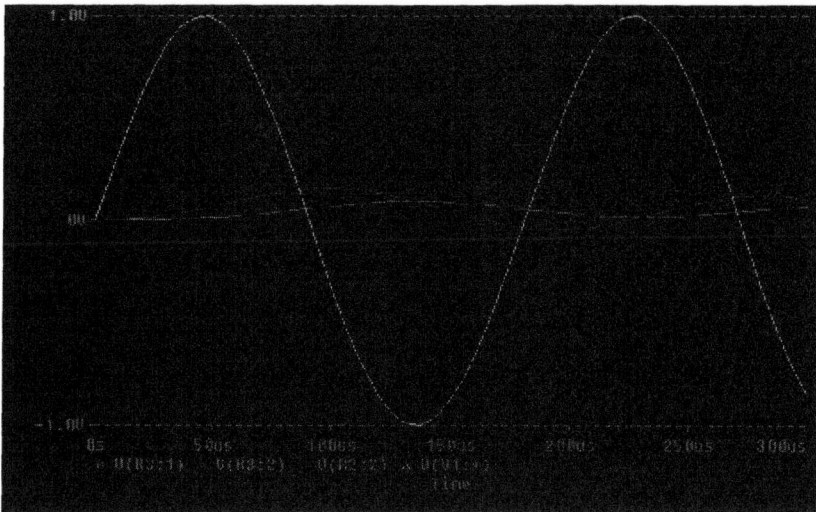

Fig. 11.13. Waveforms in the *R–C* Ladder

will be checked by the collector current reaching zero or the collector voltage reaching the base voltage and the output waveform then contains some linear sections and is distorted. The waveforms in Fig. 11.13 show the amplitude and phase relationships at all of the nodes in an *R–C* ladder. The output waveform is the smallest, and shows a phase shift of 180° at one frequency only.

A convenient value for the frequency-determining resistors is $4.7\,k\Omega$ because this permits a collector current of 1 or $2\,mA$ with a low supply voltage. To give oscillation at say 1 kHz the capacitance required can be calculated from Eqn 11.7 thus

$$C = \frac{1}{2\pi\sqrt{6}f\,R}$$

$$= \frac{1}{6.284 \times 2.45 \times 10^3 \times 4.7 \times 10^3}\,F$$

$$= 0.015\,\mu F \text{ approximately.}$$

In practice the frequency is unlikely to be precisely 1 kHz because the transistor output resistance is not infinite and its input resistance is not zero as assumed in the analysis.

A good way of ensuring purity of waveform in the output of such an oscillator would be to add an a.g.c. system, e.g. a shunt-diode detector fed via its series capacitor from the collector, and apply its output as positive bias to the base via a long-time-constant circuit. This would avoid the need for an adjustable emitter resistor.

The circuit of Fig. 11.10 shows the $R–C$ ladder with the resistive elements grounded. If the ladder is constructed with grounded capacitors and resistive series elements, a repeat of the mathematical analysis shows:

Attenuation when the phase shift is $180° = 1/29$

Frequency when the phase shift is $180°$

$$f = \frac{\sqrt{6}}{2\pi\,RC}$$

Wien-bridge oscillator

This is an example of an oscillator in which both sections introduce zero phase shift. The frequency-determining network employed has the form illustrated in Fig. 11.14. The upper arm consists of resistance and capacitance in series, the lower arm of an equal resistance and an equal capacitance in parallel. The network is supplied from a constant-voltage source and is terminated in an infinite impedance: the phase shift and attenuation introduced by the network can then be calculated as follows. The impedance of the parallel RC network is $RjX/(R + jX)$ and thus

$$\frac{v_{out}}{v_{in}} = \frac{\dfrac{RjX}{R + jX}}{R + jX + \dfrac{RjX}{R + jX}}$$

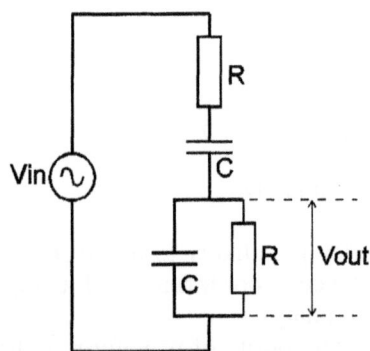

Fig. 11.14. Basic frequency-determining circuit in Wien-bridge oscillator

where $X = -1/\omega C$. Simplifying we have

$$\frac{v_{out}}{v_{in}} = \frac{RjX}{(R + jX)^2 + RjX}$$

$$= \frac{RjX}{R^2 - X_2 + 3RjX}$$

Rationalising

$$\frac{v_{out}}{v_{in}} = \frac{RjX(R^2 - X^2) + 3R^2}{(R^2 - X^2)^2 + (3RX)^2} \tag{11.9}$$

When there is zero phase shift the terms in j vanish and we have

$$R^2 = X^2$$

giving the frequency of zero phase shift as

$$\omega = \frac{1}{RC}$$

$$\therefore f = \frac{1}{2\pi RC}$$

Substituting $R = X$ in Eqn 11.9 to find the attenuation at this frequency we have

$$\frac{v_{out}}{v_{in}} = \frac{3R^2X^2}{9R^2X^2} = \frac{1}{3}$$

The maintaining amplifier thus requires a gain just exceeding 3 to sustain oscillation. Such a low voltage gain could readily be obtained from a single transistor but it is difficult in such a simple amplifier to obtain the desired terminating resistances. Usually, therefore, a two- or even three-stage amplifier is used and the gain is reduced to the required value by negative feedback.

Wien-bridge oscillators are frequently used for a.f. testing purposes. The oscillation frequency can be readily adjusted if a two-gang variable capacitor is used for C_1 and C_2 (giving a 10:1 change in frequency), the resistors R_1 and R_2 being switched in decade steps to give different frequency ranges. By this means it is possible to design an oscillator with a frequency range from say 30 Hz to 30 kHz.

The capacitors used for C_1 and C_2 must be variable and are unlikely therefore to have a maximum capacitance greater than 500 pF. For this capacitance value the values of R_1 and R_2 required to give oscillation at 30 Hz can be calculated by substitution in the above expression. We have

$$R = \frac{1}{2\pi fC}$$

$$= \frac{1}{6.284 \times 30 \times 500 \times 10^{-12}}\ \Omega$$

$$= 10\,\text{M}\Omega \text{ approximately.}$$

Fig. 11.15 gives the circuit diagram of one possible form for an oscillator of this type. The amplifier has three transistors including one common-source stage, one common-emitter stage and an emitter follower output stage. The first two stages give signal inversion and the third does not, so that the required phase characteristic is obtained. The emitter follower gives a low output resistance for feeding the frequency-determining network and the output terminals of the oscillator. As already shown the resistance in the lower arm of the frequency-determining network must be several megohms to give oscillation at low audio frequencies. To avoid resistive shunting of this arm (which would raise the oscillation frequency) the input resistance of the amplifier must be very high and a jfet is the obvious choice for TR1.

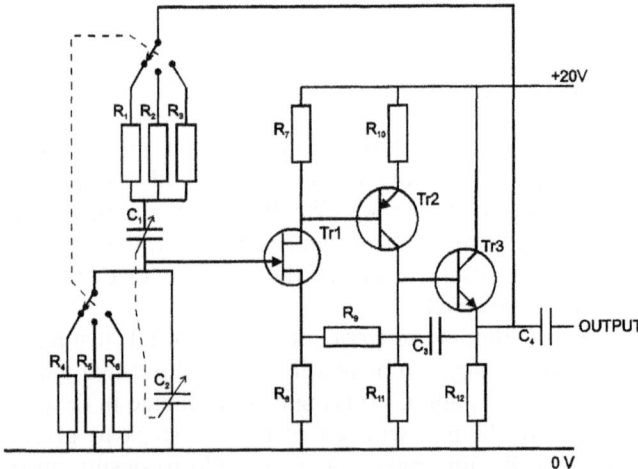

Fig. 11.15. Circuit diagram for a Wien-bridge oscillator with range switching

TR1 and TR2 can, of course, give voltage gain far greater than the value of 3 required to sustain oscillation and it is essential to reduce the gain to this value otherwise oscillation amplitude grows until limited by transistor current cut-off and the required sinusoidal waveform is not obtained. Negative feedback is used to reduce the gain to the required value and to stabilise the amplitude of the output signal.

Direct coupling is used throughout the amplifier. This gives simplicity of design and ensures a good low-frequency response. The low value of zero-frequency gain also ensures good stability of mean transistor currents.

The output amplitude of the oscillator is stabilised by making the negative feedback dependent on it. This could be achieved by using a resistor with a negative resistance-temperature coefficient (e.g. a thermistor) for R_9. R_9 is fed from the amplifier output via C_3 and the dissipation in it and hence its resistance is entirely dependent on the oscillation amplitude. Alternatively a resistor with a positive resistance-temperature coefficient (e.g. a lamp with a metal filament) could be used for R_8, R_9 being a non-temperature-dependent resistor. In both circuits any tendency for the output amplitude to increase, increases negative feedback, decreases amplifier gain and thus checks the rise in amplitude.

The circuit in Fig. 11.16 is based on an operational amplifier, but it could as easily be a bjt or fet amplifier which has been set up to provide a gain of +3. Variation of output frequency is achieved with ganged resistors and, to provide a lower limit, a separate series resistor is used.

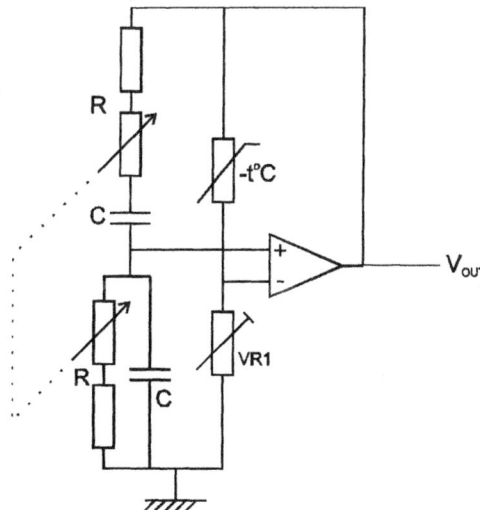

Fig. 11.16. An operational–amplifier-based Wien bridge oscillator

The idea of a resistor in series with a variable resistor is a common combination in many circuits, especially if excess current could flow when the variable resistor is reduced to zero ohms. One advantage of this circuit is that larger-value capacitors can be used because they are available up to several μF in the non-polarised form. Amplitude stability is provided by the negative-temperature-coefficient thermistor. As the output amplitude increases the thermistor resistance falls, which will reduce the gain of the amplifier and so reducing the amplitude of the output. In a bjt or jfet circuit, this would have to be as part of the negative-feedback/gain-control circuit.

Negative-resistance oscillators

Introduction

As implied at the beginning of this chapter, a resonant circuit will oscillate if connected to a suitable source of negative resistance. Such an oscillator differs from those described earlier in requiring only two connections to the frequency-determining section. Two possible shapes of negative-resistance characteristic are shown in Fig. 11.17. In both the negative-resistance region is confined to a limited range of voltage and current. For characteristic (a) a particular value of applied voltage could produce two or even three current values whilst a particular value of applied current can produce only one voltage value. Such a characteristic is termed current-controlled. Characteristic (b) – a much more familiar shape – is voltage-controlled. In both the negative resistance is a differential quantity, i.e. it is the ratio of a small change in voltage to the resulting change in current.

To use a device with such a characteristic in an oscillator the negative resistance must be sufficient to offset the positive resistance of the

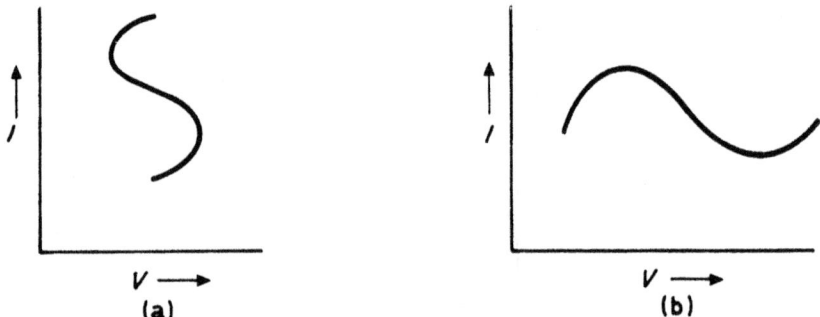

Fig. 11.17. Examples of current-voltage characteristics with a negative-resistance region, (a) current-controlled, (b) voltage-controlled

resonant circuit connected to it. Thus for characteristic (a), which is most likely to be used with a series tuned circuit, the negative resistance must be numerically greater than the positive resistance of the tuned circuit. Characteristic (b) is normally used with a parallel tuned circuit and the negative resistance must be numerically less than the parallel (i.e. dynamic) resistance of the tuned circuit. If the negative resistance is much smaller than this the oscillation amplitude grows until it occupies a voltage range greater than that of the negative-resistance region. In fact the peaks of the oscillation enter the regions of positive resistance at each end of the negative-resistance region. These apply damping to the tuned circuit, taking power from it and so limiting the amplitude of the oscillation. In this way the oscillation amplitude is limited to the value for which the average slope of that part of the characteristic which is used in the oscillation process corresponds to the dynamic resistance of the tuned circuit.

Tunnel diode

One semiconductor device which has a characteristic similar to that of Fig. 11.17(b) is a tunnel diode. This is a heavily doped pn diode

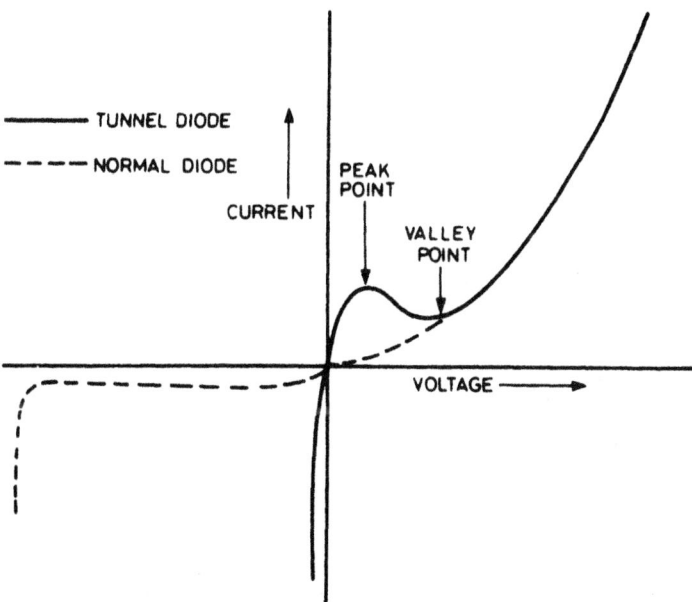

Fig. 11.18. Characteristic of a tunnel diode (solid) compared with that of a normal junction diode (dashed)

with a very thin junction region. A typical characteristic is shown in Fig. 11.18 and, for comparison, the characteristic for a normal diode with a thicker junction is shown dashed. As would be expected, breakdown for the tunnel diode occurs at a very low value of reverse bias and, in effect, there is no region of high reverse resistance. The region of negative slope which occurs at a low forward bias, typically between 0.1 and 0.3 V, was first reported by Esaki in 1958. This anomaly in the shape of the forward characteristic is caused by penetration of the potential barrier at the junction by electrons with insufficient energy to surmount the barrier. This effect – the tunnel effect – is impossible to explain in terms of classical physics but is in agreement with quantum mechanics. Tunnel diodes can be manufactured with very low capacitance and oscillators making use of the negative-resistance kink can function at frequencies as high as thousands of MHz. To obtain maximum output the quiescent point must be accurately placed at the centre of the negative-resistance region but the amplitude of the output obtainable from such an oscillator is clearly limited to a fraction of a volt.

Push-pull oscillator

A negative-resistance characteristic can also be obtained from a circuit using two transistors. The example shown in Fig. 11.19

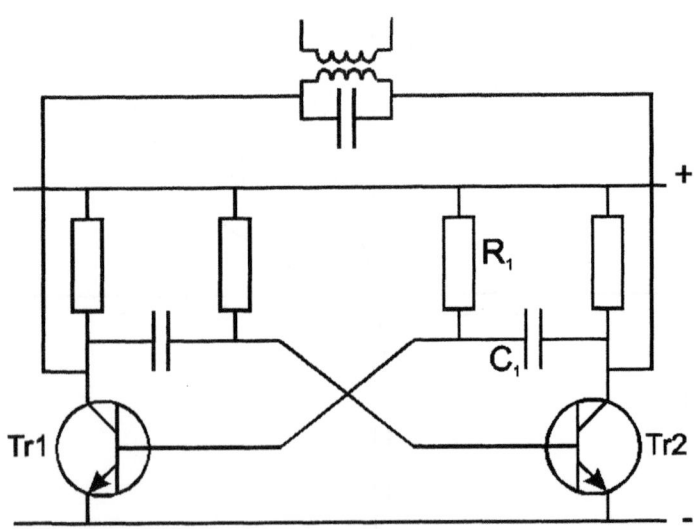

Fig. 11.19. Negative-resistance or push-pull oscillator using two bipolar transistors

consists basically of an astable multivibrator which, in the absence of the *LC* combination, would produce square-wave outputs at both collectors, the transistors switching alternately between cut-off and saturation.* The presence of the tuned circuit modifies the action because the inductor provides a low-impedance path between the collectors at low frequencies whilst the capacitor does so at high frequencies, both inhibiting normal multivibrator behaviour. Operation is confined to the resonance frequency of the tuned circuit at which it presents maximum impedance and the output of the circuit is thus sinusoidal. At this frequency the effective resistance between the collectors is approximately $-2/g_m\beta$ where g_m is the mutual conductance of the transistors and β is the attenuation of the inter-transistor coupling circuits. One such circuit is C_1R_1 but R_1 is shunted by the input capacitance of TR1 and this may affect the attenuation. For oscillation the dynamic resistance of the tuned circuit must be greater than $2/g_m\beta$.

Impatt diode

Suppose an alternating voltage is applied to a device and that electrons take an appreciable time to cross a particular region of it (known as the *drift space*). The delay is known as the *transit time* and at very high frequencies it can be a significant fraction of the period of the applied voltage. Because of the resulting phase difference between applied voltage and resulting current it is possible at a particular frequency for the current to be decreasing at the same time as the voltage is increasing, i.e. the device presents a negative resistance. This form of negative resistance is used to give oscillation in the impatt diode. This is a four-layer device of pnin construction and reverse bias causes avalanche breakdown at the pn junction. Electrons cross the i-region which acts as a drift space and the transit time is arranged to be one half the period of the required oscillation. The resulting negative resistance causes oscillation in a resonant load and power outputs up to 50 W at 10 GHz can be obtained.

Trapatt diode

If the reverse bias of a four-layer two-terminal device as described above is increased until the avalanche region fills the former drift space, operation in the Trapatt (trapped plasma avalanche transit

---

* See Chapter 13 for a detailed description of multivibrators.

time) mode occurs. The process is similar to that of a relaxation oscillator: the terminal voltage rises as the plasma region extends across the drift region and the shunt capacitance is charged. When the plasma reaches the end of the drift space a large current pulse is triggered, discharging the capacitance and destroying the plasma. The terminal voltage falls suddenly, and then begins to build up in the next cycle of oscillation.

Chapter 12

Modulators, demodulators, mixers and receivers

Introduction

The first step in communicating by radio is to establish a carrier-signal link between the two points. One characteristic of this signal is then varied to send information over the link. In radio and television broadcasting the amplitude or the frequency of the carrier is varied in accordance with the instantaneous value of the audio or video signal, a process known as modulation. Alternatively the carrier can be amplitude-modulated by a pulse train and a characteristic of the pulses can be varied to achieve modulation. For example, the pulse amplitude can be varied but there are other methods of modulation in which the amplitude is kept constant and the information is transmitted by varying the frequency, the width or the position of the pulses or by the use of pulse code modulation (see Chapter 15). In this chapter we are primarily concerned with the way in which transistors are used for modulation and demodulation of sinusoidal carriers.

Amplitude modulators

One way in which amplitude modulation could be achieved is by introducing the modulating signal into the base circuit of a bipolar-transistor r.f. amplifier as indicated in the simplified diagram of Fig. 12.1. The modulating signal must have sufficient amplitude to cause significant variations in the mutual conductance of the transistor. Thus the gain to which the carrier signal is subjected depends on the instantaneous value of the modulating signal causing the carrier amplitude at the amplifier output to vary in accordance with the waveform of the modulating signal. The gain is unlikely to be linearly related to the modulating-signal amplitude and a degree of distortion is

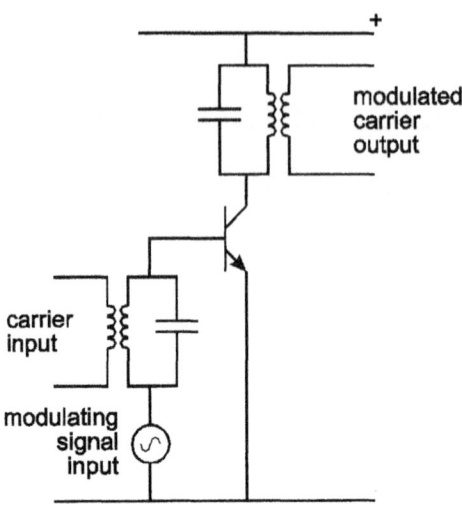

Fig. 12.1. One possible circuit for achieving amplitude modulation

inevitable in the modulation process but the circuit illustrates one principle which can be used to effect amplitude modulation.

Frequency modulators

Frequency modulation by a speech signal could be achieved very simply by connecting a capacitor microphone across an oscillating tuned circuit. But practical frequency modulators need a better degree of linearity and provision for controlling the centre (unmodulated) frequency. A varactor diode can be used as a modulator and in one arrangement two such diodes are connected across the frequency-determining circuit. The modulating signal is applied to one diode and a control signal is applied to the other to stabilise the centre frequency. The control signal is derived by comparing a sample of the frequency-modulated output with a stable frequency source and is a measure of the frequency error.

A transistor can be used as a frequency modulator and one basic circuit is shown in Fig. 12.2. Feedback from collector to base of the common-emitter stage is applied by $R_1 C_1$ and C_1 is small so that its reactance at the operating frequency is large compared with R_1. The current i_{fb} in $R_1 C_1$ therefore leads the collector alternating signal voltage v_c by $90°$ and is given approximately by

$$i_{fb} = v_c j\omega C_1$$

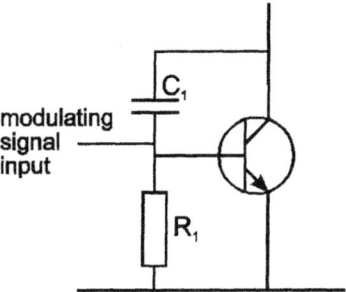

Fig. 12.2. Basic circuit of a bipolar transistor used as a reactance modulator

The signal voltage generated across R_1 by this current is thus

$$i_{fb}R_1 = v_c j\omega C_1 R_1$$

and this voltage is applied between base and emitter, generating a collector current i_c given by

$$i_c = g_m j\omega v_c R_1 C_1$$

which also leads the collector voltage by 90°. The impedance presented at the collector is hence given by

$$\frac{v_c}{i_c} = \frac{1}{g_m j\omega R_1 C_1}$$

This is a capacitive reactance and corresponds to a capacitance of $g_m R_1 C_1$. The capacitance can therefore be varied by alteration of g_m which, in turn, depends on the base-emitter voltage. Thus by connecting the collector and emitter across an oscillating tuned circuit and by applying the modulating signal between base and emitter, frequency modulation can be achieved.

If the positions of R_1 and C_1 are reversed and if the reactance of C_1 is small compared with R_1, a repeat of the above argument shows that the collector circuit now has an effective inductive reactance corresponding to an inductance value of $R_1 C_1/g_m$, again controllable by alteration of g_m.

Circuits of this type are known as *reactance modulators* and are used in f.m. transmitters. To obtain the required degree of linearity of modulation two such circuits are used in a balanced push-pull arrangement and modulation-frequency negative feedback is applied from the output of a high-grade demodulator fed from the output of the transmitter. A second feedback loop is used to stabilise the centre frequency.

An example of a u.h.f. TV picture and sound modulator is given in Chapter 16.

A.M. detectors

The function of the detector in a receiver is to abstract information from a modulated input. The information may be a replica of the original modulating waveform and the detector is then alternatively known as a *demodulator*; or the information abstracted may be a signal suitable for a.g.c., a.f.c. or some other purpose.

Series-diode detector

A semiconductor diode is commonly used as an a.m. detector and the basic circuit for a series-diode detector is given in Fig. 12.3. In this basic diagram it is assumed, for simplicity, that the diode is ideal and starts to conduct at zero bias. In practice, as we have seen, a silicon junction diode needs a forward bias of approximately 0.7 V to promote conduction and a later circuit diagram shows one way in which such bias can be applied.

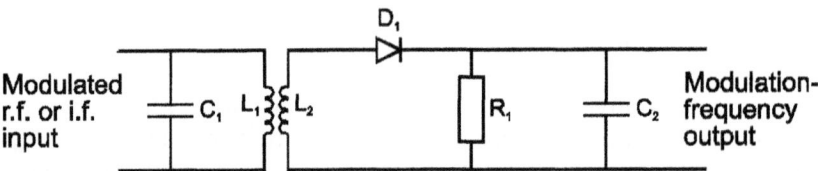

Fig. 12.3. Basic form of series-diode detector circuit

The diode D1 conducts during positive half-cycles of modulated-r.f. input and charges the reservoir capacitor C_2 to a voltage approximately equal to the peak value of the signal across L_2. During negative half-cycles D1 is cut off and C_2 discharges through the load resistor R_1. The time constant $R_1 C_2$ however is large compared with the period of the r.f. input and C_2 loses very little of its voltage during the non-conduction period. The period of conduction is only a small fraction of the r.f. period and for most of each cycle D1 is cut off, thus isolating the load circuit from the source of r.f. input.

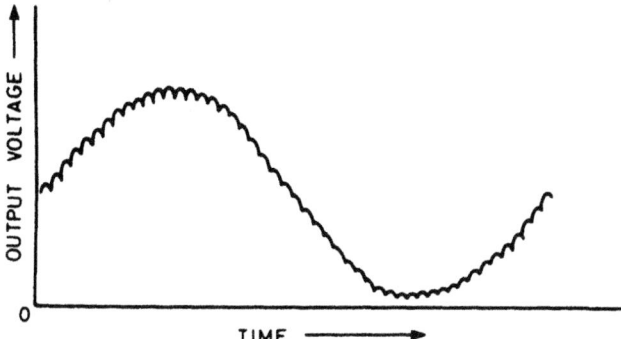

Fig. 12.4. Output waveform from a series-diode detector for a sinusoidally-modulated input signal

Detection by this circuit can be regarded as a *sampling* process: the circuit samples each positive peak of r.f. input and impresses its value on C_2. Thus a voltage is developed across C_2 which rises and falls in sympathy with the modulation waveform but never reverses in polarity as shown in Fig. 12.4. This illustrates sinusoidal modulation and the voltage output is at all times positive. In practice the frequency of the r.f. ripple is very much greater than is suggested in this diagram.

Choice of time constant

The time constant $R_1 C_2$ is important: if it is too small C_2 discharges too rapidly and the detector output is unnecessarily low; if it is too large C_2 cannot discharge rapidly enough to follow the most rapid falls in the modulation waveform and distortion results. The optimum value of the time constant is the largest which just fails to cause distortion: this value can be calculated in the following way.

If the detector is perfect, an unmodulated carrier of amplitude v_{in} gives a steady voltage of v_{in} across the load. In practice the output voltage has a ripple made up of a succession of exponential rises and falls, the rises being governed by the time constant $r_d C_2$ (r_d is the diode forward resistance) and the falls by time constant $R_1 C_2$. Normally R_1 is large compared with r_d and the falls are slower than the rises. The ripple is usually eliminated by a simple RC filter in the detector output circuit.

If the carrier input is sinusoidally modulated to a depth m, a sinusoidal signal of mv_{in} amplitude is superimposed on the steady voltage v_{in} across the detector load and this too is composed of exponential rises and falls. To avoid a distorted output the exponential falls must be rapid enough to follow the sinusoidal form. If the modulation waveform has the equation

$$v = mv_{in} \sin \omega t$$

the rate of change is given by

$$\frac{dv}{dt} = mv_{in}\omega \cos \omega t$$

The greatest value this can have occurs when $\cos \omega t = 1$ and is equal to $mv_{in}\omega$: the sine wave has this slope where it crosses the datum line.

The voltage at the datum line is v_{in} and the slope of the exponential fall which commences at this point is given by v_{in}/R_1C_2 as shown on page 276. Equating slopes we have

$$mv_{in}\omega = \frac{v_{in}}{R_1C_2}$$

$$\therefore R_1C_2 = \frac{1}{m\omega} = \frac{1}{2\pi fm}$$

The time constant is thus inversely proportional to frequency and we must choose a value which will not cause distortion at the highest modulating frequency: it will then automatically be suitable for lower frequencies.

In a 625-line television receiver the upper limit of the video band is approximately 5 MHz and signals at this end of the band can be 100 per cent modulated. Thus $f = 5 \times 10^6$, $m = 1$ and we have

$$R_1C_2 = \frac{1}{6.284 \times 5 \times 10^6} \text{ s}$$

$$= 0.032 \, \mu s$$

In high-quality sound broadcasting the upper frequency limit is 15 kHz and if signals at this extreme were 100 per cent modulated the required time constant would be given by

$$R_1C_2 = \frac{1}{6.284 \times 15 \times 10^3} \text{ s}$$

$$= 10.6 \, \mu s$$

However in sound broadcasting the high modulating frequencies are harmonics and have amplitudes small compared with those of lower-frequency fundamental signals. Moreover in a.m. receivers the i.f. amplifier usually restricts the upper frequency limit to 3 or 4 kHz by sideband cutting. Thus it is quite permissible to increase $R_1 C_2$ in an a.m. sound receiver to 50 μs or more.

When the time constant has been decided we can consider the individual values of R_1 and C_2. R_1 must not be too small otherwise C_2 cannot charge up to the peak value of the input signal and the output of the detector is unnecessarily low: to minimise this loss R_1 should be large compared with r_d the forward resistance of the diode. On the other hand R_1 must not be too large otherwise the shunting effect of the following stages becomes serious: moreover in television receivers large values of R_1 can lead to impossibly low values for C_2. A compromise value of R_1 commonly adopted in sound receivers is 5 kΩ but in television receivers slightly lower values such as 3.3 kΩ were sometimes used.

If R_1 for a television receiver is 3.3 kΩ we can calculate C_2 from the time constant thus

$$C_2 = \frac{0.032 \times 10^{-6}}{3.3 \times 10^3} \text{ F}$$

$$= 10 \text{ pF}$$

If R_1 for a sound receiver is 5 kΩ we can calculate C_2 from the time constant thus

$$C_2 = \frac{50 \times 10^{-6}}{5 \times 10^3} \text{ F}$$

$$= 0.01 \text{ μF}$$

Input resistance

If the detector is 100 per cent efficient C_2 is charged up to the peak value $v_{in(pk)}$ of the input signal. Thus the output power delivered by the detector to the load resistor is given by $v_{in\,(pk)}{}^2/R_1$. If the input resistance of the detector is r_i the input power supplied to the detector by the signal source is given by $v_{in(rms)}{}^2/r_i$. If no power is lost in the diode we can equate these two expressions thus

$$\frac{v_{in(rms)}{}^2}{r_i} = \frac{v_{in(pk)}{}^2}{R_1}$$

$$\therefore r_i = \frac{v_{in(rms)}^2}{v_{in\ (pk)}^2} \cdot R_1$$

For a sinusoidal signal

$$v_{in(rms)} = v_{in(pk)}/\sqrt{2}$$

and thus

$$r_i = \frac{R_1}{2}$$

In practice some power is inevitably lost in the forward resistance of the diode and the output voltage is less than expected: r_i then is larger than $R_1/2$ and a value of $3\,\mathrm{k\Omega}$ is commonly assumed when R_1 is $5\,\mathrm{k\Omega}$.

A.G.C. provision

The voltage required for a.g.c. may be positive or negative depending whether npn or pnp transistors are to be controlled and on whether forward or reverse control is required. Either polarity can be obtained by suitable design of the detector circuit. For example Fig. 12.5 gives the complete circuit of a detector for a sound receiver. $R_2 C_3$ attenuates i.f. ripple in the detector output and the diode load R_3 functions also as the volume control. $R_4 C_4$ is an a.f. filter which minimises a.f. content in the

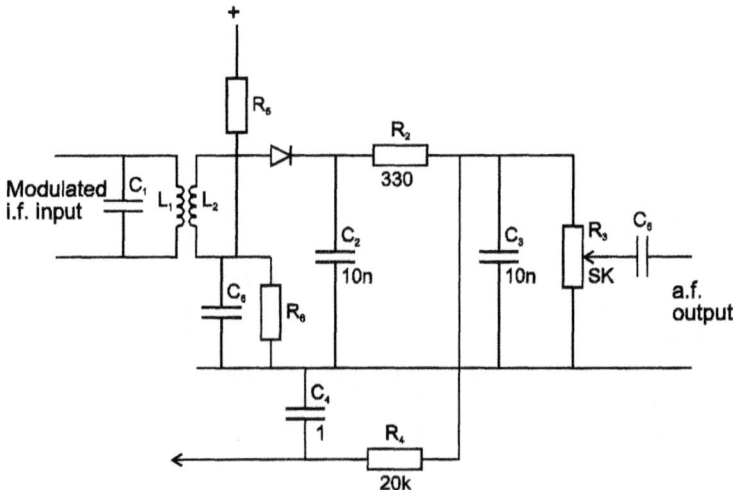

Fig. 12.5. Complete circuit diagram for a series-diode detector in a sound receiver

a.g.c. voltage which in this circuit is positive-going. The potential divider R_5R_6 provides the 0.7 V forward bias required by a silicon junction diode and C_6 decouples R_6.

Shunt-diode detector

In Fig. 12.3 the output is taken from C_2 but it could alternatively be taken from D1, the circuit being rearranged as shown in Fig. 12.6 to

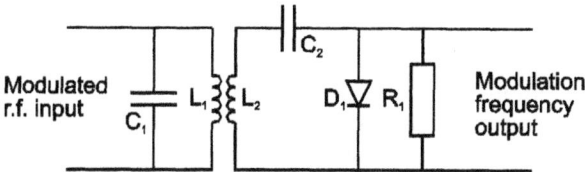

Fig. 12.6. Basic form of shunt-diode detector. Again an ideal diode is assumed

enable one terminal of the output to be earthed. The output signal is now made up of the modulation-frequency signal from C_2 plus the modulated-r.f. signal across L_2 and hence has a greater r.f. ripple content than for the series-diode circuit. As before, the diode is driven into conduction for a brief fraction of each cycle by positive peaks of the r.f. input and at these instants the diode effectively short-circuits the detector output. Thus the waveform of the output is similar to that of the modulated-r.f. input with each r.f. cycle displaced vertically so that all positive peaks are aligned or clamped at zero volts as shown in Fig. 12.7: shunt-diode detectors are sometimes called *clamping detectors*.

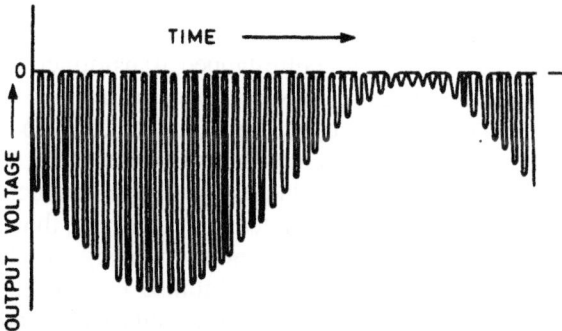

Fig. 12.7. Output waveform from a shunt-diode detector for a sinusoidally-modulated input signal

By reversing the diode all negative peaks can be clamped at zero volts.

The time constant $R_1 C_2$ should have the same value as for the series-diode circuit but the input resistance of the shunt-diode circuit is lower (approximately $R_1/3$) because the load resistor R_1 is effectively connected across the input terminals for the whole of each r.f. cycle.

An advantage of the shunt-diode circuit is that C_2 provides d.c. isolation between the input-signal source and the load circuit.

Synchronous detectors

In an a.m. transmission (Fig. 12.8a) the carrier-frequency component is of constant amplitude and independent of the modulation. It can therefore be suppressed and the effect of so doing is shown at (b): the shape of the envelope of the waveform is completely changed. Simple detectors of the type described above which respond to the peak values of the waveform would clearly give very distorted results if used to demodulate a carrier-suppressed signal.

Synchronous detectors overcome this difficulty by use of a second input consisting of a constant-amplitude sinusoidal signal synchronised with the (suppressed) carrier frequency of the modulated r.f. signal to be demodulated. The second input determines the instants at which the detector samples the amplitude of the modulated r.f. signal. These instants, shown by vertical dashed lines in Fig. 12.8, occur strictly at carrier-frequency intervals and thus the detector samples the positive peaks during one half-cycle of the modulating signal and negative peaks during the other half-cycle, thus correctly reconstituting the waveform of the modulating signal. The output has positive and negative swings and, for a symmetrical modulating signal such as a sine wave, has a mean value of zero, i.e. there is no d.c. component as in the output of a simple diode detector.

One possible circuit for a synchronous sampling detector is given in Fig. 12.9. The single series diode of the prototype a.m. detector is replaced by two diodes and a centre-tapped transformer. Both diodes conduct together to produce the low-impedance path which connects the source of modulated r.f. to the capacitor C_4. When the diodes are non-conductive the path is open-circuited and C_4 retains its charge. The diodes must be driven into conduction and non-conduction by the carrier input and not by the modulated r.f. input and thus the carrier input must be large compared with the other input signal. The balanced form of the carrier circuit is adopted to minimise any carrier component which may reach C_4. The time-constant circuits $R_1 C_1$ and $R_2 C_2$ are included as diode loads to ensure that the diodes conduct for only a small fraction of each cycle, i.e. when sampling is required.

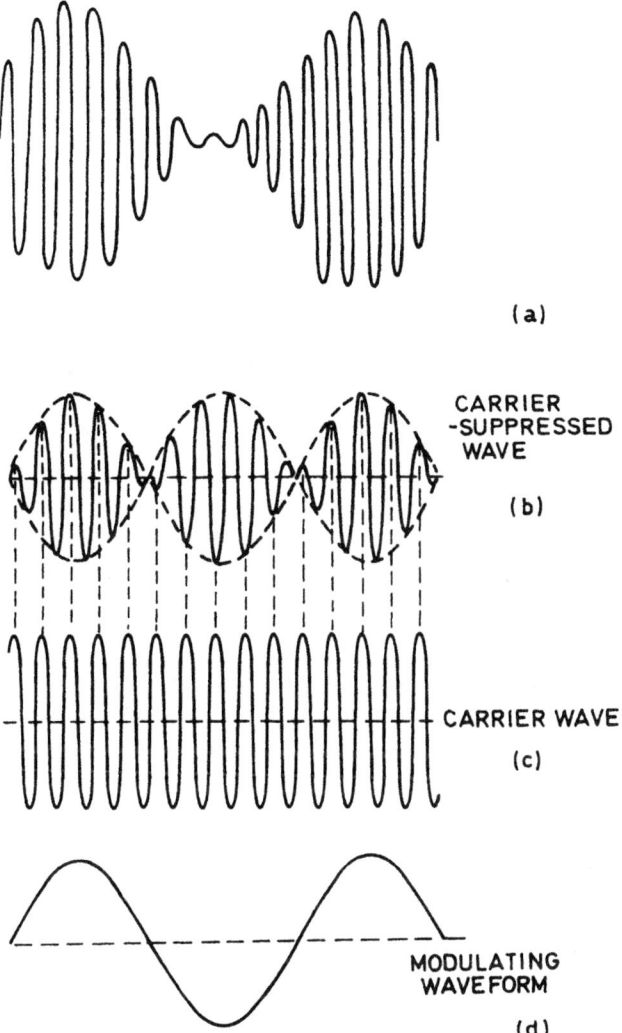

Fig. 12.8. An amplitude-modulated wave (a), and the corresponding carrier-suppressed wave (b). The action of a synchronous detection is shown at (c) and (d)

For some applications the components R_1C_1 and R_2C_2 can be omitted. The diodes then connect C_4 to the source of modulated r.f. for the whole of one carrier half-cycle.

This type of circuit can be used to demodulate the quadrature-modulated colour-difference signals in a colour television receiver. Here the modulated signal has two carrier components in quadrature, each

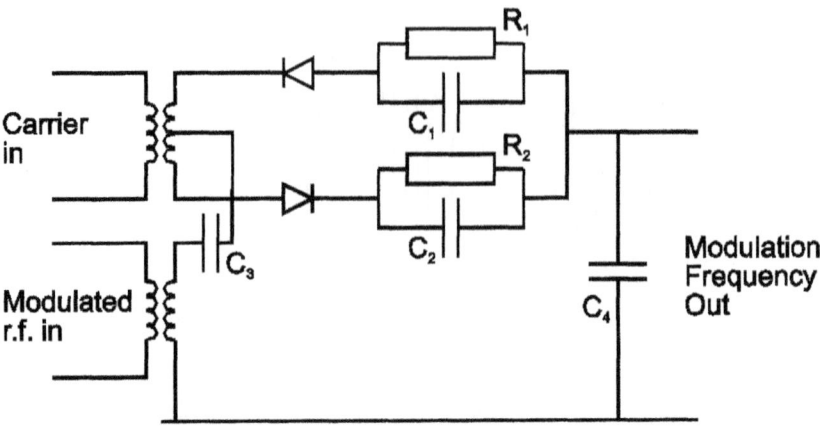

Fig. 12.9. Synchronous sampling detector using a diode bridge

amplitude modulated by a different signal. The circuit of Fig. 12.9 can demodulate one of these signals without interference from the other because, during the time it is sampling the peaks of one signal, the other is passing through zero and so has no effect on the detector output. A second detector with its carrier input in quadrature with that of the first is required to demodulate the second colour-difference signal.

F.M. detectors

There are many types of f.m. detector and we shall describe two which are in common use, namely the ratio detector (generally constructed of discrete components) and the balanced product or quadrature detector (usually forming part of an i.c.)

Ratio detector

Many f.m. detector circuits make use of the phase angle between the primary and secondary voltages of a transformer when the secondary winding is tuned to the centre frequency, e.g. the midband frequency of the i.f. amplifier. The phase angle is 90° at the resonance frequency of the secondary winding, is greater at higher frequencies and smaller at lower frequencies. By injecting a sample of the primary voltage via L_3 into the centre tap of the secondary winding as shown in Fig. 12.10, the voltage at the ends of the secondary winding are made dependent on this phase difference and hence on signal frequency. As frequency increases from the centre value the voltage increases at one secondary terminal and decreases at the other: these changes are reversed if frequency decreases from the centre value.

In the ratio detector the secondary winding is connected to two series-aiding diodes with individual reservoir capacitors C_1 and C_2 and a common load resistor R_1. As frequency increases from its centre value the voltage across one reservoir capacitor (say C_1) increases whilst that across C_2 decreases. When frequency decreases from the centre value the voltage across C_1 falls and that across C_2 increases. The voltage across either capacitor thus varies with modulation and can be used as the modulation-frequency output. In Fig. 12.10 the voltage across C_2 is used as output.

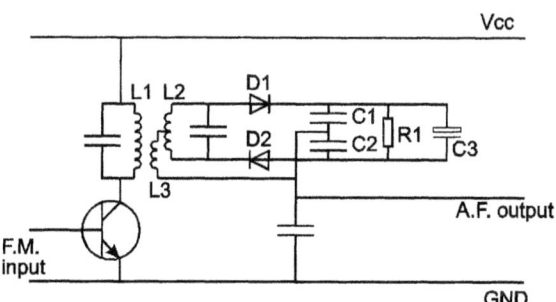

Fig. 12.10. One circuit for a ratio detector

The sum of the voltages across C_1 and C_2 tends to be constant during modulation and in the ratio detector is stabilised by the inclusion of the capacitor C_3 across R_1, the combination having a long time constant.

One of the reasons for the popularity of the ratio detector is that it can be made relatively insensitive to amplitude modulation of the input signal. This is achieved by making the detector act as a *dynamic limiter*. R_1 is made small so that the secondary winding is heavily damped. By normal diode detector action the voltage across $R_1 C_3$ is equal to the peak value of the input signal but, because of the long time constant, the voltage cannot adjust immediately to any sudden change in the amplitude of the input signal although it can follow slow changes. Any momentary increase in input-signal amplitude, e.g. a noise 'spike', causes the diodes to conduct more heavily than normal, thus increasing the damping on the secondary winding and momentarily reducing the gain of the previous stage, so offsetting the effect of the spike. Any momentary fall in input-signal amplitude, e.g. a negative spike, causes the diodes to be cut off, thus removing the damping of the tuned circuit and increasing the gain. Thus the effects of sudden increases and decreases in input-signal amplitude can be reduced.

Phase detector or comparator

A type of f.m. detector favoured by i.c. manufacturers consists in essence of two transistors connected in series across the supply. Signals from the i.f. amplifier (usually included in the i.c. also) are limited to form pulses and are applied to the base of one transistor. They are also applied to a network, the phase shift of which is dependent on frequency and is 90° at the centre frequency. An example of such a network is shown dashed in Fig. 12.11 in which L_1C_1 is resonant at the centre frequency, C_2 and C_3 being small capacitances. The output of this quadrature network is applied to the base of the second transistor. The polarity of the pulses is such that the current through the transistors is increased when the pulses overlap and as the period of overlap varies with frequency modulation the mean current through the transistors

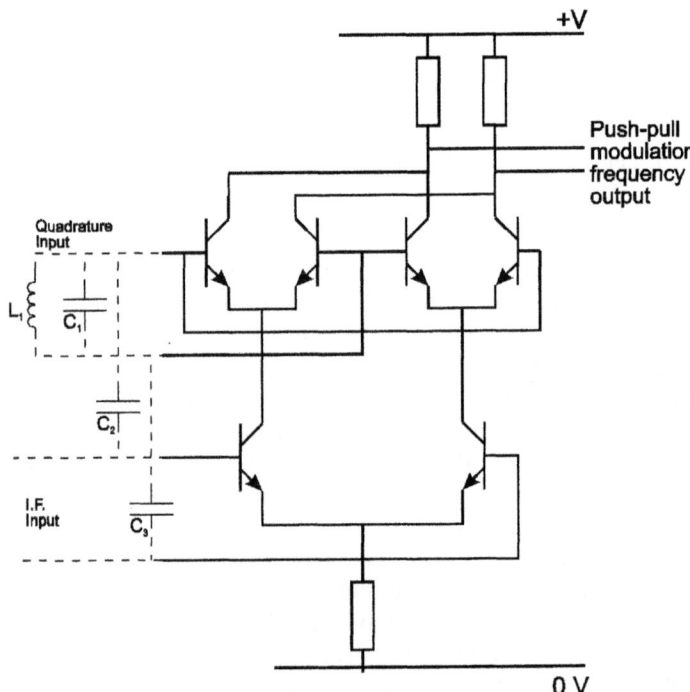

Fig. 12.11. Simplified circuit diagram of a phase detector or comparator of the type used for f.m. detection

varies correspondingly and so has a strong modulation-frequency component which can be extracted as output.

In practice the circuit is more complex than this, the lower transistor being replaced by a long-tailed pair (see page 160), the two collector circuits each containing another such pair as shown in Fig. 12.11. Pulses from the i.f. output are applied in push-pull to the lower pair and pulses from the quadrature network are applied to the other pairs in a parallel push-pull arrangement. This balanced circuit is adopted to minimise output at the pulse frequency and drift of operating currents and voltages. For simplicity, means for stabilisation of operating currents are omitted from Fig. 12.11.

Phase-locked-loop detector

A type of f.m. detector often used in i.c. form is the phase-locked-loop illustrated in the block diagram of Fig. 12.12. It consists of a phase detector (with a circuit diagram similar to that of Fig. 12.11) which compares the frequency and phase of the frequency-modulated input with those of a voltage-controlled oscillator (v.c.o.). The d.c. output of the phase detector is used to control the v.c.o. so as to minimise any

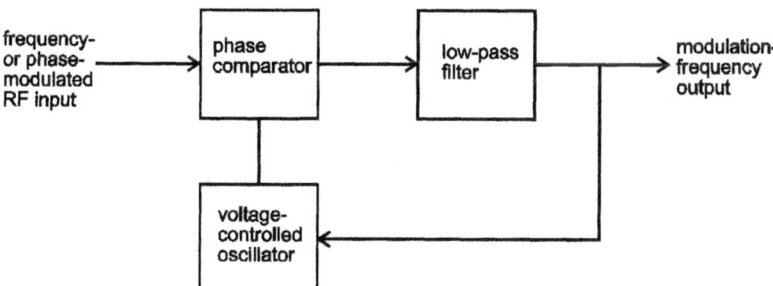

Fig. 12.12. Block diagram of a phase-locked loop

difference in frequency or phase between the two inputs. In this way the v.c.o. accurately tracks the variations in frequency and phase of the f.m. input and the control voltage for the v.c.o. is a faithful reproduction of the modulation waveform. The low-pass filter is used to suppress all the components of the phase-detector output except the modulation frequency and the d.c. output. Many types of oscillator can be used for the v.c.o. but one well suited for incorporation in an i.c. is an astable multivibrator (see page 269).

Mixers

In a superheterodyne receiver the carrier and sidebands constituting the received signal are in effect translated in frequency to give a new signal with a carrier at the intermediate frequency. This is achieved in a mixer stage in which the received signal is combined with the output of a local oscillator, thereby producing a resultant signal at the sum or difference of the received carrier and oscillator frequencies.

If two signals at frequencies f_1 and f_2 are connected in series or parallel and are applied to a device with a linear input-output characteristic the output consists only of signals at f_1 and f_2 (of larger amplitude than at the input if the device amplifies). No new frequencies are introduced. But if the device has a non-linear characteristic the output has components at several frequencies, known as combination frequencies, which can be represented as $(mf_1 \pm nf_2)$ where m and n are the integers 1, 2, 3 etc. One of these components is the difference frequency $(f_1 - f_2)$ required in superheterodyne reception. Mixers operating on this principle are termed *additive* and the most obvious device to use as an additive mixer is a diode. Another possibility is a transistor biased near cut-off to give the required non-linearity.

In a second principle which can be used in a mixer the two signals are in effect multiplied together. This is achieved practically in an active device by arranging for one input to control the gain to which the other input is subjected. For example the two inputs can be applied to the two gates of a dual-gate mosfet. No non-linearity is required in this form of mixer and only the sum and difference frequencies are generated as shown in the identity:

$$2 \cos \omega_1 t \cos \omega_2 t = \cos (\omega_1 + \omega_2)t + \cos (\omega_1 - \omega_2)t$$

Mixers using this principle are termed *multiplicative*.

Fig. 12.13 gives an example of a mixer circuit using a diode (a Schottky type is usually chosen). The modulated-r.f. and oscillator inputs are connected in parallel for application to the diode and interaction between the tuned circuits providing the inputs is minimised by using series inductance and series capacitance.

As mentioned earlier, a bipolar transistor can be used as a mixer, the oscillator input being provided by a second transistor, the combination forming a two-transistor frequency changer. It is, however, more usual to use a single transistor to perform both functions and such self-oscillating mixers can give a frequency-changing efficiency little short of that obtainable from the two-transistor circuit. The bias for a self-oscillating mixer must be chosen with care. The mutual conductance must be sufficient to sustain oscillation over the frequency bands to be

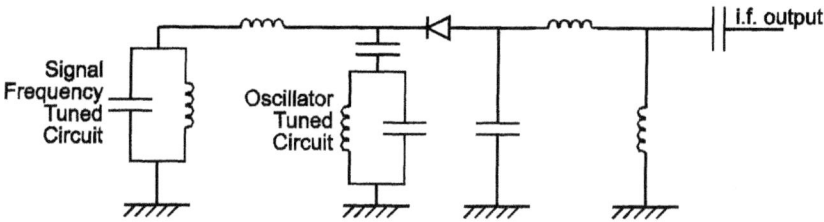

Fig. 12.13. An example of a diode mixer circuit

covered and this requires a reasonable collector current. On the other hand the bias must be sufficiently near cut-off to provide the non-linearity required for efficient mixing. Clearly a compromise is called for. A typical circuit diagram for a self-oscillating mixer is given in Fig. 12.14.

This may be regarded as a Reinartz oscillator of the type shown in Fig. 11.6 in which the base circuit is tuned to accept the signal-frequency input and the output circuit is tuned to select the difference-frequency output. The oscillator signal and the signal-frequency input are connected in series between base and emitter to enable the transistor to operate as an additive mixer. Circuits of this type are commonly employed in transistor superheterodyne receivers for a.m. reception.

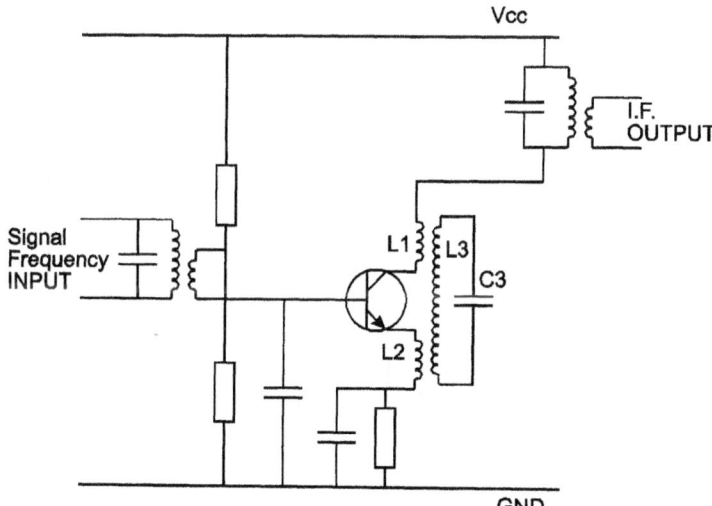

Fig. 12.14. One possible circuit for a self-oscillating mixer

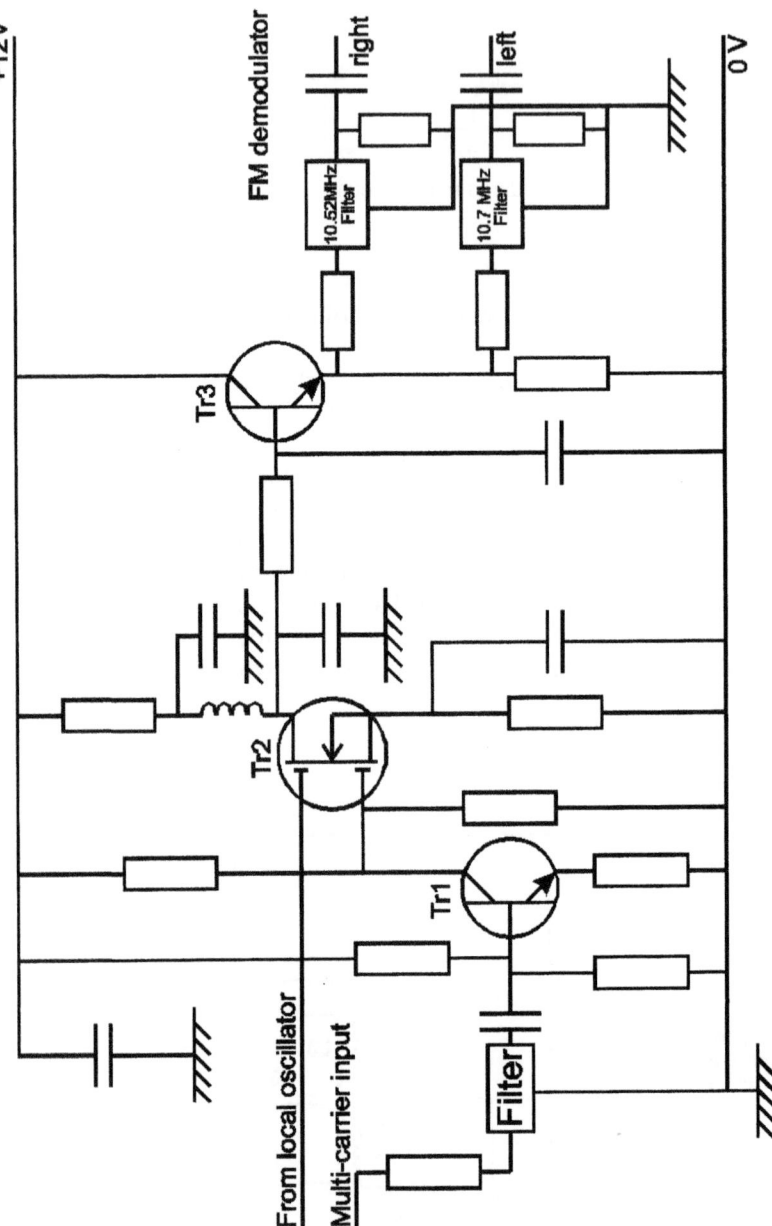

Fig. 12.15. Mosfet mixer used for stereo carrier tuning in a satellite receiver

Dual-gate mosfet mixer

An example of a multiplicative mixer-circuit is given in Fig. 12.15. The lower gate of the fet TR2 is fed with a wide spectrum of f.m. sound carriers at frequencies ranging from 6.5 MHz to 10 MHz derived from the baseband video signal in a satellite receiver. The upper gate is supplied with a sinusoidal signal from a local oscillator, the frequency of which is governed by the user's sound channel selector. The mixer output contains all the sound carriers, converted to frequencies in the 8–13 MHz range. Those centred on 10.52 and 10.7 MHz are selected by narrow-band ceramic filters for subsequent demodulation and amplification to form the right- and left-hand channels of the stereo sound output.

By suitable choice of the local-oscillator frequency, any pair of sound carriers 180 kHz apart can be brought into line with the filters and demodulators: in this f.m. transmission system the L and R sound carriers are always broadcast 180 kHz apart, though several pairs can accompany each vision carrier, the 'spares' carrying alternative languages or radio programmes.

U.H.F. television tuner

Fig. 12.16 gives the circuit diagram of a commercial television tuner covering Bands IV and V (470 to 860 MHz). Basically it comprises a dual-gate fet (TR1) as r.f. amplifier, a bipolar transistor (TR2) as oscillator, a Schottky diode (D2) as mixer and two bipolar transistors (TR3 and TR4) in cascade forming an i.f. pre-amplifier. The tuner is mounted on a printed wiring card contained within a metal housing measuring approximately 80 mm by 50 mm by 18 mm.

The aerial input circuit is not tuned but is designed as a high-pass filter. The gain of the r.f. stage TR1 can be controlled by a voltage applied to gate 2. This voltage, which is normally supplied by an external a.g.c. circuit, has a maximum range of 0 to 10 V and can vary the gain by 30 dB.

TR1 feeds a bandpass r.f. filter incorporating the two inductors L_{11} and L_{12}. These are not conventional coils but lengths of transmission line. A $\lambda/4$ length of line, if short-circuited at one end, presents a high impedance at the other. This is a resonant condition and the open-end impedance is analogous to the dynamic resistance of a parallel-tuned circuit at resonance. If the line is shortened to less than $\lambda/4$ the open end becomes effectively inductive and a capacitor can be connected across it to give resonance. Transmission lines can take many forms. In some tuners a box chassis is divided by partitions into rectangular sections, each of which can act as the outer conductor of a trough line, a stout

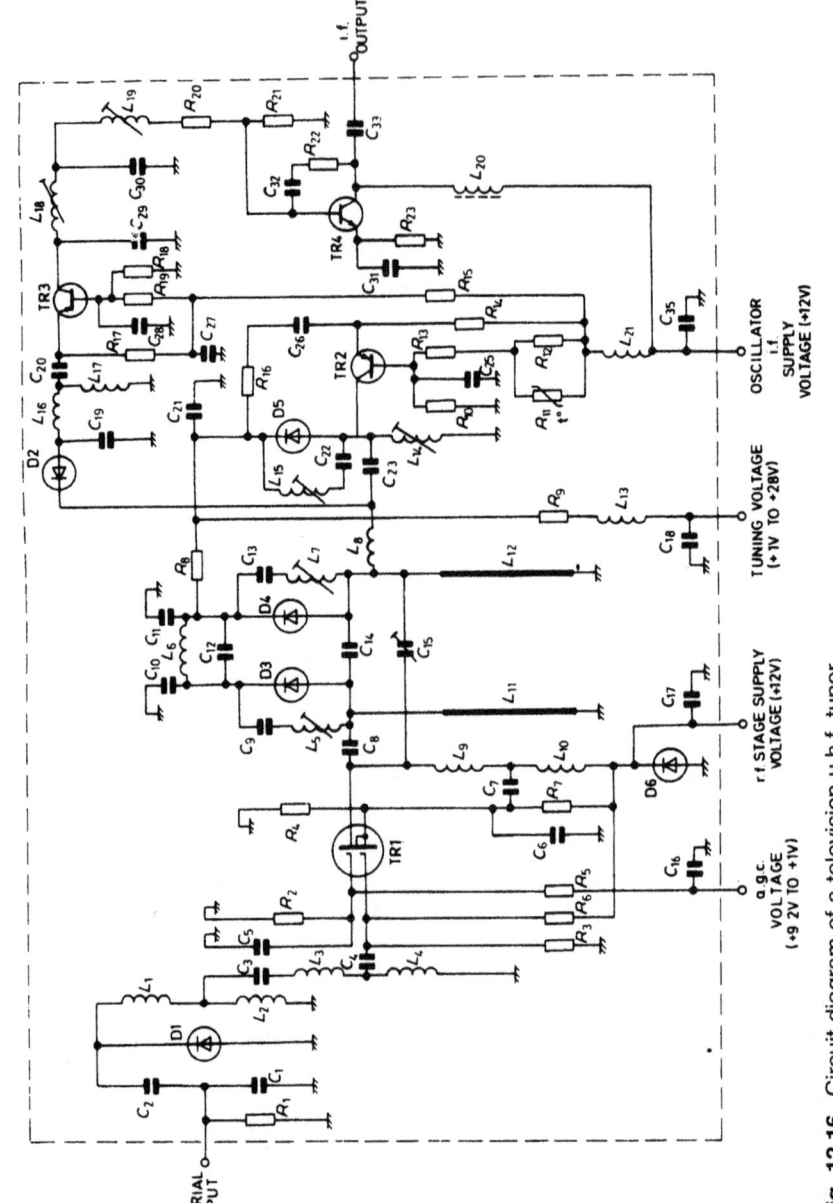

Fig. 12.16. Circuit diagram of a television u.h.f. tuner

wire forming the centre conductor. In others the lines are printed directly onto the glass fibre substrate on which the tuner is based, making for a compact and stable unit needing no alignment in production or subsequent service. In this particular tuner the lines are copper strips shaped like goalposts standing on the printed card on either side of a vertical earthed plane. The two adjustable inductors L_6 and L_7 effectively in parallel with L_{11} and L_{12} are used to align the resonance frequencies of the two tuned circuits. L_{11} and L_{12} are tuned by capacitance diodes D3 and D4 which are adjusted by a common tuning voltage which also controls the oscillator frequency. The coupling between L_{11} and L_{12}, which determines the passband of the double-tuned circuits, is controlled by L_6, C_{10}, C_{11} and C_{12} the values of which are chosen to minimise variation in the passband over the tuning range. C_{15} provides a dip in the r.f. response of the tuner at about $79\,\text{MHz}$ above the wanted vision carrier to minimise image frequency interference.

TR2 is a common-base stage acting as a Colpitts oscillator, internal capacitances providing the positive feedback. The inductor L_{14} is conventional but has only $1\frac{1}{2}$ turns and a diameter of a few mm. It is tuned by capacitance diode D5 and there is a parallel-connected coil L_{15} which can be adjusted (in addition to L_{14} itself) to give correct tracking between oscillator and r.f. circuits. The mixer diode D2 is fed with r.f. signal via L_8 and oscillator output via C_{23}. D2 output, after r.f. filtering, is applied to the emitter of the common-base stage TR3. TR3 is coupled to the common-emitter stage TR4 via L_{18} and L_{19} which together constitute an i.f. bandpass filter, the shunt coupling capacitor C_{30} determining the width of the passband. The output of TR4 is choke-capacitance coupled to the i.f. output pin of the tuner. R_{22} and C_{32} apply negative feedback to TR4 to stabilise its gain.

In all three bipolar-transistor stages the mean-emitter current is stabilised by the technique described in Chapter 6 in which a potential divider fixes the base voltage, the value of the emitter resistor then determining the mean current through the transistor. D.C. stability of the oscillator circuit is further improved by using a temperature-dependent resistor for R_{11}.

Satellite low noise block

It is not practical to convey a 12-GHz satellite carrier signal directly from the receiving dish to the indoor receiver so the frequency of the signal is downconverted in a dish-mounted *low noise block* (l.n.b.). The l.n.b. contains an amplifier, oscillator and mixer which together transfer the whole spectrum of received carriers to first intermediate frequencies

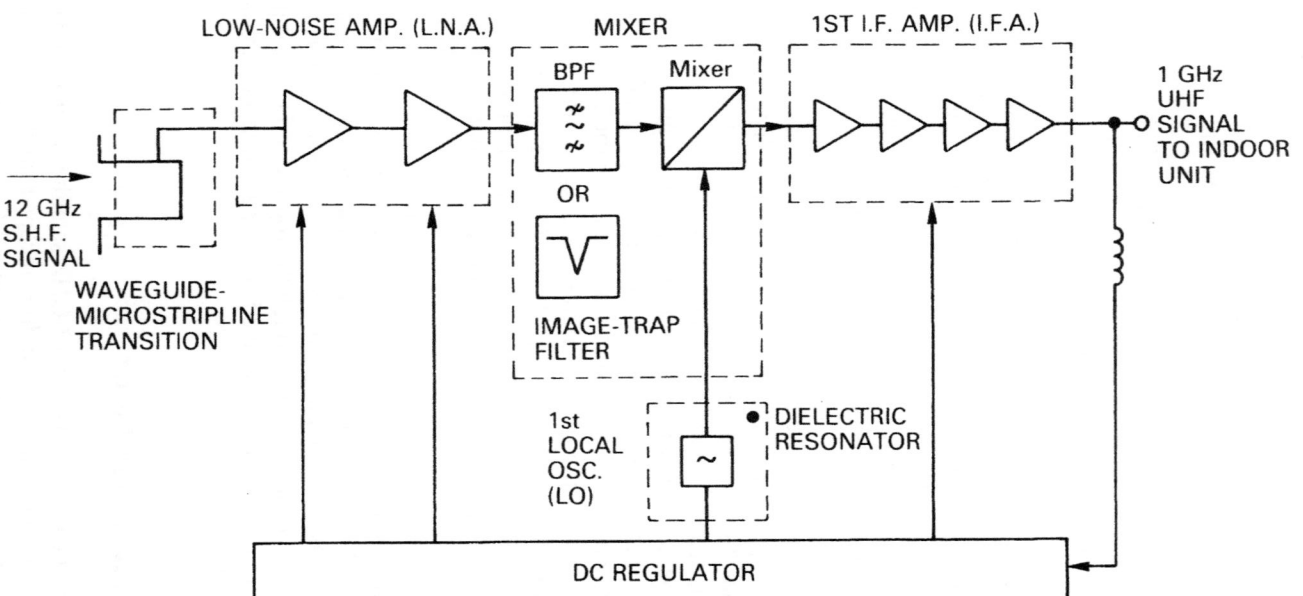

Fig. 12.17. The essentials of a head-end unit for satellite reception

in the range of 950 MHz to 1.7 GHz, which can be conveyed without excessive loss in a high-quality coaxial downlead to the indoor tuner unit.

Fig. 12.17 gives an overall block diagram of a typical l.n.b., which follows conventional superhet practice albeit at very high frequency. The low-noise r.f. amplifier depends on the waveguide or other input coupler to provide image rejection of over 50 dB. It uses two or three GaAs fets or hemts to provide an overall gain of about 20 dB to bring the signal up to a level suitable for presentation to the mixer, a Schottky-type diode of silicon or GaAs.

As with a conventional u.h.f. tuner the local oscillator must generate as little noise as possible: any noise superimposed on the oscillator output is added to the wanted i.f. signal. The oscillator transistor is again a GaAs fet, operating in conjunction with a stripline and dielectric resonator as described in Chapter 11 and Fig. 11.9. The required mixer output (1st i.f., 0.95–1.7 GHz) is selected by a bandpass filter and passed to a four- or five-stage low-noise amplifier with a gain of about 35 dB for launch into the transmission cable, which also carries the d.c. operating power up to the l.n.b. unit.

S.H.F. amplifier

An arrangement of low-noise amplifier for use at 12 GHz in a l.n.b. is shown in Fig. 12.18; it is fabricated by printed-circuit techniques

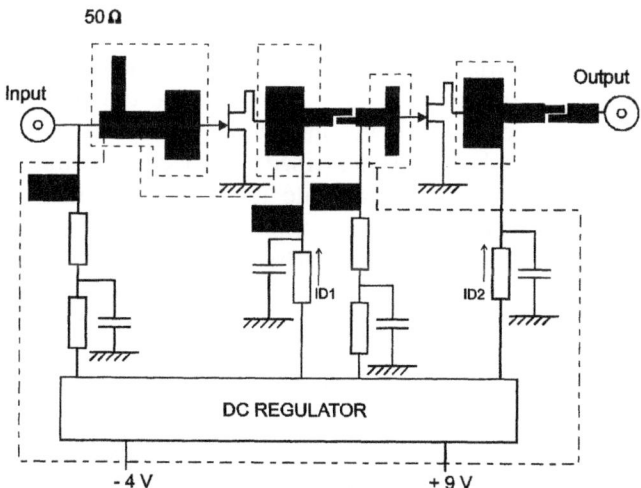

Fig. 12.18. Two-stage low-noise amplifier working at 11–12 GHz

on a low-loss substrate of glass fibre or similar material. The signal collected by the receiving dish is reflected through a waveguide and impedance-matching feedhorn to a pick-up probe consisting of a metal pin or a micropatch on the l.n.b. substrate. Printed micro-patches and stubs are used for impedance matching between the stages of the low-noise amplifier and to form couplers and bandpass filters.

The l.n.b. is followed by a filter and a mixer, both of which introduce some signal loss, but the overall noise figure is largely governed by the noise factor of the first two amplifier stages, which provide sufficient gain to override noise in the following circuits and to overcome losses in the transmission cable from dish to indoor tuner. Fig. 12.19(a) shows the layout of a commercial l.n.b. for consumer use. The square micropatch on the right intercepts the microwave signal, whence it* is transferred to one or other of the adjacent (90° apart) striplines, depending on the polarisation of the required signal. Each stripline has its own GaAs fet pre-amplifier (selected by the viewer) passing its output to the further two transistor amplifiers visible at the bottom of the picture and shown enlarged in Fig. 12.19(b). The slanted and stepped microstrips near the left of Fig. 12.19(a) form the bandpass filter for the s.h.f. signal on its way to the mixer at the far left. The oscillator can be seen at bottom left with its circular dielectric resonator while the large three-legged heat-sinked device near top centre is a series-regulator i.c. which stabilises the supply voltages to the l.n.b. circuits.

Indoor tuner

The block-converting action of the outdoor l.n.b. produces a spectrum of transmissions differing only in carrier frequency from those arriving at the receiving dish. They must now be tuned to select the required programme and the first indoor stage consists of a selector very similar to the u.h.f. tuner described earlier and illustrated in Fig. 12.16. A block diagram of a satellite tuner is given in Fig. 12.20, showing the same gain-control and stripline tuning techniques as used in tuners for u.h.f. Bands IV and V. The blocks at centre right form a prescaler and phase-lock-loop, part of the microprocessor-controlled tuning system currently fitted to all satellite tuners. Also incorporated in this tuner module is the bandpass s.a.w. filter and the second i.f. amplifier, so that the emerging signal is ready for demodulation in a p.l.l. detector similar to that shown in Fig. 12.12. In satellite receivers the second i.f. frequency is typically 480 MHz and careful design and shielding of the demodulator v.c.o. is necessary to avoid radiation and interference effects.

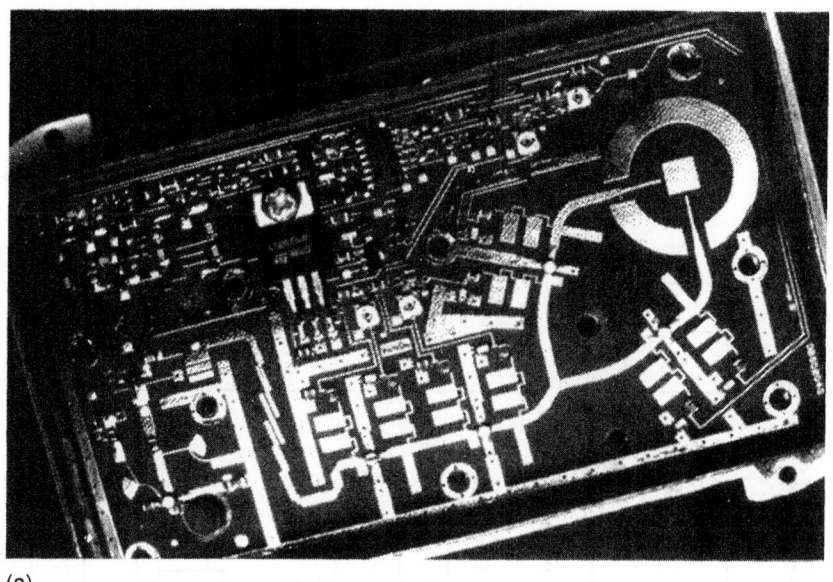

(a)

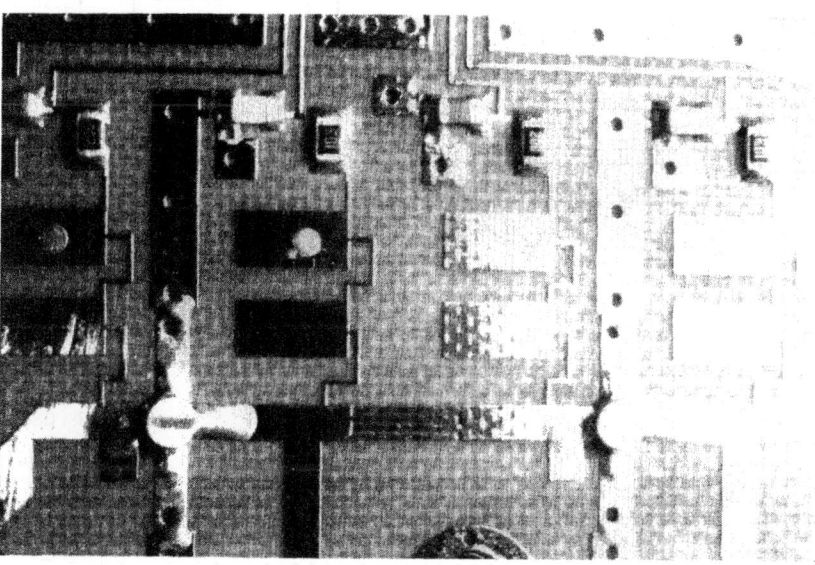

(b)

Fig. 12.19. (a) Inside an l.n.b. for domestic satellite broadcast reception; (b) enlarged view of tuned s.h.f. amplifier

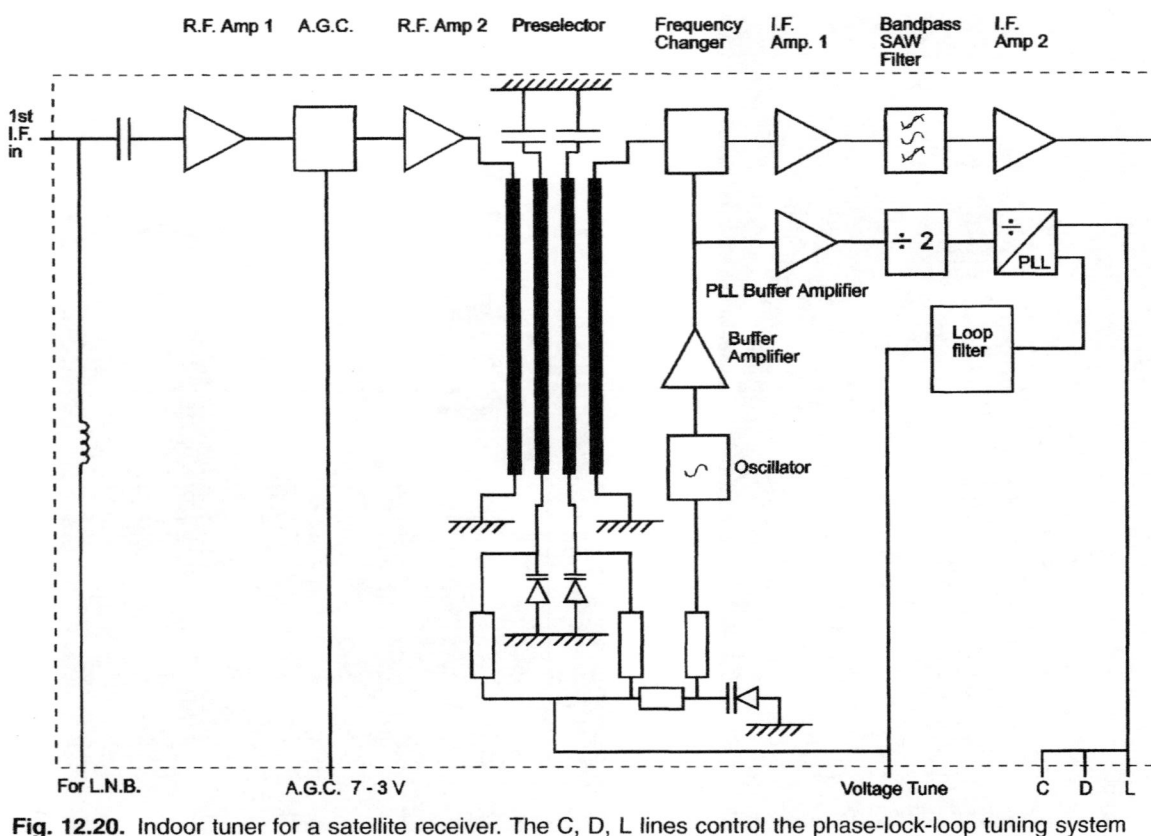

Fig. 12.20. Indoor tuner for a satellite receiver. The C, D, L lines control the phase-lock-loop tuning system

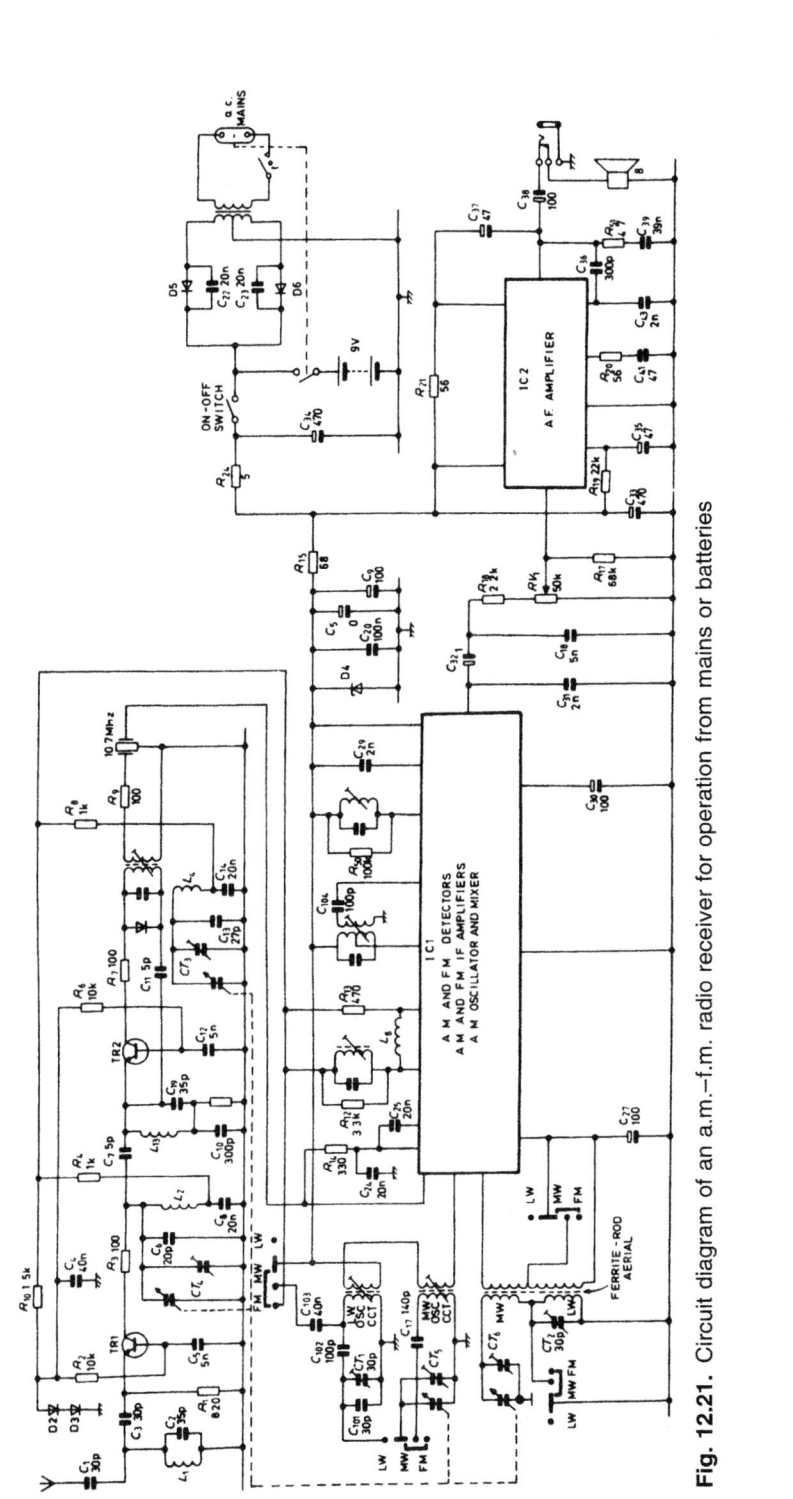

Fig. 12.21. Circuit diagram of an a.m.–f.m. radio receiver for operation from mains or batteries

Complete receivers

We have now considered how transistors can be used to perform many of the functions required in radio and television receivers. In modern equipment much of the circuitry is embodied in i.c.s. This is illustrated in Fig. 12.21 which gives the circuit diagram of an f.m.–a.m. radio receiver designed for operation from batteries or mains. The only functions not carried out by i.c.s are r.f. amplification and frequency-changing for f.m. reception. The design of the r.f. amplifier circuit TR1 was described earlier (page 185) and TR2 is a self-oscillating mixer. It is basically a Colpitts oscillator, the two fundamental capacitors providing positive feedback being C_{11} between collector and emitter (the tuning capacitor of the i.f. transformer primary winding has negligible reactance at oscillation frequency) and C_{19} between base and emitter (C_{10} and C_{12} having very low reactance). The r.f. input is injected into the emitter circuit and the i.f. output is taken from the i.f. transformer via a ceramic filter. On mains operation the 9-V supply is derived from a full-wave rectifying circuit so arranged that insertion of the mains plug into the receiver disconnects the internal batteries.

The complete circuit diagram of a colour television receiver would be formidably complicated, particularly if it had provision for remote control, teletext, display of information on screen or LED panel and stereo sound. But most of the complexities are embodied in i.c.s some of which, as mentioned earlier, can contain more than 100,000 components. If the i.c.s are shown as block symbols the diagram is considerably simplified, containing a few inter-connected i.c.s and a little discrete-transistor circuitry. Much the same can be said of the circuit diagram for a video cassette recorder. This, too, contains a colour television receiver and a display of information.

Chapter 13

Pulse generators

Introduction: the transistor as a switch

So far in this book we have discussed the use of transistors as amplifiers of pulses or sinusoidal signals. In class-A and class-B amplifiers the transistor is used as a linear amplifier in which every change in input signal brings about a corresponding change in output signal. In this type of operation (typical of analogue equipment) the shape of the input–output characteristic is all-important and any deviations from linearity must be compensated by negative feedback or other means to give an acceptable performance.

In pulse or digital circuits, however, transistors are used in a completely different type of operation in which the shape of the input–output characteristic is of minor significance. The transistor is, in fact, used as a switch with only two states usually termed *on* and *off*. In the *on* state the transistor is conductive, having a significant collector (or drain) current, the collector-emitter (or drain-source) voltage being very low. In the *off* state the transistor is non-conductive, having negligible collector (or drain) current, the collector-emitter (or drain-source) voltage being practically equal to that of the supply. In both states dissipation in the transistor is low, in the *on* state because of the low collector (or drain) voltage and in the *off* state because of the low collector (or drain) current. In the *on* state, however, there is appreciable dissipation in the base circuit of a bipolar transistor for this is usually fed via a series resistor. There is no corresponding dissipation in the jfet which gives it one advantage over the bipolar transistor in digital applications.

Stable and unstable states

A state may be such that a transistor switched into it by an external signal remains in that state indefinitely provided that the supply to the transistor is maintained: such a state is termed *stable*. On the other hand there are states which will not persist. When a transistor is put into one of these by an external signal it remains in it for a limited period only and then reverts automatically to its original state without assistance from external signals. The temporary state is termed *unstable*.

So far we have spoken of the two possible states of a single transistor but a circuit containing several transistors may also possess stable and/or unstable states. In both states some of the transistors may be on whilst others are off but it is still true that the states of the circuit as a whole may be stable or unstable.

Bistable, monostable and astable circuits

In some pulse circuits both states are stable. Such a circuit is termed *bistable* and it requires an external triggering signal to compel it to leave one state and enter the other. A further external signal is then required to compel the circuit to return to the original state again. Two triggering signals are thus necessary for each cycle of operation of the circuit which therefore has applications as a counter or frequency divider.

In other circuits one state is stable and the other unstable. Such a circuit is termed *monostable* and it always reverts to its stable state. An external signal is needed to make the circuit enter the unstable state but, after such stimulus, the circuit automatically reverts to the stable state without need of external signals. The time taken to reach the stable state after triggering can be given any desired value within wide limits by appropriate choice of time constants. Circuits of this type are often used to give accurately-timed delays and are sometimes called *delay generators*.

In a further class of circuits both states are unstable. Such a circuit is termed *free-running* or *astable* and will not remain permanently in either state. Without need of external signals an astable circuit automatically switches between the two states at a frequency determined by the time constants of its circuitry. Although an astable circuit does not need external signals to make it operate, it can readily be synchronised at the frequency of a recurrent external signal.

Some circuits can be made bistable, monostable or astable by simple changes in circuitry, sometimes even by altering bias conditions. One circuit of this type is the multivibrator.

Multivibrator: bistable

The basic circuit for a collector-coupled bistable multivibrator is given in Fig. 13.1. This shows two transistors with resistive collector loads, each collector being connected to the other base by a resistive potential divider. The arrangement is such that when one transistor is conducting, its collector potential ensures that the other transistor is cut off. The two possible states of the circuit are thus:

(1) TR1 on and TR2 off.
(2) TR2 on and TR1 off.

If the circuit is initially in state (1), an external signal will cause it to enter state (2) and the change of state is accomplished with great rapidity. The speed of transition is due to the positive feedback inherent in the circuit: this feedback is obvious if the circuit is regarded as that of a two-stage direct-coupled amplifier with the output signal returned to the input. A second external signal applied to the circuit will cause it to return to state (1) again.

During these changes of state the collector potentials of TR1 and TR2 alternate between a positive value when the transistor is cut off and almost zero when it is fully conducting.

If a recurrent triggering signal is applied to the circuit, the voltage waveform generated at each collector is approximately rectangular in form. Moreover, the frequency of the rectangular signal is half that of the triggering signal, enabling such circuits to be used in frequency dividers or binary counters.

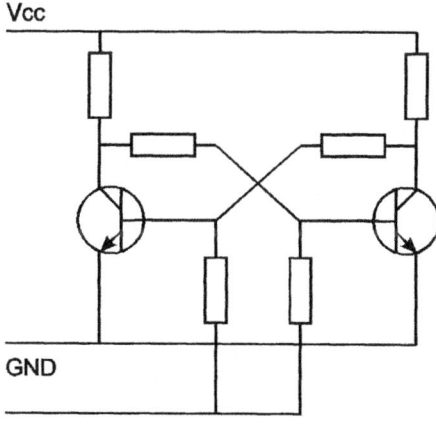

Fig. 13.1. Basic circuit of bistable multivibrator using bipolar transistors

Speed-up capacitors

In the simple form illustrated in Fig. 13.1 the waveforms generated at the collectors during changes of state have poor rise times. This is because the circuit coupling each collector to the other base consists of a series resistor feeding into the parallel RC combination of the transistor input impedance. Such a network has a response which falls as frequency rises and thus the positive feedback in the circuit is more marked at low than at high frequencies. To achieve steep edges (small rise times) in the output signals the degree of feedback should be independent of frequency: thus the coupling circuits should not have a high-frequency loss. The coupling circuits can be made aperiodic (i.e. non-frequency-discriminating) by shunting the series resistors with capacitors as shown in Fig. 13.2 and the condition for an aperiodic response is that the time constant of the coupling resistor-coupling capacitor combination should equal that of the transistor input impedance. If a smaller value of capacitor is used the rise time of the multivibrator output signals is not improved as much as is possible: if a larger value of capacitor is used the capacitance placed in parallel with the collector load becomes significant and it slows up the speed with which the collector potential approaches the negative supply potential when the collector current is cut off: in other words the rise time of the positively-going collector potential changes is impaired.

To avoid the necessity for the separate negative supply for the bases shown in Fig. 13.1, the two emitters may be bonded and connected to the earth line via an RC circuit as shown in Fig. 13.2. In this circuit the emitter current of the conductive transistor ensures that the other is cut off by making its emitter potential more positive than its base potential.

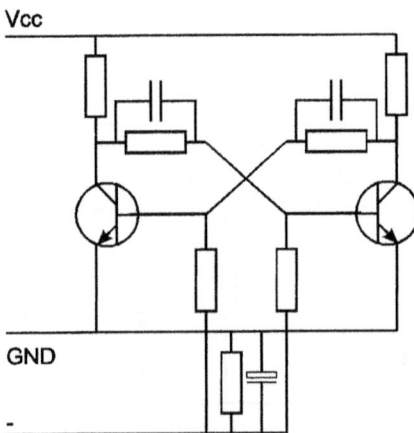

Fig. 13.2. Transistor bistable multivibrator with speed-up capacitors

It is advisable to decouple the common-emitter resistance to preserve steep edges in the output waveform.

The triggering signals needed to bring about a change of state in a multivibrator can be applied to the base of either transistor and can be positive or negative so that the transistor is cut off if conductive or switched on if non-conductive. It is preferable, however, to apply the signals to the base of a conductive transistor rather than that of a non-conductive one. This is because a smaller trigger amplitude is required, there being no bias to overcome, and also because the conductive transistor amplifies the trigger so applying a much larger trigger to the non-conductive transistor.

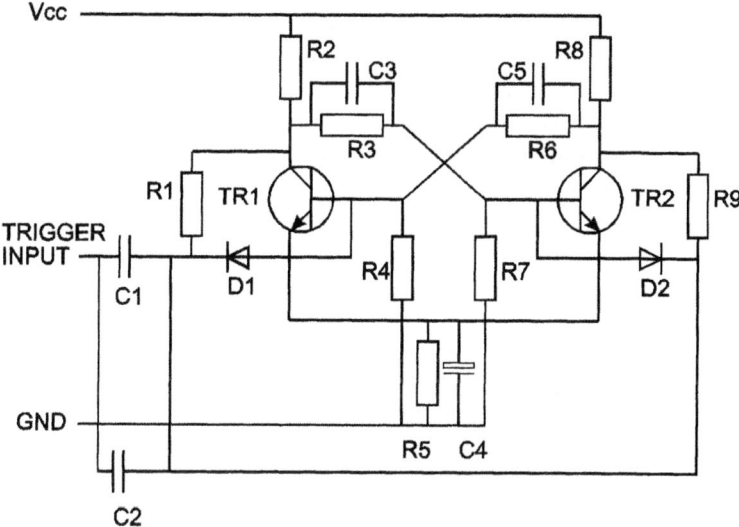

Fig. 13.3. The circuit of Fig. 13.2 with diode triggering gate

In counter circuits it is usual to inject a series of unidirectional pulses into a bistable circuit and some means is required of directing the pulses alternately to the base circuits of the two transistors. One method of achieving this is by the use of the diode gate circuit illustrated in Fig. 13.3. The input signals are applied to the two capacitors C_1 and C_2 and then to the two base circuits via diodes D1 and D2. The switching action of the diodes can be followed from the following section.

Diode gate circuit

Suppose TR1 is conductive. The collector current is a maximum and the collector potential is very low so that there is little difference between the collector and base potentials. The diode D1 is connected between collector and base (via R_1) and there is little potential difference across the diode. The diode is so connected that any negative-going edge applied to capacitor C_1 makes the diode conduct, i.e. low-resistance, and the edge is conducted to the base of TR1. The edge cannot, however, reach the base of TR2: this transistor is cut off and its collector current is a minimum. The collector potential of TR2 is very high and nearly equal to the supply voltage. There is thus a maximum potential difference across the diode D2 which is connected between collector and base (via R_9). Moreover this potential difference biases D2 in the reverse direction so that it does not conduct when a negative-going edge is applied to C_2 unless this edge has sufficient voltage to offset the static bias (and the amplitude of the edge must be controlled to avoid this).

A negative-going edge applied to C_1 and C_2 thus reaches the base of TR1 but not that of TR2. This edge initiates the regenerative change of state characteristic of multivibrators which ends with TR1 cut off and TR2 conducting. A second negative-going edge applied to the circuit is now conducted to the base of TR2 but not to that of TR1. This triggers off a second change of state, hastened by regeneration, which ends with TR2 cut off and TR1 conductive. This is the state originally postulated.

Waveforms

The waveforms for a triggered bistable multivibrator are given in Fig. 13.4. At t_1 TR1 is conductive and its base and collector potentials are slightly positive with respect to emitter potential. At the same instant TR2 is cut off: its base potential is considerably negative with respect to emitter potential and the collector potential is almost at supply potential $+V_{cc}$. When $t = t_2$ a triggering pulse is received by TR1 and interchanges the states. At $t = t_3$ a second triggering edge is received (by TR2 this time) and interchanges the states again. Two triggering signals are thus required to produce one complete cycle of rectangular wave from TR1 or TR2 collector circuit.

Complementary bistable multivibrator

The transistors in a multivibrator must provide the two essentials of signal inversion and gain. A bistable circuit can therefore consist of a pnp and an npn transistor. Such a complementary multivibrator has the unusual feature that in one state both transistors are on and in the other both are off. Direct coupling can be used between collectors and bases and as a result the circuit is particularly simple containing only five resistors as shown in Fig. 13.5(a): here the output is taken from the

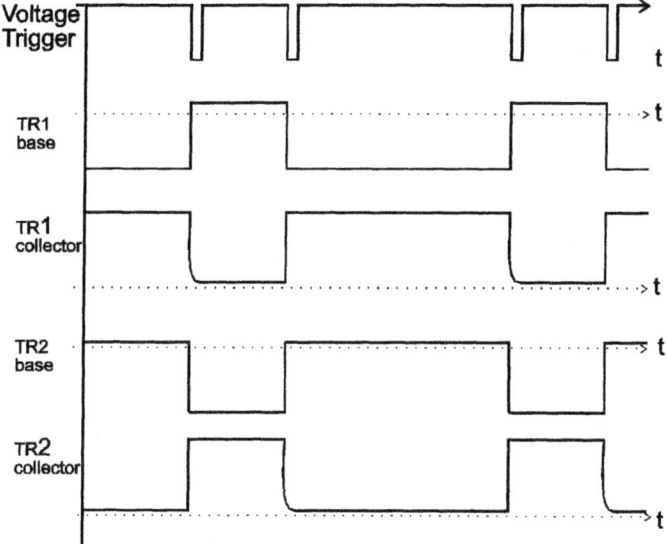

Fig. 13.4. Base and collector waveforms for the bistable multivibrator of Fig. 13.3

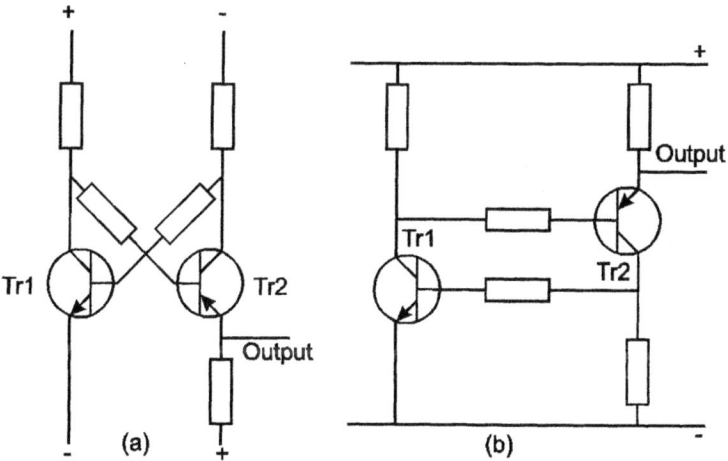

Fig. 13.5. Two ways of representing the circuit diagram of a complementary bistable multivibrator

emitter circuit of the npn transistor. It is perhaps easier to visualise operation of the circuit by drawing it as shown in Fig. 13.5(b) although this loses the familiar crossed lines of the multivibrator circuit.

Multivibrator: monostable

The bistable circuit of Fig. 13.2 can be made monostable by replacing one of the direct inter-transistor couplings by a capacitance coupling and by suitable choice of bias. One possible circuit for a monostable collector-coupled multivibrator is that shown in Fig. 13.6. TR1 is biased

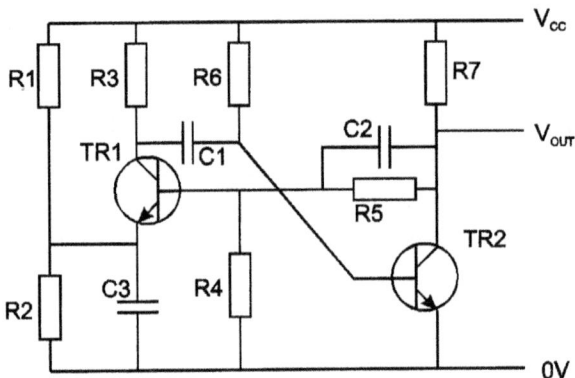

Fig. 13.6. Transistor monostable multivibrator

by the potential divider R_1R_2 but TR2 has no bias components other than R_6 which connects the base to the positive supply line. Clearly any negative potential applied to TR2 base (to cut this transistor off) cannot remain there indefinitely but will leak away as C_1 discharges and thus the circuit will always revert to the state in which TR2 is on (and TR1 therefore cut off). Thus we have

stable state: TR1 off and TR2 on

from which it follows that

unstable state: TR2 off and TR1 on.

The operation of the circuit is illustrated by the waveforms of Fig. 13.7. Before time t_1 the circuit is assumed here to be in the stable state.

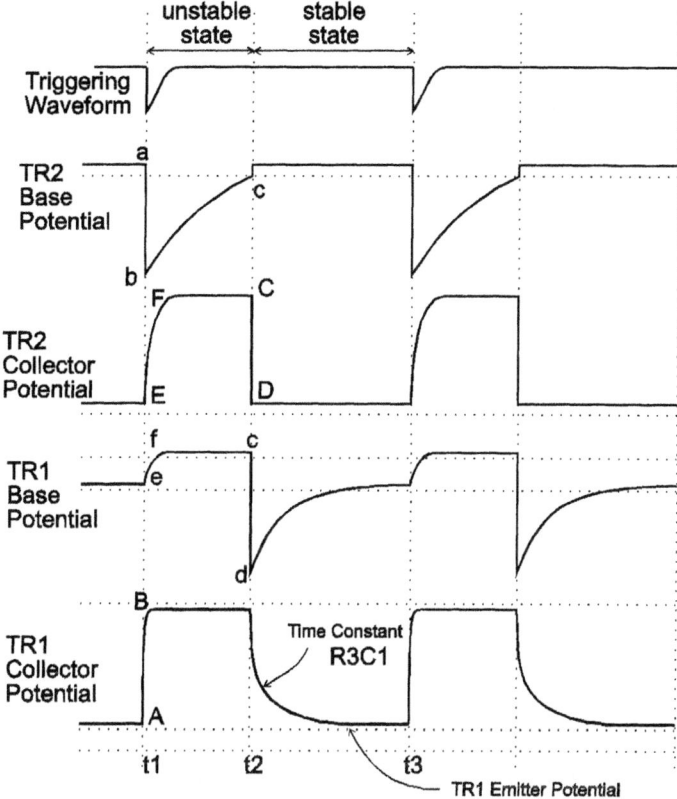

Fig. 13.7. Base and collector potential waveforms for the multivibrator of Fig. 13.6

At t_1, TR1 is made conductive by applying a positive-going edge to its base or preferably (as shown in Fig. 13.7) by applying a negative-going edge to TR2 base. The collector potential of TR1 makes a negative excursion (*AB* in Fig. 13.7). The voltage across a capacitor cannot be changed instantaneously and this edge is transferred without loss in amplitude by C_1 to the base of TR2 (*ab*), cutting TR2 off. The circuit is now in the second of its two possible states (TR1 conductive and TR2 cut off), but it cannot remain in it permanently because C_1 begins immediately to discharge through R_6 and the output circuit of TR1. As the discharge proceeds the potential at the base of TR2 becomes less negative (*bc*) until a point (*c*) is reached at which TR2 begins to conduct. This causes the collector potential of TR2 to move negatively, generating an edge *CD* which is transferred simultaneously, without change in amplitude, to TR1 base (*cd*) cutting TR1 off. This in turn causes the collector potential of TR1 to move positively, which accelerates the onset of conduction in TR2.

The process becomes regenerative, in fact, as soon as the overall gain of the two transistor stages exceeds unity, and the change of state is then very rapid. After the change, TR1 is off and TR2 on. The potential at TR1 base now falls to near zero as C_2 discharges through R_4, R_5 and the output circuit of TR2. The potential at TR1 collector moves exponentially towards $+V_{cc}$ as C_1 charges through R_3. When these two changes have ceased the circuit rests and no further action occurs unless further triggering signals are received: in other words this is the stable stage again.

When a monostable multivibrator is used as a delay generator, the delay is equal to the period of the unstable state and we thus need to be able to calculate the component values needed to produce a wanted delay. This can be achieved as follows.

Duration of unstable state

The duration of the unstable state is equal to the time taken for the potential at the base of the non-conductive transistor to fall to just beyond cut-off value.

First consider the conditions in the circuit in the stable state: TR1 is cut off and TR2 conductive. TR1 collector potential is equal to that of the supply $+V_{cc}$. TR2 is conductive and its base potential is at the cut-off voltage $+V_{be}$. Thus the voltage across C_1 is equal to $(V_{cc} - V_{be})$, the right-hand plate being positive with respect to the other.

The unstable period begins when TR1 is suddenly switched on. The collector current rises at this instant to a value which is sufficient with proper design to bring the collector potential very nearly to TR1 emitter potential V_{e1}. The change in TR1 collector potential is thus $(V_{cc} - V_{e1})$ and this is transferred by C_1 to TR2 base, causing the potential here to rise from $+V_{be}$ to $(V_{cc} - V_{be1} - V_{be})$. C_1 now begins to discharge through R_6 and the output circuit of TR1 but the resistance of this circuit (with TR1 conductive) is small compared with R_6 and can be neglected. If the discharge were completed TR2 base potential would fall from its initial (negative) value of $(V_{cc} - V_{e1} - V_{be})$ to zero and would then reverse in polarity and reach the (positive) supply voltage of V_{cc}. However, as soon as TR2 base potential reaches V_{be}, the cut-off value, TR2 begins to conduct and the unstable period is abruptly terminated.

The rate of discharge of C_1 is determined by the discharge current. This flows in R_6 and thus the current can be measured by the voltage across R_6. At the beginning of the discharge TR2 base has a potential of $(V_{cc} - V_{e1} - V_{be})$ and as the supply voltage is $+V_{cc}$, the voltage across R_6 is $(2V_{cc} - V_{e1} - V_{be})$. At the moment when TR2 becomes conductive the voltage across R_6 is $(V_{cc} - V_{be})$. The duration of the unstable state is approximately equal to the time taken for this fall of voltage to occur in a circuit of time constant R_6C_1. In general the fall of voltage is given by

$$V_t = V_o e^{-t/R_6C_1}$$

where V_t is the voltage after a time t and V_o is the initial voltage. This may be written in the form

$$\log_e \frac{V_o}{V_t} = \frac{t}{R_6C_1}$$

from which

$$t = R_6C_1 \log_e V_o/V_t$$

Substituting for V_o and V_t

$$t = R_6C_1 \log_e (2V_{cc} - V_{e1} - V_{be})/(V_{cc} - V_{be})$$

As typical practical values we may have $V_{cc} = 12$ V, $V_{ce} = 3$ V and $V_{be} = 0.7$ V. For these

$$t = R_6C_1 \log_e 1.8 = 0.587 R_6C_1$$

to obtain approximate answers it is often possible to neglect V_{e1} and V_{eb} in comparison with V_{cc}: we then have

$$t = R_6C_1 \log_e 2 = 0.6931 R_6C_1$$

Thus to obtain a duration of, say, 1 ms

$$R_6C_1 = \frac{t}{0.6931}$$

$$= \frac{1}{0.6931} \text{ ms}$$

$$= 1.44 \text{ ms}$$

Any values of R_6 and C_1 would thus appear to be suitable provided their product (time constant) is 1.44 ms. In fact there are limitations to the values of R_6 and C_1 which can be used.

For example, R_6 supplies the base current to TR2 when TR2 is conductive and this sets an upper limit to the value of R_6 which can be used. If TR2 is to take 2 mA collector current and if β is 50, the base current is given by 2/50 mA, i.e. 40 μA. If the supply voltage is 12 V R_6 is given by $11.3/(40 \times 10^{-6})$, i.e. 280 kΩ. This is the maximum permissible value and it is advisable to use a smaller value, say, 200 kΩ to ensure that the collector potential is near zero when TR2 conducts. Thus the minimum value of C_1 which can be used is given by

$$C_1 = \frac{1.44 \times 10^{-3}}{R_6}$$

$$= \frac{1.44 \times 10^{-3}}{200 \times 10^3} \text{ F}$$

$$= 0.007 \text{ μF}$$

The end of the unstable period is marked by the abrupt start of the collector current in TR2 and the sudden cut-off of collector current in TR1. At this instant, therefore, a negative-going step is generated at TR2 collector and a positive-going step at TR1 collector. Either of these steps could be used as the output of the circuit but the output at TR2 collector is, in general, preferred because it has a shorter rise time. This is borne out by the following calculations.

When TR2 is turned on, its collector potential goes negative and cuts TR1 off by driving its base negative. There is little change in voltage across the coupling capacitor C_2 and the only capacitance which has to be charged or discharged by the collector current of TR2 is the output capacitance of TR2 with possibly a contribution of capacitance from TR1 input. The relationship between the voltage across a capacitor and the current flowing into it is given by

$$V = \frac{Q}{C} = \frac{1}{C} \int i \cdot \mathrm{d}t$$

from which

$$\frac{\mathrm{d}V}{\mathrm{d}t} = \frac{i}{C}$$

Suppose the capacitance at TR2 collector is 20 pF and that the collector current of TR2 is 2 mA. We have

$$\frac{dV}{dt} = \frac{i}{C}$$

$$= \frac{2 \times 10^{-3}}{20 \times 10^{-12}}$$

$$= 10^8 \; V/s$$

$$= 100 \; V/\mu s$$

If the supply to TR2 is 12 V the collector potential excursion is limited to this value and occupies only approximately 0.1 μs. The rise time of this potential step is thus of the order of 0.1 μs. This is perhaps an optimistic value because we have neglected any effect due to TR2 collector resistor R_4 which tends to slow up the discharge of TR2 collector capacitance by charging it from the supply.

Alternatively, of course, the potential step generated at TR1 collector may be used as output. When TR1 is cut off at the end of the unstable period, the collector potential approaches that of the positive terminal of the supply. This tends to drive the base of TR2 positive but as soon as TR2 starts to conduct, its input resistance becomes very low and its base potential is thus effectively stabilised near emitter potential. Thus TR1 collector potential can only go positive as C_1 charges from the supply through R_3. The voltage–time relationship is exponential and the time taken for the voltage to change from 10% to 90% of its final value (i.e. the rise time) is 2.2 times the time constant RC governing the change. For an approximate calculation it is often taken as twice the time constant. Thus in this particular example the time constant is R_3C_1 and the rise time is approximately $2R_3C_1$. A suitable value for R_3 is 5 kΩ because this permits a collector current of 2 mA with a supply of 12 V (2 V being assumed lost across TR2 emitter resistor). C_1 has already been calculated as 0.007 μF and thus we have

$$\text{rise time} = 2R_3C_1$$

$$= 2 \times 5 \times 10^3 \times 0.007 \times 10^{-6} \; s$$

$$= 70 \; \mu s$$

which is considerably greater than the rise time of the step at TR2 collector.

Hole storage

The difference between the rise times of the voltage steps at TR1 and TR2 collectors is in practice likely to be greater than suggested by the above calculations. This is because of an effect, known as hole storage, which tends to degrade the rise time of the collector-potential step when a transistor is cut off. Hole storage is an effect, internal to the transistor, which causes the collector current to persist for a few microseconds after the emitter current has been cut off. This occurs in transistors which have been driven hard into conduction so that the collector-emitter potential is nearly zero. In these circumstances the emitter injects more charge carriers (holes in a pnp transistor) into the base region than are required to give the collector current, and the excess carriers are stored ready to be swept into the collector region to prolong the collector current for a short period after the emitter current has been cut off. This hole-storage effect can be avoided by so designing the circuits that the collector-emitter potential does not approach zero but is limited to a value such as 1 V: this can be achieved by the use of diodes known as catcher diodes because they catch the collector voltage and hold it at a value above saturation.

Control of duration of unstable state

In the deduction of an expression for the duration of the unstable state it was assumed that R_6 was returned to the positive supply line (Fig. 13.6). When TR1 is suddenly switched on C_1 causes TR2 base voltage to jump instantaneously to $- V_{cc}$ approximately. At this instant the voltage across R_6 is approximately $2 V_{cc}$ and this determines the rate of discharge of C_1. As C_1 discharges, TR2 base voltage moves towards $+ V_{cc}$ (supply voltage) but never reaches it because TR2 starts to conduct when its base voltage reaches $+0.7$ V approximately. Thus only about one half of the possible extent of the voltage change at TR2 base is in fact realised. This accounts for the factor 0.6931 in the expression $0.6931 R_6 C_1$ for the duration of the unstable state. 0.6931 is $\log_e 2$.

Now suppose R_6 is returned to a source of negative voltage equal to one half the supply voltage, i.e. $+ V_{cc}/2$. TR2 base voltage still jumps to approximately $- V_{cc}$ when TR1 is abruptly switched on but the instantaneous voltage across R_6 is now only $1.5 V_{cc}$ approximately, implying a slower rate of discharge of C_1. As C_1 discharges TR2 base voltage moves towards $+V_{cc}/2$. Again it never reaches this value because TR2 starts to conduct before the discharge is completed. But now two thirds of the discharge is achieved and, as the time constant is still $R_6 C_1$, this takes longer than before. In fact the expression for the duration of the unstable state is now approximately $1.1 R_6 C_1$, the natural logarithm of 3 being 1.0986.

As the voltage to which R_6 is returned is moved towards zero more of the discharge of C_1 is achieved and the longer is the duration of the unstable state. If R_6 is returned to the slider of a potentiometer across the supply, this control can in fact be used as a means of adjusting the duration to a desired value.

Multivibrator: astable

The bistable circuit of Fig. 13.2 can be made astable by replacing both of the direct inter-transistor couplings by capacitance couplings and by suitable choice of bias. One possible circuit for an astable collector-coupled multivibrator is given in Fig. 13.8.

No emitter bias is used and both bases are returned via resistors to the positive supply line. The behaviour of the circuit will be illustrated by the waveforms shown in Fig. 13.9. At t_1, TR1 is cut off by a negative signal at the base and the collector potential is a positive maximum. At the same time, TR2 is conducting, having minimum positive values of base and collector potentials. TR2 is in a stable state but TR1 is not, because the potential at its base is moving positively as C_2 discharges through R_2. At t_2, the potential at TR1 base reaches the value at which TR1 starts conducting. As explained for the monostable multivibrator, this starts a regenerative action which causes TR1 to become abruptly conductive. The resultant steep negative-going voltage (*AB*) at TR1 collector is transferred by C_1 to the base of TR2 (*ab*), cutting TR2 off. There now follows the exponential change in base potential (*bc*) previously described. It keeps TR2 cut off for a period given approximately by $0.69R_3C_1$, which terminates at time t_3 when TR2 becomes abruptly conductive. TR2 is now in a stable state but TR1 is not, being cut off by a negative edge (*de*) which decays exponentially (*ef*). This keeps TR1 cut off for a period given approximately by $0.69R_2C_2$. This is the circuit condition assumed initially. Thus the cycle continues, the period of oscillation T being equal to the sum of the durations of the two unstable periods

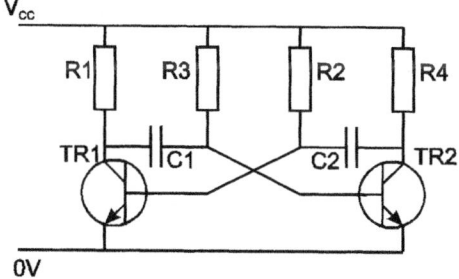

Fig. 13.8. Transistor astable multivibrator

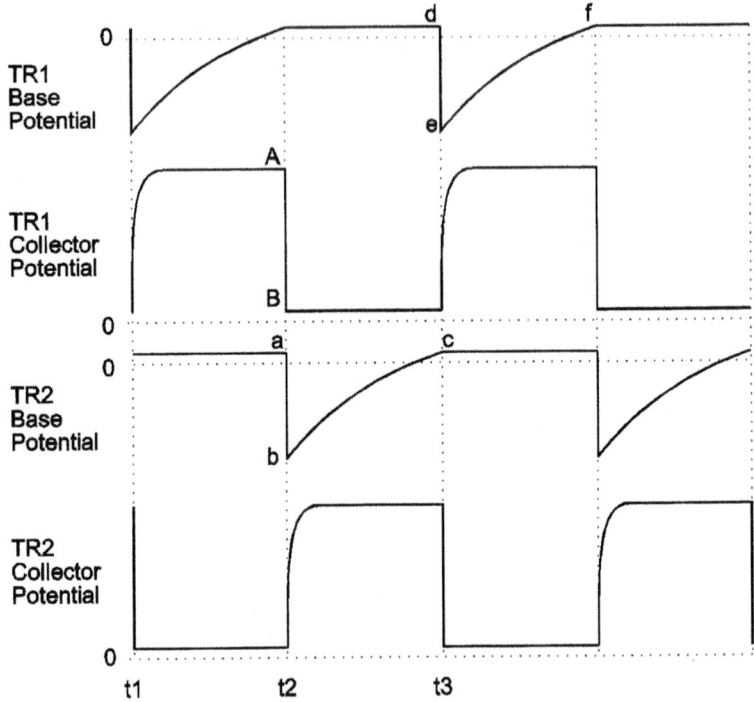

Fig. 13.9. Base and collector waveforms for the astable multivibrator of Fig. 13.8

$$T = 0.69(R_3C_1 + R_2C_2)$$

If $R_2 = R_3 = R$ and $C_1 = C_2 = C$ the multivibrator is symmetrical and generates square waves (equal mark space ratio) at both collectors. The period of oscillation is given by

$$T = 1.38\, RC$$

The free-running frequency of such a multivibrator is given by

$$f = \frac{1}{T}$$

$$= \frac{1}{1.38RC}$$

A multivibrator required to be synchronised at the line frequency (approximately 15 kHz) of a 625-line television system would be designed to have a natural frequency somewhat lower than this – say 10 kHz. We have already seen that suitable values for the load resistors and base resistors are 5 kΩ and 200 kΩ: we have now to determine a suitable value for the coupling capacitors. From the above expression we have

$$C = \frac{1}{1.38Rf}$$

$$= \frac{1}{1.38 \times 2 \times 10^5 \times 10^4} \, F$$

$$= 360 \, pF \text{ approximately}$$

Control of free-running frequency

In Fig. 13.8 the base resistors R_2 and R_3 are shown returned to the positive supply line but either or both could alternatively be returned to a source of voltage nearer zero and, as explained for the monostable circuit (page 268), this has the effect of increasing the duration of the unstable period(s). For an astable multivibrator this implies a decrease in the free-running frequency. In this way the frequency of operation can be controlled by adjustment of the voltage to which the base resistors are returned. The astable multivibrator can thus be used as a voltage-controlled oscillator and has applications in phase-locked-loop circuits (page 241).

Synchronising of multivibrators

Bistable and monostable multivibrators are triggered by signals which cause the circuit to leave a stable state. Such a technique cannot be applied to astable circuits because these have no stable states. The synchronising signals injected into an astable circuit are designed to terminate the unstable periods earlier than would occur naturally. It follows that the natural frequency of the circuit must be lower than the frequency of the synchronising signals.

To terminate the unstable periods unnaturally early the synchronising signals can take the form of negative-going pulses applied to one or both of the bases of npn transistors. The amplitude of the edges is important. If the natural frequency of the multivibrator is small compared with that of the synchronising signals, the amplitude of these signals, if small, may be such that every fifth one terminates an unstable period. The multivibrator then runs at one fifth of the sync-pulse frequency. Increase in the amplitude of the sync signal may cause every fourth signal to trip the multivibrator so that its frequency is now one quarter that of the sync signals. Further increase in sync-pulse amplitude may cause the multivibrator frequency to jump to one third the synchronising frequency. Thus a synchronised multivibrator can be used as a frequency divider but close control over the sync-pulse amplitude is necessary to obtain a consistent large division ratio.

Emitter-coupled multivibrator

All the multivibrators described above have employed two collector-to-base coupling circuits. One of these can be replaced by an emitter-to-emitter coupling without significantly altering the principles of operation and Fig. 13.10 gives the circuit diagram of a bistable emitter-coupled multivibrator. When TR2 is turned on, its emitter current, in flowing through the common-emitter resistor R_e, makes the emitter of TR1 positive with respect to the fixed base potential and thus cuts TR1 off. The effect of R_e is hence similar to that of a direct coupling between TR2 collector and TR1 base.

One advantage of this circuit over the collector-coupled version is that the collector of TR2 plays no part in the mechanism of operation. It is free and can provide an output, the loading on which has no significant effect on the performance of the circuit.

Collector-coupled multivibrators can be symmetrical but emitter-coupled types cannot. This can easily be seen for in one of the two possible states of the circuit the voltage across R_e must be less than that across R_2 to give the positive base-emitter voltage which keeps TR1 conductive. In the other state, however, the voltage across R_e must be greater than that across R_2 to give the negative base-emitter voltage

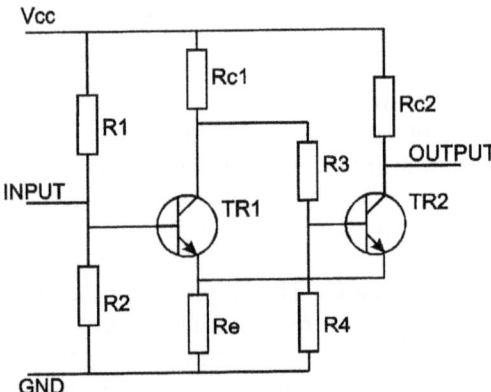

Fig. 13.10. A bistable emitter-coupled transistor multivibrator

which keeps TR1 cut off. In the first state the only current in R_e is that due to TR1 whereas in the second state the current in R_e is due entirely to TR2. It follows that TR2 must take a larger collector current (when conductive) than TR1 (when conductive) and the circuit is hence asymmetrical.

The circuit can be triggered by positive-going or negative-going signals applied to TR1 or TR2 base and is sometimes used as a limiter by applying, say, a sinusoidal input to TR1 base. The output is then rectangular at the frequency of the input with an amplitude substantially independent of that of the input provided this exceeds a certain minimum value.

A monostable version of the emitter-coupled multivibrator of Fig. 13.10 can be obtained by replacing one of the direct inter-transistor couplings by a capacitance coupling. This can be achieved by replacing R_3 in Fig. 13.10 by a capacitor and by returning R_4 to a source of positive voltage such as the supply line.

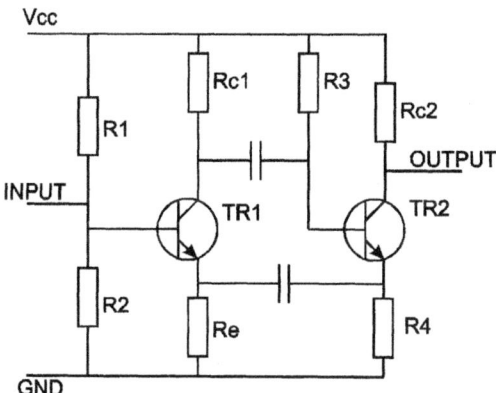

Fig. 13.11. An astable emitter-coupled transistor multivibrator

An astable version of the multivibrator is obtained if capacitors are included in both inter-transistor couplings. This may be achieved by modifying the collector-to-base coupling as for the monostable circuit and in addition using individual emitter resistors coupled by a capacitor as shown in Fig. 13.11.

Blocking oscillator

As we have seen, a multivibrator consists essentially of two transistors so connected that the output of each is coupled to the input of the other.

Each transistor introduces a signal inversion, a condition essential to give the positive feedback which accelerates the changes of state. If one of the transistors is replaced by a transformer connected so as to give signal inversion, the resulting circuit will still generate pulses. The circuit derived in this way is termed a blocking oscillator and is extensively employed as a pulse generator and as a sawtooth generator. Two forms of blocking oscillators are in common use, an astable or free-running circuit and a monostable circuit: a bistable circuit is not possible because the transformer cannot maintain a signal across the windings indefinitely.

Astable circuits

One form of transistor blocking oscillator is illustrated in the circuit diagram of Fig. 13.12. The primary winding of the transformer is included in the collector circuit and the secondary winding in the base circuit. The black dots indicate points of similar instantaneous polarity and show the winding sense required for the required signal inversion in the transformer. Base bias is provided by the resistor R_1 and in this circuit R_1 is returned to a source of positive voltage: the transistor is a npn type and this bias voltage ensures that the circuit will always revert to the state in which the transistor is conductive. As soon as conduction starts, however, positive feedback causes oscillation at the resonance frequency of the primary winding (assumed the larger of the windings). The design of the circuit is such that the oscillation amplitude builds up very rapidly and as it does so the capacitor C_1 is charged by the base current in the transistor. The voltage generated across C_1 by this current is such as to bias the base of the transistor negatively with respect to the emitter. Such a polarity cuts an npn transistor off, of course, and one of the aims in the design of blocking oscillators is to cut the transistor off in the first half-cycle of oscillation.

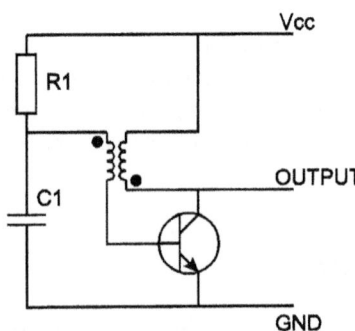

Fig. 13.12. An astable blocking oscillator

After cut-off there is a period of relaxation in the circuit whilst C_1 discharges through R_1. As soon as the voltage across C_1 has fallen to a value at which the transistor starts to conduct, oscillation begins again and the cycle is repeated. Thus the circuit is astable (free-running) and takes simultaneous bursts of base and collector current at a repetition frequency governed by the time constant $R_1 C_1$, the duration of the burst being determined by the resonance frequency of the transformer primary winding. Provided the burst of collector current is large enough to bring the collector-emitter voltage down to nearly zero, pulses of substantially rectangular waveform can be obtained from the collector terminal as suggested in Fig. 13.12.

Very close coupling is desirable between the two windings of the transformer to give a large degree of positive feedback and heavy damping of the windings is sometimes arranged to discourage ringing at the resonance frequency of the primary winding. A practical precaution which is usually desirable is to connect a diode across the primary winding as shown in Fig. 13.13. This has the effect of eliminating the large overshoot which would otherwise be generated across the primary winding when the collector current is suddenly cut off. Such overshoots could cause the collector-emitter voltage to exceed the safe rating for the transistor.

We can obtain an estimate of the natural frequency of the blocking oscillator in the following way. If the collector supply voltage is V_{cc} then a voltage with a peak value of nearly V_{cc} is generated across the primary winding when the transistor is suddenly turned on. We will assume the transformer to be a step-down type with a turns ratio of $n{:}1$. The voltage generated across the secondary winding when the transistor is turned on is then approximately V_{cc}/n. This is also the voltage generated across the

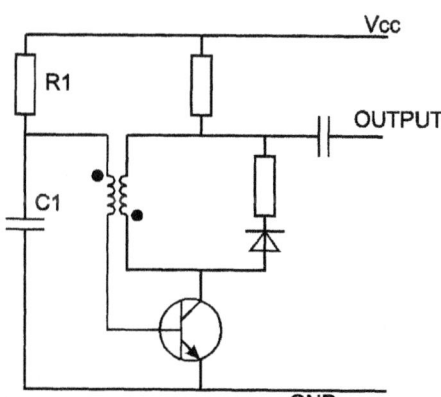

Fig. 13.13. An astable blocking oscillator with a diode to suppress overshoot

capacitor C_1 because the base-emitter circuit of the blocking oscillator behaves in a manner similar to that of a diode detector and a capacitor in the circuit is charged up to the peak value of the applied signal. Initially the voltage across C_1 biases the base negatively but C_1 is connected via R_1 to a source of positive voltage V_{bb}. Thus in the absence of the transistor the voltage across C_1 would initially fall to zero and then change sign, ultimately reaching the value V_{bb}, the curve following an exponential law. However, the moment the voltage reaches the cut-off value V_{be} (approximately +0.7 V for npn silicon transistor) the transistor begins to conduct. Oscillation then occurs and C_1 is again charged to V_{cc}/n volts. Thus only the initial part of the exponential change in voltage across C_1 is therefore achieved and the interval between successive bursts of oscillation (which, of course, determines the natural frequency of the blocking oscillator) is equal to the time taken for a voltage initially equal to $(V_{cc}/n + V_{bb})$ to fall by $V_{cc}/n + V_{be}$, the time constant being $R_1 C_1$.

Now the initial part of an exponential curve is nearly linear and has a slope given by $V_o/R_1 C_1$ where V_o is the initial voltage. Thus the fall in voltage in a time t is given by

$$\left(\frac{V_{cc}}{n} + V_{bb} \right) \cdot \frac{t}{R_1 C_1}$$

and if we equate this to $V_{cc}/n + V_{be}$ we can obtain an estimate of the natural frequency of the oscillator. We have

$$\frac{V_{cc}}{n} + V_{be} = \left(\frac{V_{cc}}{n} + V_{bb} \right) \cdot \frac{t}{R_1 C_1}$$

giving

$$\text{natural frequency} = \frac{1}{t}$$

$$= \frac{(1 + nV_{bb}/V_{cc})}{(1 + nV_{be}/V_{cc})} \cdot \frac{1}{R_1 C_1}$$

Control of natural frequency

If R_1 is returned to the collector supply voltage, $V_{bb} = V_{cc}$: if, in addition, V_{cc}/n is large compared with V_{be} the natural frequency is given by

$$f = \frac{n + 1}{R_1 C_1}$$

The natural frequency is now independent of the supply voltage and, for a given oscillator with a fixed value of n, can be controlled by adjustment of R_1 or C_1.

As a numerical example suppose a blocking oscillator is required to operate at 10 kHz. Such an oscillator is required in a 625-line television receiver, the synchronised frequency being 15 kHz. A typical value for the turns ratio of the transformer is 5:1 and rearranging the above expression we have

$$R_1 C_1 = \frac{n + 1}{f}$$

$$= \frac{6}{10^4} \text{ s}$$

$$= 600 \, \mu s$$

If we wish the collector current to peak to 5 mA, the base current should preferably be not less than, say, 0.3 mA. R_1 must then be 80 kΩ to give such a base current with a 24-V supply. C_1 is then given by

$$C_1 = \frac{\text{time constant}}{R_1}$$

$$= \frac{600 \times 10^{-6}}{80 \times 10^3} \text{ F}$$

$$= 0.008 \, \mu F$$

It is probably better, however, to have separate supplies for collector and base as shown in Fig. 13.13 and to control the natural frequency by adjustment of V_{bb} using fixed values of R_1 and C_1. The way in which this controls the natural frequency of oscillation is discussed for monostable and astable multivibrators on pages 264 and 268. Such control has the advantage that the natural frequency can be adjusted by a potentiometer (as shown in Fig. 13.14) which need carry only direct current and can therefore be situated at some distance from the blocking oscillator itself. This could be useful, for example, in the design of a television receiver.

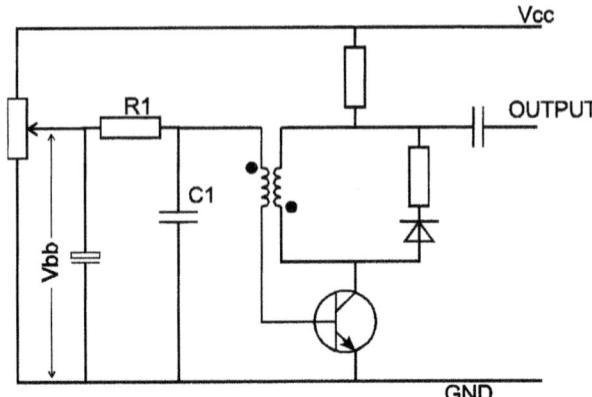

Fig. 13.14. Control of natural frequency of a blocking oscillator by adjustment of V_{bb}

Control of pulse duration

The duration of the burst of conduction is governed by the resonance frequency of the primary winding of the transformer, being approximately equal to half the period of oscillation. For example, if the winding resonates at 50 kHz the period is given by

$$T = \frac{1}{f} = \frac{1}{50 \times 10^3} \text{ s} = 20 \, \mu\text{s}$$

and the burst of conduction is approximately 10 μs duration. Often blocking-oscillator transformers are designed to have a primary inductance and self-capacitance which will give the required duration. Alternatively precautions can be taken to keep self-capacitance to a minimum and the required duration can be obtained by adding a physical component of the correct capacitance across the primary winding.

Output terminals of the blocking oscillator

The output of a blocking oscillator can be taken from the collector terminal as suggested in Fig. 13.12 but external loads added at this point could affect operation of the oscillator and it is better to include a resistor in the collector circuit as shown in Figs. 13.13 and 13.14. This gives negative-going output pulses (if the transistor is an npn type). If positive-going output pulses are required, a pnp transistor can be used, the supply voltage being negative.

Synchronisation of blocking oscillators

Blocking oscillators can readily be synchronised at the frequency of any recurrent signal. Synchronisation is achieved by terminating the non-conductive period earlier than would occur naturally. Thus, the natural frequency must be lower than the frequency of the sync pulses. To terminate the non-conductive period, a positive-going signal can be applied to the base of the transistor (assumed npn) or a negative-going signal can be applied to the collector. Alternatively if the transistor is a pnp type, a negative-going signal is required at the base or a positive-going signal at the collector. The precise waveforms of the sync pulses are not important provided they have a steep leading edge.

Monostable blocking oscillator

The blocking oscillator circuits so far described are essentially astable because the transistor always reverts to the conductive state as C_1 discharges and in this state the transistor bursts into oscillation because of the inherent positive feedback in the circuit. If, however, R_1 is returned to a point negative with respect to the emitter potential, when the circuit is triggered into oscillation, e.g. by a positive-going signal applied to the base, it will revert to the state where the base is negative with respect to the emitter. In this state the transistor is cut off and the circuit will remain indefinitely in this state unless triggered out of it by another sync signal. This is therefore now a monostable circuit which requires a triggering signal to initiate each output pulse.

Chapter 14

Sawtooth generators

Introduction

Sawtooth or ramp waveforms are extensively employed in electronic equipment. Two obvious applications are as sources of deflecting signals in oscilloscopes, television cameras and television receivers but they are also used in radar, in test equipment and in analogue-to-digital converters.

Usually sawtooth generators are classified as *free-running* (astable) which do not require external signals (but can readily be synchronised by such signals) and *driven* (monostable) which will not operate without external signals.

We can distinguish three principles which can be used to generate a sawtooth waveform. The first uses the initial part of an exponential curve as an approximation to the required linear change. The second principle is a more fundamental solution to the problem: if a constant current is applied to a capacitor the rise in voltage across it is linear. The third principle can be derived from the second by using the concept of duality,* i.e. by substituting voltage for current, current for voltage and inductance for capacitance: if a constant voltage is applied to an inductor the current in it grows linearly with time. All three principles are used in practice as the following examples of sawtooth generators illustrate. In some of these transistors are used as switches and in others as linear amplifiers.

Simple discharger circuit

It was mentioned in the previous paragraph that the initial part of an exponential curve is almost linear and in many sawtooth generators this

---

* The inductive circuit is said to be a dual of the capacitive one and the concept of duality can be applied generally to electronic circuits.

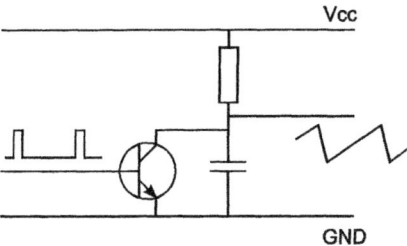

Fig. 14.1. A simple transistor discharger

part of such a curve is used as the working stroke, a faster exponential change providing the rapid flyback which terminates this stroke.

One form of sawtooth generator operating on this principle is illustrated in Fig. 14.1. TR1 has an input consisting of widely spaced positive-going pulses which turn the transistor on to short-circuit (discharge) the capacitor. In the intervals between the pulses, TR1 is cut off and C_1 charges through R_1 to provide the working stroke of the output. The discharge period is also called the flyback.

This is essentially a driven circuit because there is no output in the absence of input pulses. Increase in R_1 or C_1 reduces the rate of charge of C_1 with the result that a smaller voltage builds up across C_1 during the interval between successive pulses: in other words increase in the time constant $R_1 C_1$ reduces the amplitude of the sawtooth output. Similarly reduction in the time constant increases the output amplitude but, if good linearity of forward stroke is required, the output should be kept small compared with the supply voltage to ensure that only the initial part of the exponential curve is used as output.

As an illustration the departure from linearity is 5 per cent if the output swing is restricted to 10 per cent of the supply voltage. The time constant of an RC combination is, by definition, the time taken for the voltage across a capacitor to rise to 63 per cent of the final value. The time taken for the voltage to rise to only 10 per cent of the final value is approximately 0.1 RC. Expressed differently the time constant of the network required in a simple discharger should be approximately 10 times the duration of the forward stroke. Thus if the forward stroke is 50 μs, a time constant of 500 μs is required.

There is, of course, an unlimited number of combinations of resistance and capacitance with this value of time constant but there are limitations on the values which can be successfully used and a decision can usually be made by considering the charging and discharging currents. The capacitor is charged via the resistor during the forward stroke and is discharged by the transistor during flyback. The ratio of average discharge current to average charge current is thus equal to the ratio of forward stroke period to flyback period and may be 10 or 20:1 in a practical circuit. Suppose the forward stroke is to occupy $50\,\mu\text{s}$ and the flyback $3\,\mu\text{s}$. It is convenient to take the average discharge current as $20\,\text{mA}$ for this is easily within the capability of even a small transistor. The sawtooth amplitude is to be 10 per cent of the supply voltage and we will assume an amplitude of 2 V. We can now calculate the capacitance required from the relationship.

$$C = \frac{Q}{V} = \frac{it}{V}$$

Substituting for i, t and V for the discharge of C

$$C = \frac{20 \times 10^{-3} \times 3 \times 10^{-6}}{2}\,\text{F}$$

$$= 0.03\,\mu\text{F}$$

The average charging current is 3/50 of $20\,\text{mA}$, i.e. $1.2\,\text{mA}$. The supply voltage is 20 V and the average value of voltage across the charging resistor during the forward stroke is 19 V. The current is $1.2\,\text{mA}$ and the resistor value is clearly $16\,\text{k}\Omega$ approximately.

We can check the validity of this calculation from the time constant of the combination. This is $16 \times 10^{3} \times 0.03 \times 10^{-6}$, i.e. $480\,\mu\text{s}$, approximately 10 times the duration of the forward stroke as required.

The simple discharger circuit has a rectangular input signal from which it develops a sawtooth output signal of the same frequency. Mathematically the output waveform is the time integral of the input waveform and the circuit may therefore be described as an integrator.

The collector current of TR1 is, of course, turned on at regular intervals by the input signal and it follows that any circuit in which current flows as a series of recurrent bursts can be connected to a resistor and capacitor combination (an integrating circuit, in fact) such as R_1 and C_1 in Fig. 14.1 to produce a sawtooth output. For example, an RC combination can be added to the output circuit of a multivibrator. Fig. 14.2 gives the circuit diagram of a monostable emitter-coupled multivibrator modified in this way to produce a sawtooth output. The time constant $R_3 C_1$ determines the duration of each forward stroke of the sawtooth output whilst the time constant $R_4 C_2$ determines the amplitude of the output.

Fig. 14.3 gives the circuit of an astable blocking oscillator used as a discharger. This is a free-running circuit, the natural frequency of which is determined by the time constant $R_1 C_1$. The output amplitude is

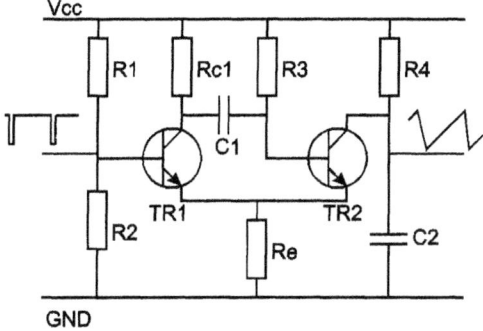

Fig. 14.2. A driven multivibrator sawtooth generator

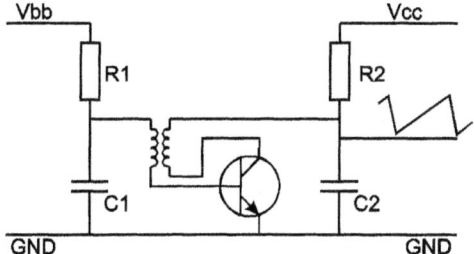

Fig. 14.3. Basic circuit of a blocking oscillator sawtooth generator

determined by both R_1C_1 and R_2C_2 and the ratio of these quantities should be kept constant if the output amplitude is required to be constant when the natural frequency is varied. This circuit can, however, be synchronised at the frequency of any recurrent signal applied to it as explained in the previous chapter.

Miller-integrator circuit

Perhaps the most serious disadvantage of the simple discharger circuit is that the output obtainable with good linearity is limited to a small fraction of the supply voltage. Thus to give a reasonable output with good linearity a large supply voltage is required. One circuit which is free from this limitation is the Miller integrator which can give a sawtooth output nearly equal to the supply voltage and with good linearity.

As shown in Fig. 14.4 the circuit is similar to that of the simple discharger but the capacitor C_1 is returned to the base instead of to the emitter. This introduces negative feedback and makes the operation of

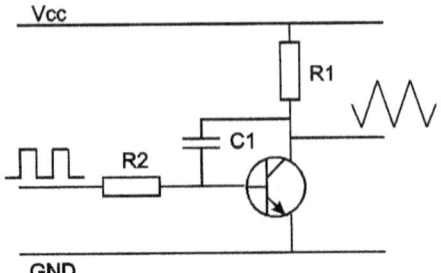

Fig. 14.4. Fundamental circuit of a transistor Miller integrator

this circuit quite different from that of the simple discharger. For example the transistor in the Miller integrator is conductive throughout the cycle and is used as a class-A amplifier: in fact the transistor is used as an operational amplifier and this approach is described on page 167. Suppose that C_1 is fully charged and that the transistor (assumed npn) has suddenly been turned on by a positive-going pulse applied to the base. The immediate tendency is for the collector potential to approach the emitter potential as C_1 begins to be discharged. As the collector potential changes, however, the change is transferred by C_1 to the base. The resulting feedback reduces the collector current and slows up the discharge of C_1. In fact the effective time constant is increased by the feedback from $R_2 C_1$ to $AR_2 C_1$ where A is the voltage gain of the transistor with R_1 as load. In addition to slowing up the discharge of C_1, the feedback exerts a powerful linearising effect and the sawtooth generated at the collector terminal is a much better approximation to the ideal than that obtained from the simple discharger circuit.

The reduction in the speed of the collector-voltage change is undesirable during the flyback period, which is usually required to be as short as possible, and a common technique is to include a second transistor in practical Miller-integrator sawtooth generators to interrupt the feedback circuit for the duration of the flyback period. One circuit of this type is illustrated in Fig. 14.5. The two transistors TR1 and TR2 are connected in series as a cascode but the input is applied to TR2 base, not TR1 base as in r.f. applications. The overall gain is approximately equal to that of a single transistor and there is also a signal reversal between the signals at TR1 base and TR2 collector. Thus the fundamental components R_1, R_2 and C_1 can be connected as shown, these corresponding with those of the previous diagram. During the period when the base is biased positively by the input, both transistors conduct and the circuit operates in the manner already described for Fig. 14.4, negative feedback slowing down the discharge of C_1 and linearising the sawtooth generated at TR2 collector. When, however,

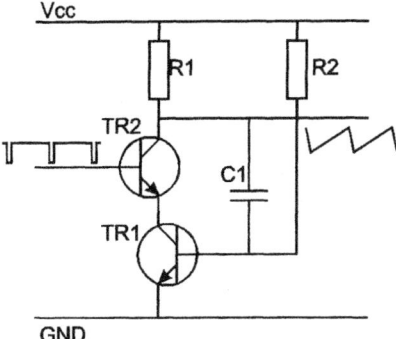

Fig. 14.5. Transistor Miller-integrator circuit with fast flyback

TR2 base is held negative by the input signal, TR2 is cut off. This removes negative feedback from the circuit and C_1 now charges through R_1 and the base-emitter junction of TR1 at a rate probably 40 or 50 times that which applies with feedback. In this way, therefore, a rapid flyback is obtained.

Constant-current charger

If the base voltage of a transistor is held at a fixed and stable voltage, its emitter takes up a corresponding voltage less the drop in the base-emitter junction. The current flowing in the external emitter circuit then depends entirely on the value of resistance it contains. The emitter current also flows in the external collector circuit so long as its resistance is low enough to sustain it. Thus the collector represents a *constant-current source*.

A linear rise of voltage takes place across a capacitor when it is charged from a constant-current source. In the circuit of Fig. 14.1, R_1 can be replaced by a constant-current generator as shown in Fig. 14.6. TR2 generates a current determined by ZD1 and R_e; it flows to charge C until TR1 switches on to discharge the capacitor. When TR1 turns off again another linear positive-going ramp is generated, and so on. This principle is often used in i.c. field scan generator circuits, with switch TR1 being closed momentarily once per 20 ms by the field oscillator, itself locked to incoming field sync pulses.

A simple constant current charger for C can also be constructed from a jfet as in Fig. 14.7. When V_{gs} is zero, the current which flows through the jfet is relatively constant at I_{DSS}, the saturated drain-source current. This is a basic parameter, but the drain current can be adjusted with the potentiometer as shown.

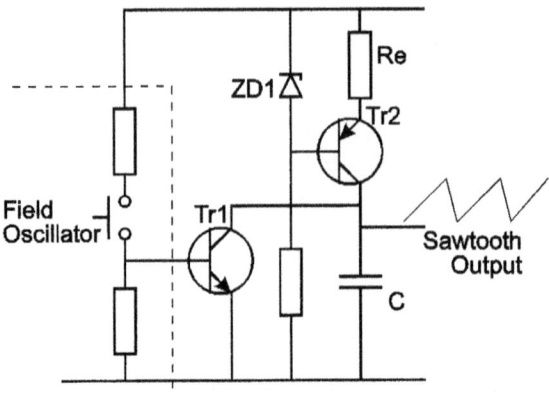

Fig. 14.6. Generation of a linear sawtooth by constant-current charging of a capacitor. This circuit is similar to that of Fig. 14.1, with TR2 taking the place of R_1

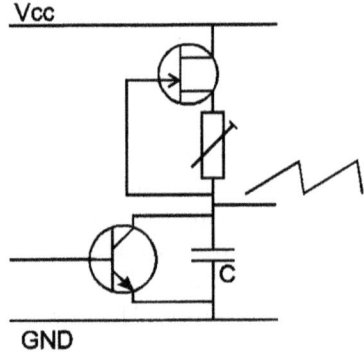

Fig. 14.7. A simple constant-current charger for C, constructed from jfet

Line output stage for television receiver

The principles of the method of generating a sawtooth current commonly used in television receivers are quite different from those so far described. At frequencies of the order of 15 kHz the scanning coils are predominantly inductive and are represented by L_1 in Fig. 14.8. The parallel capacitor is chosen to resonate with L_1 at a frequency such that the period is double the required flyback time.

When S is closed (t_1 in Fig. 14.9) it connects L_1 across the supply. Current in L_1, initially zero, starts to grow and rises linearly with time:

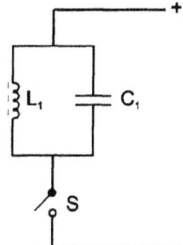

Fig. 14.8. Basic circuit for line output stage

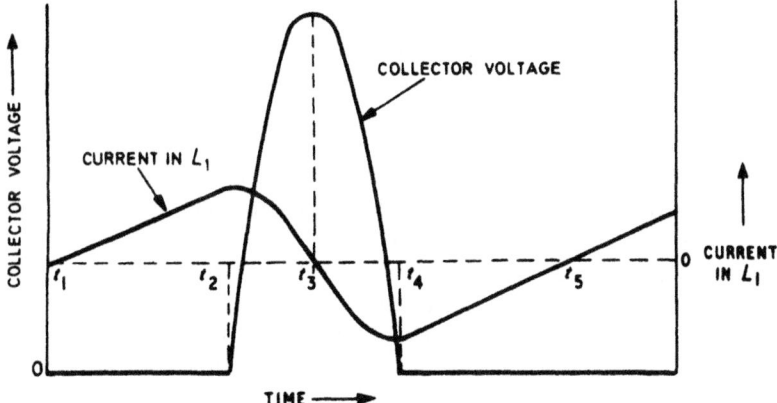

Fig. 14.9. Voltage and current waveforms for the circuit of Fig. 14.8

the voltage across an inductor is related to the rate of change of current in it according to the expression

$$V = -L\frac{di}{dt}$$

V in this circuit is the supply voltage and is constant: L_1 is also constant. Thus di/dt is constant, i.e. the current changes linearly with time as required for the working stroke of the scanning system. When the working stroke is completed, S is opened (t_2), isolating $L_1 C_1$ from the supply. $L_1 C_1$ now begins to oscillate freely and the following voltage and current changes occur.

Current continues to flow in L_1 but is now taken from C_1 instead of from the supply. This discharges C_1 reducing the voltage across L_1 and thus reducing the rate of rise of current in L_1. When C_1 is completely discharged, the current in L_1 has zero rate of rise, being now at a maximum. As current continues to flow C_1 begins to charge up with the opposite polarity. This causes current in L_1 to decrease and it falls to zero and builds up in the opposite direction. At the instant (t_3) when the current is zero, the voltage across C_1 is a maximum. As soon as the current has reached a value equal to the previous maximum, S is closed again (t_4). At this instant the voltage across C_1 is equal to the supply voltage. Current now flows from L_1 into the supply and because the supply voltage is constant the current falls linearly with time to zero (t_5). This completes the first half of the working stroke and current now begins to increase linearly, as described initially, to complete the second half of the stroke. The current and voltage waveforms are given in Fig. 14.9: the voltage waveform is of course proportional to the rate of change of the current waveform.

Fig. 14.10 shows, in basic form, a practical line output stage for a television receiver. The switch has been replaced by an npn power transistor TR1 and the inductor L by the parallel combination of line output transformer (l.o.p.t.) primary winding L_1 and scanning coil L_2: large capacitors C_2 and C_3 may be regarded as short-circuits at line-scanning frequency. At about the middle of each scanning stroke TR1 is switched on by the line drive waveform at its base, grounding the bottom end of L_1 to give a linear build-up of current within it as already described. Line flyback is initiated by the sudden turn-off of TR1, whereupon the magnetic field collapsing about L_1/L_2 causes an immediate reversal of the direction of magnetic flux and coil current, which now flows into tuning capacitor C_1, charging it. At a time determined by the period of resonant circuit LC $(t_4$ in Fig. 14.9) one half-cycle of oscillation has taken place and the energy stored in C_1 has

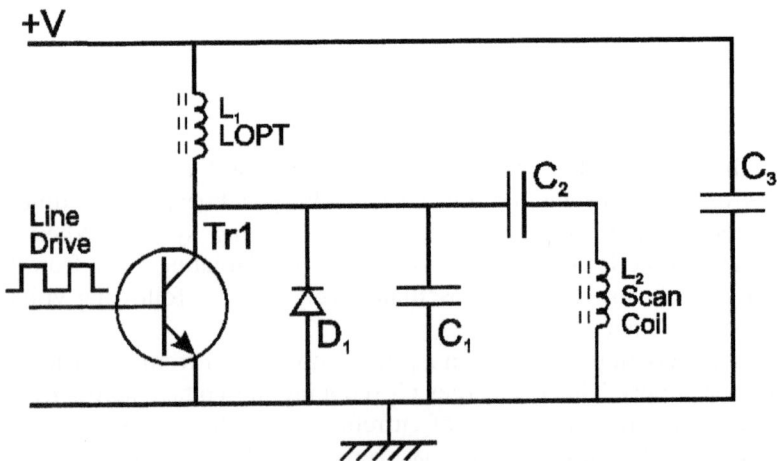

Fig. 14.10. Skeleton circuit of a television receiver line output stage

been transferred back into L_1/L_2. Further oscillation would involve the bottom end of L_1 moving negatively with respect to ground to recharge C_1 but this is prevented by the action of clamp or efficiency diode D1 which conducts at t_4 to clamp the bottom end of L_1 to ground. The magnetic energy stored in the inductors now decays linearly to zero during the period t_4-t_5, feeding energy back into reservoir capacitor C_3 and providing deflection energy for the first half of the active scanning line. At t_5 the circuit is at rest with no energy left in C_1 or L_1/L_2, and the scanning spot has once again reached screen centre, corresponding to the situation at t_1, and at this point TR1 switches on again to repeat the cycle. In this way a sawtooth current is built up in the l.o.p.t. L_1 and scan coil L_2; in practice L_2 consists of a pair of coils in a saddle-wound ferrite deflection yoke.

The voltage generated across an inductor is proportional to the rate of change of current within it. Thus the voltage across the l.o.p.t. primary winding and scan yoke is steady and relatively low during the forward scanning stroke but high during the flyback period when the rate of change of magnetic flux is high and varying, as shown by the collector voltage waveform of Fig. 14.9. Thus the voltage waveform across the l.o.p.t. and scan coil is a series of pulses about 12 μs wide recurring at 64-μs intervals. As is now plain, the flyback period is governed by the LC product of the components in the line output stage rather than by any characteristic of the line drive waveform. The currents in the transistor and the diode charge and discharge the decoupling/storage capacitor equally and there would appear to be no net current drawn from the supply line + V. In practice the losses contributed by coil resistance, the finite switching time of TR1 and D1, and the extra loads imposed on the l.o.p.t. by the various auxiliary power feeds taken from it create a considerable current demand from the power supply.

The amplitude of the sawtooth current (which determines the picture width) is directly dependent on the supply voltage and is normally controlled by adjustment of this voltage.

So far, for the purpose of circuit explanation, we have assumed scan-coil coupling capacitor C_2 to be infinitely large. In practice, in addition to providing d.c. isolation of the yoke coil, it performs the function of S-correction. In scanning a large flat screen a linear deflection current would result in a picture cramped in the centre and stretched at the ends of the scan. This is caused by the variation in distance between the deflection centre and the picture-tube screen. Critical selection of the value of the coupling capacitor C_2 has the compensating effect of reducing scan speed at the start and end of each sweep as shown in Fig. 14.11.

The line output transistor is normally a large high-voltage silicon type with high I_c and V_{ce} ratings: peak collector voltage during flyback can

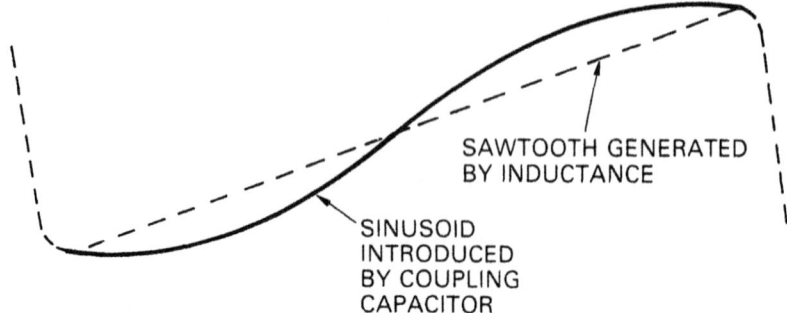

SAWTOOTH GENERATED BY INDUCTANCE

SINUSOID INTRODUCED BY COUPLING CAPACITOR

Fig. 14.11. S-correction introduced by the yoke coupling capacitor

exceed 1,000 V. The construction of such a transistor requires a large base region to carry the heavy current. To overcome the inherent hole-storage effect (see page 267) the drive waveform must remove the charge built up in the current-saturated base junction very quickly. This is usually achieved by using a transformer to drive the line output transistor base. The negative voltage pulse at the instant of switching off removes the charge from the transistor base. Even so there is a delay of about 8 μs from the start of the turn-off pulse to the time when the transistor starts to cut off and this must be allowed for in the phasing of the line drive pulse relative to that of the picture signal. Some receivers, generally small-screen models, use a compound (e.g. Darlington-type) transistor for line deflection, rendering a driver transformer unnecessary. In some transistors designed for 90° line deflection the clamp/efficiency diode is encapsulated with the transistor in the same package.

Practical design

Mention has already been made of the auxiliary uses to which the l.o.p.t. is put in a television receiver. In the complete circuit diagram of Fig. 14.12 the elements already described can readily be seen: they are the line output transistor TR11; the line output transformer T5 (with primary winding 13–15); tuning capacitor C_{34}/C_{35}; clamp diode D33/D34; and scan coil with S-correction coupling capacitor C_{38}, here inserted in the earthy side of the yoke circuit. Also in series with the deflection coils is L_4, the saturable magnetic core of which is biased by a permanent magnet for correction of horizontal scanning linearity. Windings 8–11 and 4–12 on the l.o.p.t., with D39 and D40, provide ±12-V power lines for the field timebase, whilst winding 1–3 generates a 6.3-V r.m.s. supply for the heaters of the picture tube. The other secondary windings on T5 provide high voltages for beam acceleration

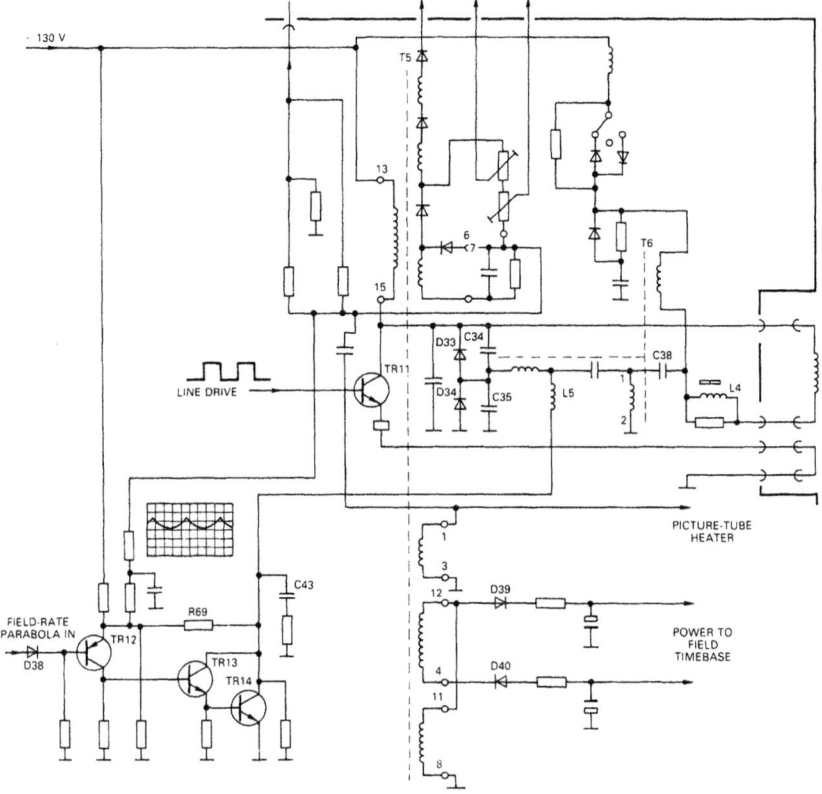

Fig. 14.12. Complete circuit diagram of commercially produced line-scanning stage

and focus electrodes in the tube. The winding connected to pin 7 feeds the anode of the (in built) diode which provides a 7-kV voltage for focus control, adjustable by the potentiometer built into the l.o.p.t. In its bottom leg is a second moulded-in potentiometer from which is tapped off about 1 kV for the first anode of the picture tube. The diode conducts on the flyback pulses and the associated reservoir capacitance is formed by the inter-layer capacitance of the l.o.p.t. windings. The components connected to transformer pin 6 provide a control voltage for beam-current limiting via the luminance drive circuit. Further diodes leading upwards from pin 7 provide a 25-kV supply for the final anode of the tube, each separate coil/diode combination forming, in conjunction with the carefully designed inter-winding-layer capacitance, a voltage-generating 'cell', three of which are stacked in series to form a voltage multiplier.

Diode modulator

In the circuit of Fig. 14.12 both the tuning capacitor and clamp diode are split to facilitate the insertion of a diode modulator circuit which amplitude-modulates the line scan current to compensate for the pincushion distortion inherent in wide-angle picture tubes. A 50-Hz parabolic waveform derived from the field timebase is fed via D38 (bottom left-hand side of diagram) to a direct-coupled class-A amplifier made up of TR12, TR13 and TR14 with feedback stabilisation via R_{69}. TR14 is a medium-power transistor, usually heat-sinked. It acts as a variable load which governs the distribution of current in the yoke, part of tuned circuit L_5C_{34}, and winding 1–2 on the EW transformer T6, which is part of tuned circuit T6 (1–2) C_{35}. Injection coil L_5 and capacitor C_{43} decouple modulator transistor TR14 from line pulses. In addition to dynamic modulation of scan amplitude, picture width can be controlled by adjustment of the d.c. voltage on which the parabolic waveform sits at D38 anode.

Chapter 15

Digital circuits

Introduction

As mentioned in the previous two chapters, transistors in multivibrators and other pulse circuits are used as switches, i.e. except for the brief periods during which the transistors are changing state they are either *on* (taking considerable current) or *off* (non-conductive). For most of the time therefore the collector (or drain) voltage has one of two possible values namely a low value near emitter or source potential (when the transistor is on) and a high value near the supply potential (when the transistor is off).

A vast range of circuits has been developed during the past decade or two in which diodes and transistors are used as switches and in which the signal paths have at all times one or other of two possible voltage levels. Such circuits are used to perform mathematical and logical operations on signals in computers and similar equipment. Circuits used in this manner are known as digital (strictly binary digital) or logic circuits and the principal types are described briefly in this chapter.

Logic levels

The two significant values of voltage on the signal-carrying lines are referred to as logic level 0 and logic level 1.* If level 1 is more positive than level 0 the circuit is said to use the *positive logic convention* and if level 1 is more negative than level 0 the circuit is said to use the *negative logic convention*. The distinction is important because circuits can behave differently according to the logic convention chosen. This is illustrated below but it is worth stressing now that the logic convention chosen should always be stated or implicit on diagrams of logic circuits. The tendency in the design of digital equipment is to favour positive

---

* The two levels could alternatively be values of current or more generally of air pressure in pneumatic systems or fluid pressure in hydraulic systems.

logic and if no logic convention is indicated on a diagram it can normally be assumed that positive logic is used. Positive logic is used in all the diagrams in this chapter.

Binary scale

The advantage of labelling the two significant voltage levels 0 and 1 is that it simplifies the process by which logic circuits are able to carry out mathematical and other operations. Arithmetical operations, for example, can be performed by using the binary scale of numbers which has only two digits 0 and 1.

Conventional counting uses the scale of 10 (the decimal scale) and in numbers the digits are arranged according to the power of 10 they represent. For example the number 4721 (four thousand, seven hundred and twenty-one) means, if written out in full:

$$4 \times 10^3 + 7 \times 10^2 + 2 \times 10^1 + 1 \times 10^0$$

i.e. $\quad 4000 \quad + \quad 700 \quad + \quad 20 \quad + \quad 1$

$\quad = 4721$

Similarly in the binary scale of counting, the digits (0 or 1) in a number are arranged according to the power of 2 they represent. For example the binary number 110101 means, if written out in full:

$$1 \times 2^5 + 1 \times 2^4 + 0 \times 2^3 + 1 \times 2^2 + 0 \times 2^1 + 1 \times 2^0$$

i.e. $\quad 32 \quad + \quad 16 \quad + \quad 0 \quad + \quad 4 \quad + \quad 0 \quad + \quad 1$

$\quad = 53$

The first nine numbers in the binary scale are as follows:

binary number	decimal equivalent
1	1
10	2
11	3
100	4
101	5
110	6
111	7
1000	8
1001	9

It is common practice, however, with very large numbers to translate each digit of the decimal number separately into the binary scale. This is known as the *binary-coded decimal* system and in it the number 4721 would be coded as follows:

100 : 111 : 10 : 1

i.e.   4 : 7 : 2 : 1

This system has the advantage that after a little experience binary-coded numbers can be translated into decimal form on inspection. It also simplifies the design of equipment for coding decimal numbers into binary form and for decoding and displaying binary-coded numbers in decimal form.

Logic gates

A simple logic circuit employing two diodes is illustrated in Fig. 15.1. Input A is at all times at +10 V or 0 V, these being the standard voltage levels chosen for use in this circuit. Input B is similarly at +10 V or 0 V. If either input is at 0 V the associated diode conducts and, if the forward resistance of the diode is neglected, 0 V appears at the output. A diode with an input at +10 V does not conduct and the output is isolated from this input. For two inputs there are four possible combinations of input voltage. They are shown in Table 15.1 together with the value of the output voltage. This table shows that the output is 0 V if either or both of the inputs is at 0 V and that the only way to obtain +10 V at the output is for both of the inputs to be at +10 V.

If the positive logic convention is used +10 V is logic level 1 and 0 V is logic level 0. If Table 15.1 is repeated in terms of logic levels, the

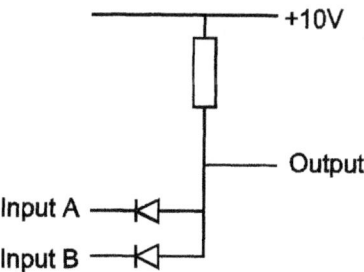

Fig. 15.1. Simple diode gate circuit. It can be an AND gate or an OR gate depending on the logic convention adopted

Table 15.1 Voltages in the circuit of Fig. 15.1

Input volts		Output volts
A	B	
0	0	0
0	+10	0
+10	0	0
+10	+10	+10

Table 15.2 Truth table for an AND gate

Input		Output
A	B	
0	0	0
0	1	0
1	0	0
1	1	1

result is as shown in Table 15.2. Such a table is known as a *truth table* and it shows that a logic 1 is obtained at the output of the circuit of Fig. 15.1 only when input *A and* input *B* have a logic 1 signal. A circuit such as this which requires a logic 1 signal at all the inputs to give a logic 1 signal at the output is known as an AND gate.

Suppose negative logic is used in the circuit of Fig. 15.1. Now +10 V represents logic level 0 and 0 V represents logic level 1. The truth table now has the form shown in Table 15.3. A logic 1 is obtained at the

Table 15.3 Truth table for an OR gate

Input		Output
A	B	
0	0	0
0	1	1
1	0	1
1	1	1

output of the gate when either input *A or* input *B* has a logic 1 signal. A gate which requires a logic 1 signal at any one input to give a logic 1 at the output is known as an OR gate.

Thus the circuit illustrated in Fig. 15.1 can behave as an AND gate or an OR gate depending on the logic convention used. In general it is not possible to state the nature of a logic circuit until the logic convention to be used with it is known.

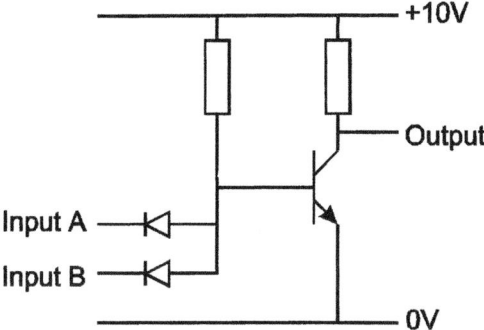

Fig. 15.2. An inverter stage following the diode gate gives a circuit which can be a NAND gate or a NOR gate depending on the logic convention adopted

Suppose a common-emitter amplifier is added after the diode gate as shown in Fig. 15.2. The transistor circuit is so designed that its collector voltage is always at one or other of the two chosen voltage levels. The signal inversion introduced by the amplifier gives yet another type of behaviour and the relationship between the inputs and output, for positive logic, is as given in Table 15.4. This shows that the only way to obtain a logic 0 at the output is for all the input signals to be at logic 1. This behaviour is not surprisingly the inverse of that of the AND gate: this circuit, for positive logic, is therefore known as a NOT-AND or more simply a NAND gate.

Table 15.4 Truth table for a NAND gate

Input		Output
A	*B*	
0	0	1
0	1	1
1	0	1
1	1	0

Table 15.5 Truth table for a NOR gate

Input		Output
A	B	
0	0	1
0	1	0
1	0	0
1	1	0

For negative logic the behaviour of the circuit of Fig. 15.2 is as illustrated in the truth table of Table 15.5. When any of the input signals is at logic 1 the output is at logic 0. The only way to obtain a logic 1 at the output is for all the inputs to be at logic 0. This behaviour is the inverse of that of the OR gate and this circuit, for negative logic, is known as a NOT-OR or NOR gate.

Gate symbols

For simplicity only two inputs are shown in Figs. 15.1 and 15.2 but there can, of course, be more. The block symbols for the four basic types of gate so far introduced are shown in Fig. 15.3 and here three inputs are shown. The circle at the output of the NOR and NAND gates indicates the inversion of the output relative to that of the OR and AND gates.

Fig. 15.3. Shows the logic symbols drawn according to the British Standard BS3939, which was adopted from the International Standard IEC 617:12. There is a further common standard in use, and this is the American Mil-Std-806, which uses the so-called 'curvy' symbols shown in Fig. 15.4

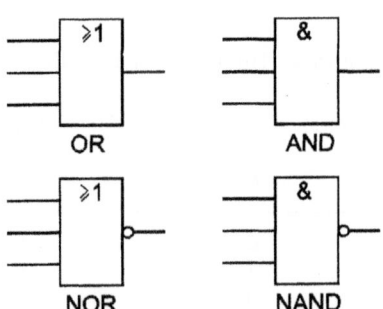

Fig. 15.3. Block (or logic) symbols for the four basic types of gate

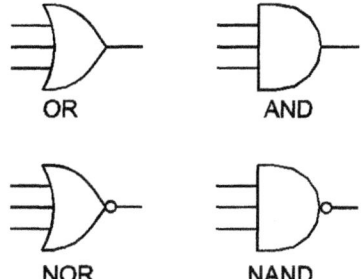

OR   AND

NOR   NAND

Fig. 15.4. 'Curvy' symbols used in the American Mil-Std-806

Integrated circuit gates

Gate circuits can be constructed of discrete components but normally they are in the form of integrated circuits. For example a single i.c. may contain three 3-input gates. To use such an i.c. in an equipment it is not necessary to know details of the circuitry of the device. All the designer needs to know to be able to use the i.c. successfully are details of input and output signal levels, polarities, impedances and supply voltages. In preparing diagrams of computers and computer-like equipments the gates and other functional units are represented by block symbols such as those given for gates in Fig. 15.3. To help in the layout of printed wiring cards and in maintenance the inputs, outputs and supply points of the gates can be identified in the block diagram by giving the pin numbers of the i.cs. Block diagrams of logic equipment are usually known as logic diagrams.

To illustrate the versatility of logic gates a number of applications will now be considered.

Gates as switches

Table 15.2 shows that when input A is at logic 0, the output is also at logic 0 (irrespective of the signal on input B) whereas if input A is at logic 1 the output signal is the same as the signal on input B. Thus an AND gate can be used as a switch, a logic 1 signal on input A allowing the signal on input B to pass through the gate, a logic 0 signal on input A blocking the signal on input B.

An OR gate can be used similarly but here a logic 0 on one input allows the signal on the other input to pass through the gate.

Inverters

Table 15.4 shows an interesting property of a NAND gate. When there is a logic 1 signal on input A the output from the gate is the inverse of

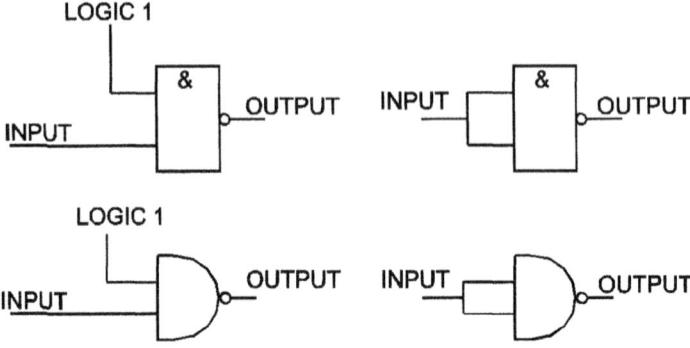

Fig. 15.5. A NAND gate used as an inverter

that on input *B*. A NAND gate is frequently so used, input *A* being connected permanently to a source of logic 1 voltage as shown in Fig. 15.5. For some types of NAND gate i.cs it is sufficient simply to leave input *A* unconnected: this has the same effect as connecting it to a logic 1 source. A gate with this property is known as a NOT gate. Another way of converting a NAND into a NOT gate (inverter) is to couple the inputs together. This is also shown in Fig. 15.5.

A NOR gate can also be used as an inverter but one input must be connected to a source of logic 0 voltage to obtain inversion of the signal on the other input. Similarly, a NOR gate can also be turned into a NOT gate by coupling the inputs together.

Inhibiting input

Reference to Table 15.5 shows that if one input is at logic 1, the output of the gate is prevented from taking up the logic 1 state irrespective of the signal on the other input. This ability of an input of a device to prevent the output taking up the logic 1 state is sometimes exploited in logic circuitry and an input so used is then known as an inhibiting input: it is indicated on logic diagrams by a short stroke across the signal line to that input.

Exclusive-OR gate

Combinations of gates can be used to perform desired operations. For example suppose a circuit is required to give a logic 1 output when either of two inputs is at logic 1 but not when both are at logic 1. Such a circuit is known as an *exclusive-OR* gate.

There are a number of possible circuits and one is given in Fig. 15.6. That this circuit gives the required performance can be checked from the

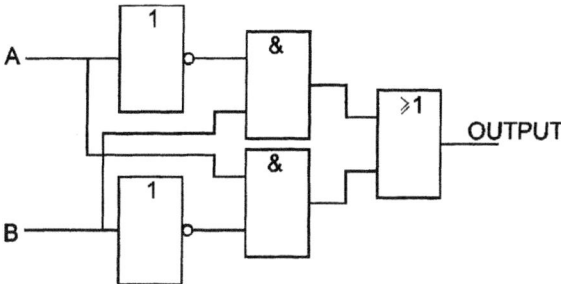

Fig. 15.6. One possible logic circuit for an exclusive-OR gate

Table 15.6 Truth table for an exclusive-OR gate

Input		Output of inverter A	Output of inverter B	Output of AND gate A	Output of AND gate B	Output of OR gate
A	B					
0	0	1	1	0	0	0
0	1	1	0	1	0	1
1	0	0	1	0	1	1
1	1	0	0	0	0	0

truth table of Table 15.6 or by the use of Boolean algebra and this second method is the one normally used in designing logic gate circuits. The exclusive-OR (ex-OR) gate is represented by the logic symbols of Fig. 15.7. The '= 1' in the BS symbol is intended to represent the fact that the output is at logic 1 whenever one, and only one, of the inputs is at logic 1. There are two other symbols to represent the exclusive-NOR gate. The output of an exclusive-NOR is 1 whenever there are two 0s or two 1s present at the input of the gate. The normal signal inversion 'o' is used on the output of the gate to imply the change from OR to NOR.

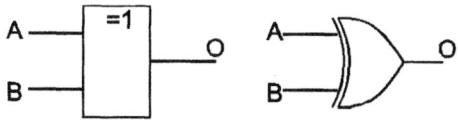

Fig. 15.7. Logic symbol for an exclusive-OR gate

Equivalence element

It is sometimes necessary to compare two binary digits and to give an indication when they are the same. A circuit used for this purpose is known as an *equivalence element* or a *comparator* and is required to give a logic 1 output when the two inputs are both at logic 1 or both at logic 0. The required output is the inverse of that of the exclusive-OR gate and can be obtained from a combination of gates similar to that shown in Fig. 15.6 but with the OR gate replaced by a NOR gate.

Practical NAND gate circuit

The circuit shown in Fig. 15.2 has a number of disadvantages. One is that the speed with which the output changes from the 0 state to the 1 state (positive logic assumed) is low compared with that of the reverse transition. This is because a bipolar transistor can be switched on quickly but the time of switch off is appreciable. The delay in switch off slows up the rate at which logic operations can be carried out. The effect can be avoided by using an output stage with two transistors in series, one of which is switched on to perform one transition, the other being switched on to give the opposite transition. The push-pull output circuit shown in Fig. 15.8 fulfils these conditions. The input diodes of Fig. 15.2 have been replaced by a common-base stage in which the transistor has a number of emitters, each providing one gate input. It is easy to fabricate such transistors using i.c. techniques but if discrete components are used one transistor would be required for each emitter. The

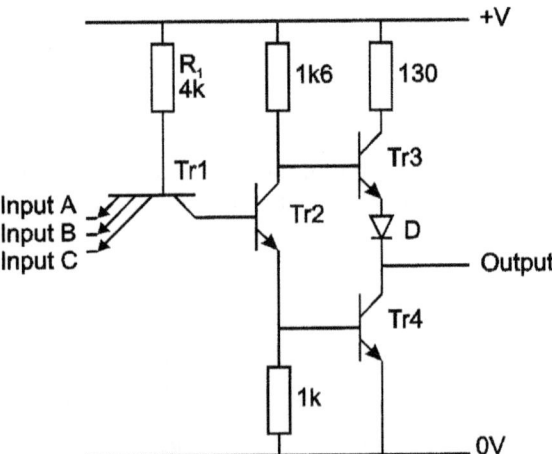

Fig. 15.8. Circuit diagram of an i.c. NAND gate

second transistor in Fig. 15.8 is a phase splitter driving the push-pull output stage. Diode D provides base bias for TR3.

For positive logic this circuit is a NAND gate. To confirm this, assume the output to be at logic 0. This requires TR3 to be off and TR4 on. This, in turn, requires TR2 to be on. TR2 must therefore have a high base potential which requires TR1 to be off. The only way for TR1 to be off is for all the emitters to be at positive supply potential, i.e. at logic 1. By definition a gate which requires all inputs to be at logic 1 to give a logic 0 output is a NAND gate.

This circuit is an example of transistor–transistor logic, usually abbreviated to TTL.

Fan in

In circuits containing numerous gates it may happen that several gates feed one particular gate. Several output circuits are then connected in parallel across one input circuit and it is essential that this loading should not affect the voltage level at the junction point, no matter whether this is the logic 1 or logic 0 voltage. There is normally a tolerance on the voltage levels but if too many circuits are connected together the junction voltage may be outside the tolerance and normal circuit behaviour becomes impossible. The greatest number of outputs which may be connected to a gate input whilst still permitting normal behaviour is known as the fan in. This may in practice be as high as 12 but in the gate circuits and symbols given earlier the number of inputs was shown for simplicity as only 2 or 3.

Fan out

A junction between input and output circuits also occurs when one gate is required to feed a number of others: one output circuit is then connected to several input circuits. Current is required to operate an input circuit using bipolar transistors. For example in Fig. 15.8, to put a logic 0 signal on an emitter of TR1, the emitter potential must be lowered to that of the supply negative voltage (positive logic is assumed). An emitter current (determined by R_1 and the supply voltage) of, for example, 1.5 mA then flows in the input circuit and the output circuit feeding it. The conductive transistor in the output circuit must be capable of passing this current without undue rise in the voltage across it. Any significant rise may cause this voltage to fall outside the range recognised as a logic 0 signal. The maximum collector current through the output transistor may be, say, 30 mA, which enables up to 20 gate circuits to be fed satisfactorily. This maximum number is known as the fan out of the circuit.

Wired-OR connection

It is sometimes useful in a circuit to achieve a logical-OR function without using a gate circuit at that point for the purpose. As an example, consider the circuit of Fig. 15.9 in which two outputs are connected directly together. The resistor is called a pull-up resistor. If neither transistor is turned on, then the output is a logic '1'. If either of the transistors turns on, then the output will be connected to the 0 V line through whichever transistor is turned on. Of course, if both turn on, then the conduction path will be through both transistors. An external circuit will not know this, the output being at logic '0' however caused.)

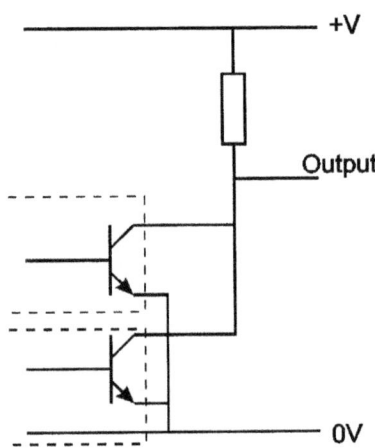

Fig. 15.9. Illustrating a wired-OR connection

Many logic devices do not use the 'totem-pole' output of Fig. 15.8, and instead they have an 'open-collector' output as shown in Fig 15.9. The BS symbol for this output is shown in Fig. 15.10(a). The Mil-Std-806 does not have a special symbol; if an open collector is indicated at all, it is with the letters 'o/c'.

Insulated-gate fet i.cs as logic gates

I.Cs using bipolar transistor techniques as so far discussed in this chapter are extensively used. They have the advantages that they operate satisfactorily with low collector voltages, switch-on times are low and they have a high current-handling capacity. Dissipation in the collector circuits is low in the off state because the collector current is nearly zero

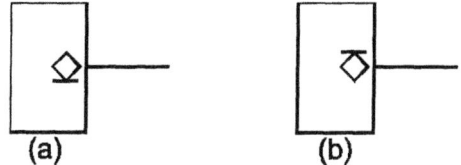

Fig. 15.10. BS logic symbol for (a) pull-down (npn) open-collector output and (b) pull-up (pnp) open-collector output

and in the on state because the collector voltage is nearly zero. However appreciable power is required in the base circuit to keep a bipolar transistor in the on state.

Insulated-gate fets have a number of advantages over bipolar transistors in logic-gate i.cs and are now being used. One advantage is that fets can be manufactured much smaller than bipolar transistors and thus greater miniaturisation is possible. Secondly because of the very high input resistance of fets it is possible to parallel a large number of gate inputs on one gate output, i.e. a very high fan out can be achieved. Thirdly there is no dissipation in a fet in the on or off state. A disadvantage of the fet is that it can easily be damaged by excess voltages and its current-handling capacity is less than that of bipolar transistors.

An example of an inverter circuit using mosfets is given in Fig. 15.11. It could hardly be simpler and no resistors (to dissipate power) are necessary. Two complementary mosfets are connected in series across the supply and the gates are connected to the input signal. A logic 1 signal (positive logic assumed) cuts TR1 off and turns TR2 on so that a logic 0 signal appears at the output. A logic 0 input turns TR1 on and

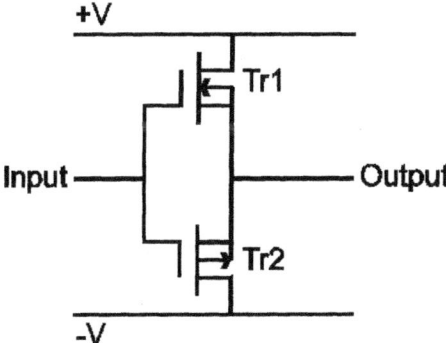

Fig. 15.11. Complementary mosfets (CMOS) as used in an i.c. inverter

cuts TR2 off so that a logic 1 output is obtained. No current is taken from the supply in either state but only at the instants of switching.

This is an example of a complementary metal-oxide-semiconductor circuit, usually abbreviated to CMOS.

Bistables

In general the output of a gate circuit may change when the inputs are removed: in other words gates have no memory. Logic operations must frequently be carried out in sequence and thus there is a need for a device with a memory, i.e. which can store logic information. Bistable multivibrators (usually abbreviated simply to bistables) are commonly used for this purpose. It was pointed out in the description of the multivibrators that the two transistors alternate between the on and off states. Thus the collector potential is either near that of the supply positive or the supply negative and if the supply potentials are taken as logic levels we can say that at any moment one collector voltage will be at logic 1 level whilst the other is at logic 0 level. After a change of state the collector potentials are reversed. The collectors provide the outputs of the bistable circuits and these are always in complementary states: the outputs are denoted by Q and $\bar{Q}$.

RS bistable

A simple form of bistable circuit is illustrated in Fig. 15.12. Outputs are taken from the collectors of the transistors and two inputs known as S (set) and R (reset) are connected to the bases via diodes arranged to

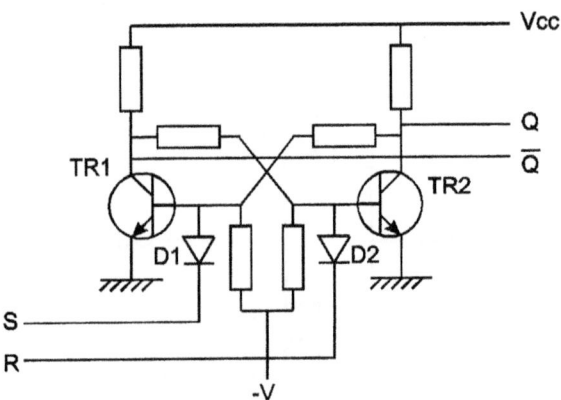

Fig. 15.12. Basic *RS* bistable circuit

conduct negative-going signals. A negative-going signal applied to the *S* input cuts off TR1 (and hence turns TR2 on) and the circuit will now remain in this state indefinitely if need be, so storing information. To remove the information and restore the circuit to the state it had originally, a negative-going signal is applied to the *R* terminal which cuts off TR2 and turns TR1 on.

If negative-going inputs are applied to the *R* and *S* terminals simultaneously there is no way of knowing what the resultant state of the circuit will be. *RS* bistables are therefore never used in circumstances where simultaneous negative-going *R* and *S* inputs are possible. Positive-going signals applied to the *R* and *S* inputs have no effect on the circuit which remains in its previous state.

Thus for three of the four possible combinations of *R* and *S* inputs, the resultant state of the bistable is predictable: for the fourth combination the output is indeterminate. This behaviour is summarised in the following table in which a logic 1 signal is taken to be a change to the more positive voltage level and a logic 0 signal is a change to the less positive level.

Table 15.7   Truth table for an
RS bistable

R input	S input	Q output
0	0	no change
0	1	1
1	0	0
1	1	indeterminate

The *RS* bistable can also be used in a clocked mode. For this purpose a diode gate (see page 259) is included and a regular train of pulses, known as clock pulses, is fed to the gate. The circuit is so arranged that the clock pulses have no effect on the circuit until an input is applied to the *R* or *S* terminals, after which the next clock pulse initiates the change of state of the bistable. The behaviour of the bistable is still as indicated in Table 15.7 but the third column should be interpreted as giving the logic level of the *Q* output after receipt of a clock pulse. Another example of a clocked bistable is given later in this chapter.

JK bistable

The *JK* bistable may be regarded as an improved form of the *RS* bistable in which there is no indeterminate state. The improvement is achieved

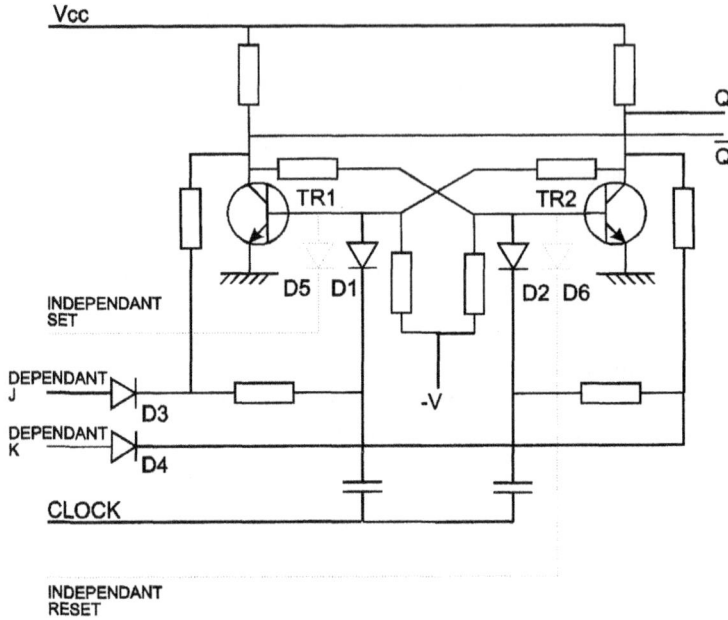

Fig. 15.13. Circuit diagram of a typical clocked *JK* bistable

by arranging for the signal inputs to operate on the gating diodes D1 and D2 to which the clock pulses are applied. A typical circuit diagram is given in Fig. 15.13. It is very similar to that of Fig. 13.3. D1 and D2 are biased by the difference between the collector and base potentials of the associated transistor. If the transistor is on, there is little potential difference and the diode can conduct negative-going clock signals to the base to cut the transistor off. When the transistor is off, the considerable difference between collector and base potentials reverse-biases D1 or D2 so that clock signals cannot reach the base. A positive signal on the *J* input forward biases D3 and reverse-biases the clock-pulse diode D1 and thus prevents TR1 being cut off by the clock pulses. Similarly, a positive signal on the *K* input forward biases D4 and reverse-biases D2, so preventing TR2 being cut off by the clock pulses. If *J* and *K* inputs are made positive simultaneously, clock pulses are prevented from reaching either base and have therefore no effect on the bistable, which remains in its former state. Negative-going signals on the *J* and *K* inputs are blocked by D3 and D4 so that clock pulses operate on the bistable and the outputs alternate between supply negative and earth at the clock frequency. The behaviour of the bistable is thus predictable for all four combinations of *J* and *K* inputs and is summarised in Table 15.8 in which the logic 1 signal is regarded as a change to the more positive

Table 15.8 Truth table for a *JK* bistable

J input	K input	Q output after receipt of a clock pulse
0	0	no effect
0	1	0
1	0	1
1	1	C

C = complement of the state before the clock pulse

level and a logic 0 signal as a change to the less positive level. A further example of a clocked *JK* bistable is given later.

The information to be stored in the bistable is fed in via the signal inputs *J*, *K*, *R* or *S* and the convention adopted is that a logic 1 signal on the *J* or *S* input gives a logic 1 output from the *Q* terminal (and hence a logic 0 output from the *Q̄* terminal) on receipt of a clock pulse. A logic 1 signal on the *K* or *R* input gives a logic 1 output from the *Q̄* terminal (and hence a logic 0 output from the *Q* terminal) when a clock pulse is received.

This convention is indicated on the block symbol used to represent a bistable on logic diagrams. As shown in Fig. 15.14 the symbol is a rectangle, the upper half of which carries the *J* and *S* inputs and the *Q* output, the lower half carrying the *K* and *R* inputs and the *Q̄* output. The clock input is applied to the midpoint of the leading vertical side. Any logic 1 signal applied to an input on the upper half of the symbol gives a logic 1 output from that half i.e. from *Q*: similarly, any logic 1 signal applied to an input on the lower half of the symbol gives a logic 1 output from that half, i.e. from *Q̄*.

Where there is more than one set of clock-dependent signal inputs and more than one clock input, the interdependence is indicated by a numerical code. For example two clock inputs are known as *C*1 and *C*2,

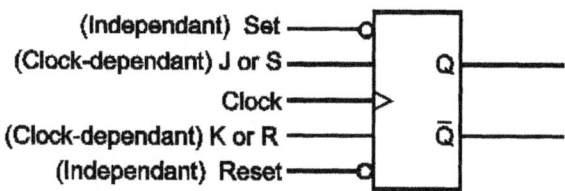

Fig. 15.14. Logic symbol for a bistable

and the inputs dependent on $C1$ are designated $1J$, $1K$, $1R$ and $1S$; those dependent on $C2$ are similarly designated $2J$, $2K$, $2R$ and $2S$.

Independent set and reset inputs

RS and JK bistables commonly have two further inputs which override the signal and clock inputs and are used to put the bistable into a desired state. These are known as the independent set and reset inputs and one way in which such inputs could be applied is indicated by dashed lines in Fig. 15.13. To give a logic 1 output at Q, TR2 must be turned on and TR1 turned off. To cut off TR1 a negative-going input is required at TR1 base and this is applied via the diode D5. To restore the circuit to its original state and give a logic 1 output from $\bar{Q}$ a negative signal is required at TR2 base and this can be applied via D6. In this example a logic 1 signal applied to the independent set input gives a logic 1 output at Q. This applies also to CMOS logic circuits and agrees with the convention for clock-dependent J, K, R and S inputs.

In other types of logic circuit, notably TTL, the independent set input requires a logic 0 signal to give a logic 1 output at Q and a logic 0 input is also required at the independent reset input to restore the circuit to its original state and to give a logic 1 output at $\bar{Q}$. To indicate when logic 0 signals are required to operate these independent inputs, a circle is placed on the independent set and reset signal lines to the block symbol as shown in Fig. 15.14.

These independent inputs were formerly known as the clear and preset inputs.

Use of gates to form bistables

The bistable circuits of Figs. 15.12 and 15.13 are suitable for construction with discrete transistors but the manufacturer of integrated circuits can make up bistables by interconnecting gates. For example, two NAND gates of the type shown in Fig. 15.7 can be interconnected as shown in Fig. 15.15(a).

Suppose a logic 0 signal is applied to input $A1$. Then, irrespective of the logic state of the signal on input $A2$, the output Ao goes to logic 1. This is also the signal on input $B2$ of gate B and we will assume the signal on input $B1$ also to be at logic 1 so that the output Bo is at logic 0. Output Bo is connected to input $A2$ so this is also at logic 0. In this state the circuit is stable and the outputs Ao and Bo remain at logic 1 and logic 0 indefinitely unless an input is applied to change the output of one of the gates. Gate A is unaffected by changing input $A1$ to logic 1 and the only way to switch the bistable is to apply a logic 0 input to $B1$. This causes output Bo to go to logic 1 and Ao to logic 0 and this again is a

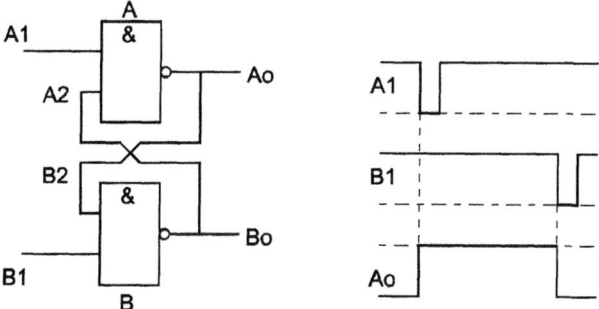

Fig. 15.15. (a) A bistable formed of two NAND gates. This requires logic 0 input signals for its operation as indicated in the idealised waveform diagram (b)

stable state which can be changed only by a further logic 0 signal applied to input $A1$. Thus this combination of gates behaves as a bistable and changes of state can be brought about by logic 0 signals applied to inputs $A1$ and $B1$. The operation of the circuit is illustrated by the waveform diagram of Fig. 15.15(b).

It is preferable to have logic 1 input signals to operate a bistable which then conforms with the signal conventions of the block symbol of Fig. 15.14. This can be achieved by adding a further pair of NAND gates D and E before A and B as shown in Fig. 15.16. This arrangement also has the advantage of permitting clocked operation. If input $D1$ and the clock pulse input on $D2$ are both at logic 1 then the output Do of gate D is at logic 0. This is also the input $A1$ and leads to an output at Ao of logic 1 as already explained. Similarly, if input $E1$ and the clock pulse input at $E2$ are both at logic 1 then output Eo goes to logic 0 causing output Bo to go to logic 1. Thus a logic 1 input to the upper or lower row

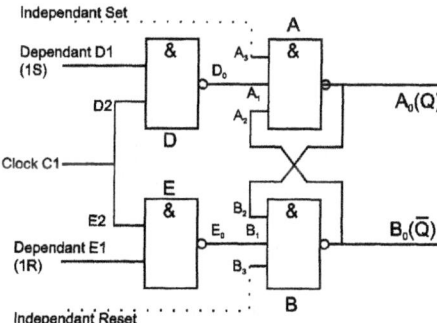

Fig. 15.16. An improved bistable with four NAND gates which operates with logic 1 signals and with provision for a clock input. The independent set and reset inputs (requiring logic 0 signals) are shown in dashed lines

of gates gives a logic 1 output from that row: each row can thus be represented by the upper or lower half of the bistable block symbol of Fig. 15.14.

The dashed lines in Fig. 15.16 indicate how independent set and reset inputs can be applied. They form third inputs $A3$ and $B3$ to the A and B NAND gates. When the output Ao of gate A is at logic 0 the two inputs $A1$ and $A2$ are at logic 1. Thus a logic 1 signal to terminal A3 from the independent set input leaves the bistable unaffected but a logic 0 input instantly changes the state of gate A to give a logic 1 output at Q and, of course, a logic 0 output from Bo ($\bar{Q}$). By symmetry, when $\bar{Q}$ is at logic 0, a logic 0 signal applied to the independent reset terminal ($B3$) switches the bistable to give a logic 1 output at $\bar{Q}$. This bistable is thus switched by logic 0 signals at the independent set and reset inputs and conforms to the block symbol of Fig. 15.14.

Master and slave bistables

As a further stage of elaboration the gate arrangement of Fig. 15.16 can be followed by an identical arrangement driven from the clock pulses via an inverter stage as shown in Fig. 15.17. When the clock input is at logic 1 a logic 1 input on $D1$ sets the first bistable (the master) so that output Ao is also at logic 1. This cannot, however, be accepted by the second bistable (the slave) because the clock input to this stage is at logic 0. But when the clock signal changes (to logic 0 for the master stage and to logic 1 for the slave stage) the logic 1 output at Ao sets the slave stage so that its Q output is at logic 1. If the clock input is in the form of rectangular pulses the signal input is accepted by the master stage on the positive-going edge of the clock pulse and is transferred to the slave stage on the negative-going edge as shown in Fig. 15.18.

A disadvantage of this circuit is that the output cannot be predicted if logic 1 signals are applied simultaneously to the two signal inputs: in other words this is an RS master-slave bistable. The difficulty can be resolved by adding the connections shown in dashed lines in Fig. 15.17. The two outputs of the bistable are, of course, complementary and this feedback therefore applies a logic 0 signal to one of the input gates so preventing it from responding to a logic 1 clock signal. At the same time the feedback applies a logic 1 signal to the other input gate, enabling this to respond to a logic 1 clock signal which therefore causes the bistable to change state. The bistable outputs now reverse and the input gate which was formerly inhibited by the feedback connection can now respond to the next clock pulse whilst the other input gate is inhibited. Thus if both signal inputs are held at logic 1 the clock pulses will cause the bistable to switch regularly between its two possible states. This feedback connection makes the circuit that of a JK master-slave bistable.

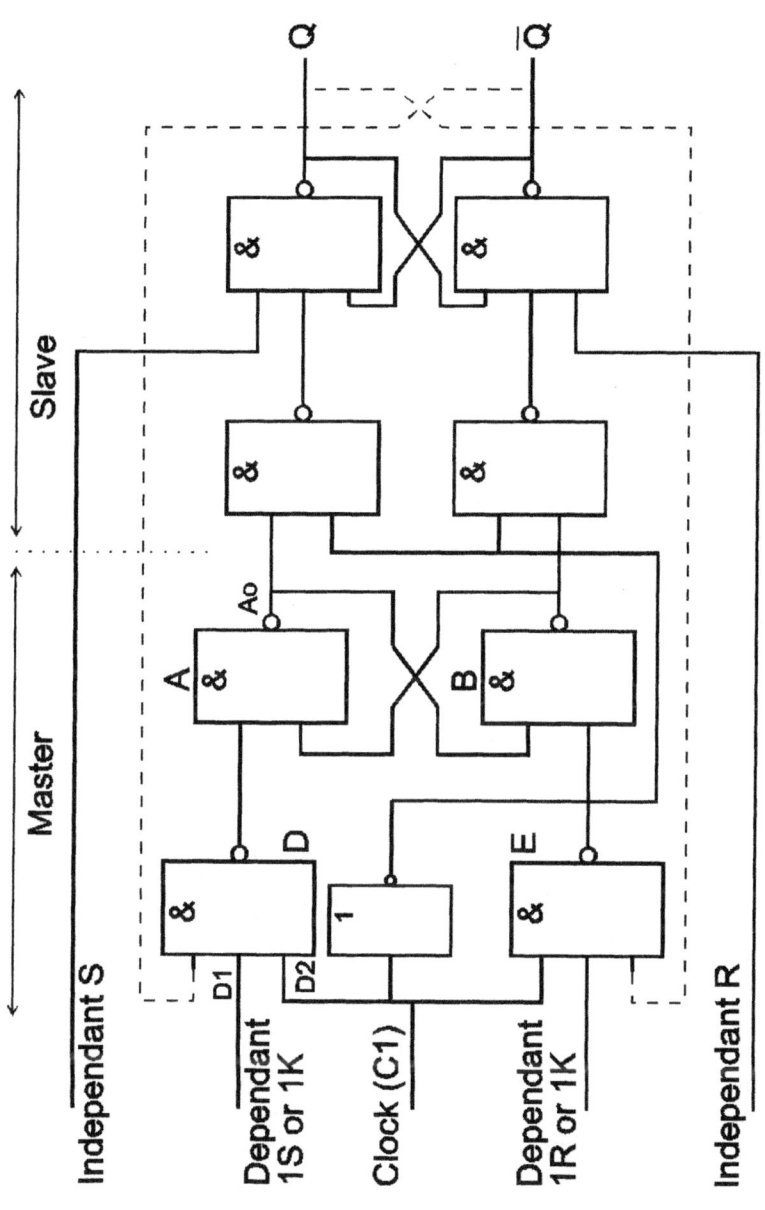

Fig. 15.17. Eight NAND gates arranged to form an *RS* or *JK* master/slave bistable. The dashed connections are required for the *JK* circuit

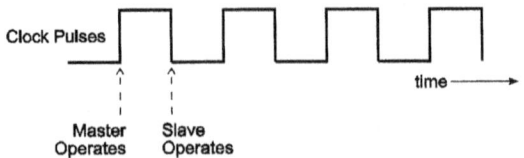

Fig. 15.18. Operation of master/slave bistables

Binary counters and dividers

Bistables give one complete cycle of output–signal variation (at Q and $\bar{Q}$) for every two clock input signals. They can therefore be used as basic elements in binary counters and dividers.

Consider a cascade of bistables A, B, C in which the Q output of each bistable is connected to the clock input of the next as shown in Fig. 15.19. Suppose that the Q outputs are all initially at logic 0: this can be arranged by use of the independent set inputs. The first clock pulse (equivalent to a logic 1 signal) applied to bistable A sets its Q output to logic 1 and this, in turn, sets the Q outputs of bistables B and C also to logic 1. So all Q outputs are at logic 1 after the first clock pulse. The changes of state which occur after the next few successive clock pulses are detailed in Table 15.9. It is assumed that the state of each bistable is switched when the clock input changes from logic 0 to logic 1 and that the changes from logic 1 to logic 0 have no effect. In other words the bistables are assumed to be switched by the positive-going edges of the clock pulses: the circuit of Fig. 15.13 is an example of such a bistable.

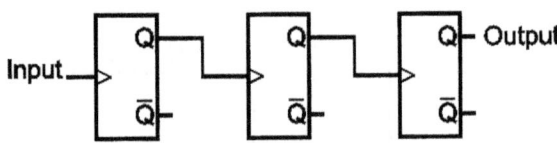

Fig. 15.19. Logic diagram of a simple cascaded counter

After the eighth clock pulse the bistables are in the states assumed initially and thereafter the cycle repeats itself indefinitely. Such an arrangement is known as a counter and this particular counter gives one complete cycle of output for every eight input pulses. If the Q outputs of the bistables are read in the order C, B, A they are in fact the numbers from 7 to 1 in the binary scale. This circuit can therefore be described as that of a down counter.

Table 15.9 also shows the way in which the $\bar{Q}$ outputs of the bistables change on receipt of the successive clock pulses. The $\bar{Q}$ outputs are, of

Table 15.9

	Q outputs			Q̄ outputs		
	A	*B*	*C*	*A*	*B*	*C*
Initial state	0	0	0	1	1	1
After 1st clock pulse	1	1	1	0	0	0
2nd	0	1	1	1	0	0
3rd	1	0	1	0	1	0
4th	0	0	1	1	1	0
5th	1	1	0	0	0	1
6th	0	1	0	1	0	1
7th	1	0	0	0	1	1
8th	0	0	0	1	1	1

course, the complements of the Q outputs. If the $\bar{Q}$ outputs are read in the order C, B, A they represent the numbers from 1 to 7 in the binary scale. These outputs could therefore be described as those of an up counter.

Some bistables change state on the negative-going edges of the clock input pulses: an example is the master-slave bistable of Fig. 15.17 with both J and K inputs held at logic 1. A cascade of bistables of this type behaves differently from the type just described. If all the Q outputs are initially at logic 0, the first negative-going clock signal gives a Q output from bistable A of logic 1. These bistables respond only to changes from logic 1 to logic 0 in the clock input and thus bistables B and C are not affected by this change in the Q output from A. The next clock pulse switches the Q output of A back to logic 0 and in so doing switches B so that its Q output goes to logic 1. The changes brought about by the next few clock pulses follow the pattern shown for the $\bar{Q}$ outputs in Table 15.9: in other words the Q outputs for a cascade of bistables operating on negative-going clock pulses form an up counter. The $\bar{Q}$ outputs therefore form a down counter.

By cascading n bistables in this manner it is possible to realise an up or down counter with a ratio of 2^n. Such an arrangement is of limited application because the count ratio can be only a power of 2 but by use of feedback any desired count ratio can be obtained.

Feedback counter

One method of modifying the count ratio is by feedback to the independent set or reset inputs. For example a divide-by-four counter incorporating two bistables can be converted to a divide-by-three

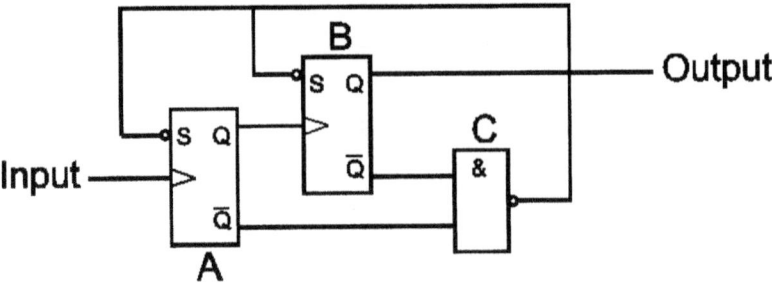

Fig. 15.20. A divide-by-four counter modified by feedback to divide by three

counter by detecting the bistable outputs which correspond to count three and then resetting the bistables to count zero. The operation of this example will now be examined in detail because, though simple, it illustrates the principles employed in this type of feedback counter.

The logic diagram is shown in Fig. 15.20. It is assumed that the bistables are triggered by the positive-going edges of the clock pulses and that counting begins when both bistables have a logic 1 output.

The changes which occur with each subsequent clock pulse are indicated in the waveform diagrams of Fig. 15.21, the upper waveforms applying in the absence of feedback and showing one complete cycle of output for every four input pulses. The lower waveforms apply when

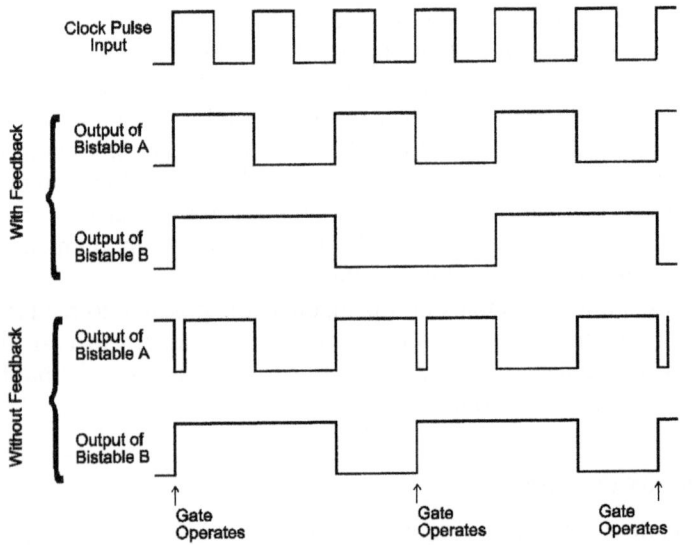

Fig. 15.21. Idealised waveforms illustrating the operation of the logic circuit of Fig. 15.20

Table 15.10

	Without feedback		With feedback	
	Output of bistable A	Output of bistable B	Output of bistable A	Output of bistable B
Initial state	1	1	1	1
After 1st clock pulse	0	1	0	1
2nd	1	0	1	0
3rd	0	0	1	1
4th	1	1		

feedback is present. When the Q outputs of both bistables are at logic 0 (representing count 3) the bistables are reset to logic 1 and the cycle recommences. There is now one output cycle for every three input pulses.

The operation of the circuit can alternatively be represented as in Table 15.10. This also shows that in the absence of feedback four input pulses are required to give one complete cycle of output whereas if feedback is present only three input pulses are required.

Possibly the simplest way of achieving the desired feedback is to use a gate to provide the signal for the independent set terminals and if we assume that a logic 0 set signal is needed then a NAND gate is necessary. This requires two logic 1 inputs to operate it and these signals can be obtained from the $\bar{Q}$ outputs of the two bistables. Thus the counter has the logic diagram shown in Fig. 15.20.

Waveform generators using counting techniques

Combinations of bistables and gates can be used to generate recurrent waveforms containing a number of transitions between the logic levels. Such waveforms are often used for control purposes in analogue and digital equipment: one example is the television sync waveform.

We will consider the generation of a simple recurrent pulse. As before the counter consists of a cascade of bistables operated from a master- or clock-pulse source (and triggering on positive-going edges), the count ratio being so chosen that the output of the cascade is at the desired pulse frequency. A divide-by-eight counter is illustrated in Figs. 15.22 and 15.23.

Suppose the transitions in the pulse to be generated are required to be at t_1 and t_2. The instant t_1 can be identified by the fact that it is the only time in the output cycle at which the clock pulse is at logic 0, the Q

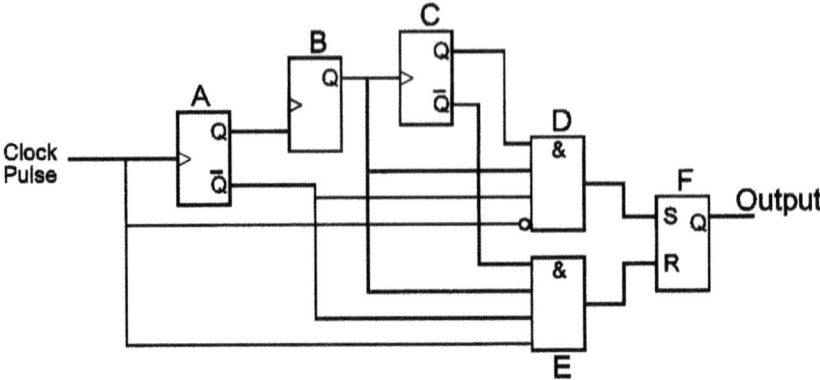

Fig. 15.22. A combination of bistables and gates suitable for generating the waveform shown in Fig. 15.23

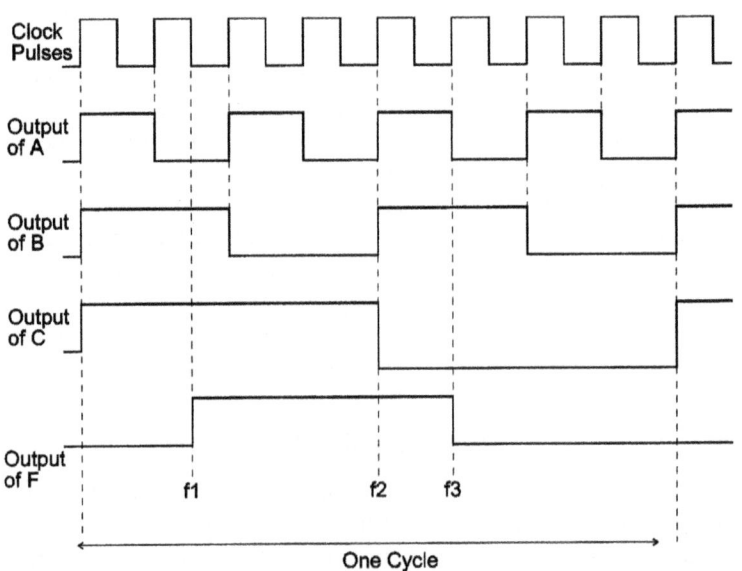

Fig. 15.23. Idealised waveforms illustrating the action of the circuit of Fig. 15.22

output of bistable A is also at logic 0 but the Q outputs of bistables B and C are both at logic 1. The AND gate D is designed to detect these levels, an inverter being used to provide the logic 1 input from the clock pulses and the $\bar{Q}$ output from bistable B providing the logic 1 output from that source. At t_2 the clock pulse and the Q outputs from bistables A, B and C are at logic 1, 0, 1 and 0 respectively. Gate E is designed to detect this combination of levels, the outputs from bistables A and C being taken from the $\bar{Q}$ terminals to give the required logic 1 inputs for the gate. The required waveform can now be generated in a bistable F which is turned on by the output of gate D and off by the output of gate E as shown in Fig. 15.22.

Delay in bistables

In the descriptions of Figs. 15.20 and 15.22 it was assumed, for simplicity, that all changes of state in the bistables occur coincidently with the edge of the clock pulse which initiates them. In practice there is inevitably a slight delay in a bistable: in other words there is a significant time lag between the triggering edge of the clock pulse and the corresponding edge in the output signal. In a cascade of bistables this delay can be appreciable and must be allowed for in the design of circuits. For example a gate responding to logic 1 outputs from the clock pulses and from all three bistables in Fig. 15.23 would operate appreciably later than the leading edge of the clock pulse because of the cumulative delays in operating bistables A, B and C. Such a gate would also operate at t_3 in Fig. 15.23 because at this instant the output of bistable C is still at logic 1 although the outputs of the other two bistables have changed to logic 1.

If the delay through bistables A, B and C is less than the duration of a clock pulse, the possibility of false outputs from gate E can be eliminated by using a circuit similar to that of gate D which is arranged to operate after the trailing edge of the clock pulse.

Shift registers

An assembly of JK bistables is often used as a temporary store for binary information. For example suppose the binary signal 11001 is to be stored. Five bistables are needed (one for each binary digit) and the information could be fed simultaneously into the bistables via the independent set and reset inputs. The information can now be stored for the necessary period and then simultaneously read out by operation of the clock inputs. This is the parallel method of feeding in (writing) and feeding out (reading) the information.

Alternatively the bistables may be connected in cascade as shown in Fig. 15.24 and the information can be fed into the input of the first bistable one digit at a time. The clock inputs of all bistables are commoned and the clock rate must be the same as the rate of receipt of the digits in the incoming information (the bit rate). On the first clock pulse the first binary digit (1) is transferred to the first bistable, its Q output registering logic 1. On the second clock pulse the first digit is

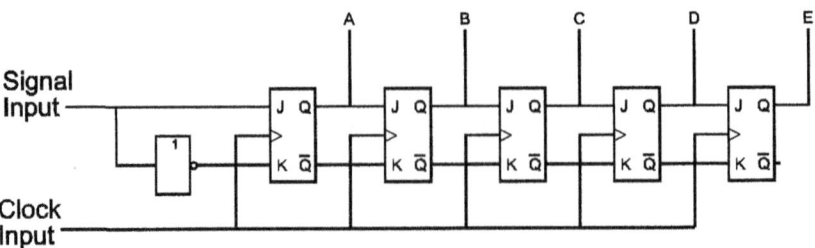

Fig. 15.24. A five-stage shift register using *JK* bistables

transferred to the second bistable and the second digit (also 1) is transferred to the first bistable. Thus the process continues, the stored information moving to the right in the cascade of bistables until, after the fifth clock pulse, the whole binary word has entered the register. The *Q* outputs, in order, now read 11001. At this point the clock pulses can be stopped and the word stored for as long as necessary. When the word is required to be read out, the clock pulses are restarted and the stored information again moves to the right at clock rate and can be read at the *Q* output of the fifth bistable. This is the serial method of storing and reading information in a shift register.

It is possible to write the information into the register by parallel methods and to read it out serially: the converse is also possible. Thus there are serial/parallel and parallel/serial shift registers.

Digital representation of information

The various processes (including transmission and storage) of data required in computers and similar equipments are carried out by binary digital techniques but operations cannot be performed on an individual bit. Bits are commonly organised into groups of 8 (known as *bytes*) and a byte is the smallest item of information which can be operated upon. Between 00000000 and 11111111 there are 2^8, i.e. 256 unique combinations of 0 and 1 in an 8-bit byte which can be used to represent different items of information. Data is commonly fed into a computer via a keyboard similar to that of a typewriter and appears on a c.r.t. screen (a *visual display unit* or v.d.u.) also used to display the computer output. The data is expressed in alphanumeric symbols and the 256 possible bit combinations available in a byte are sufficient to represent the upper-case and lower-case letters of two different alphabets, the figures from 0 to 9, punctuation marks and various control codes (such as 'start a new line' equivalent to 'carriage return' on a typewriter), still leaving combinations for other characters such as graphics (e.g. blocks with various outlines which can be used to make up diagrams). The American Standard Code for Information Interchange (ASCII) relates the bit arrangement within the byte with the symbol represented. When the arrangement of bits in a byte is recognised, the character represented is displayed on a v.d.u. as a pattern of dots on adjacent scanning lines.

Stores

The shift register is one of a number of types of store needed in a computer. We shall now consider how transistors can be used in other types.

Static random access memories (SRAMs)

When a bistable is triggered from one state to the other, the collector (or drain) voltage switches from a low value (for the 'on' transistor) to a high value (for the 'off' transistor). If these two voltage values are taken as logic levels a trigger can be said to cause one collector voltage to take up its 1-state. Moreover this state will persist indefinitely if no further triggers are received. In other words the bistable is capable of storing one binary digit. A group of 8 bistables can thus store a byte.

To achieve the storage capacity required in a digital computer, bistables are needed by the thousand and are available in the form of i.cs. Early storage i.cs held only 1,000 bistables but as technology improved and greater miniaturisation became possible, the capacity of the i.cs increased to 64,000 bistables (and more recently much greater than this). The bistables are usually m.o.s. types which are more economical of power than bipolar transistors, requiring no base current to hold them in the conductive state. The bistables are arranged in a rectangular array or matrix with leads from every row and column. A 16-kbit i.c., for example, may have 128 rows and 128 columns. A wanted bistable can be switched by signals applied to the appropriate horizontal and vertical leads. At each lead intersection there are, in addition to the bistable itself, up to four other transistors, the whole forming a group known as a *memory cell*. Fig. 15.25 gives one possible circuit diagram for a memory cell.

TR1 and TR2 form the bistable and transistors TR3 and TR4 act as drain loads. TR5 and TR6 are used to connect the bistable to the vertical lines (bit lines) which convey the input (write) signals to and the output (read) signals from the bistable. TR5 and TR6 are switched on by signals applied to the horizontal (row select) line – sometimes called the word line. In the absence of a signal on the horizontal line the bistables are isolated and so maintain their states.

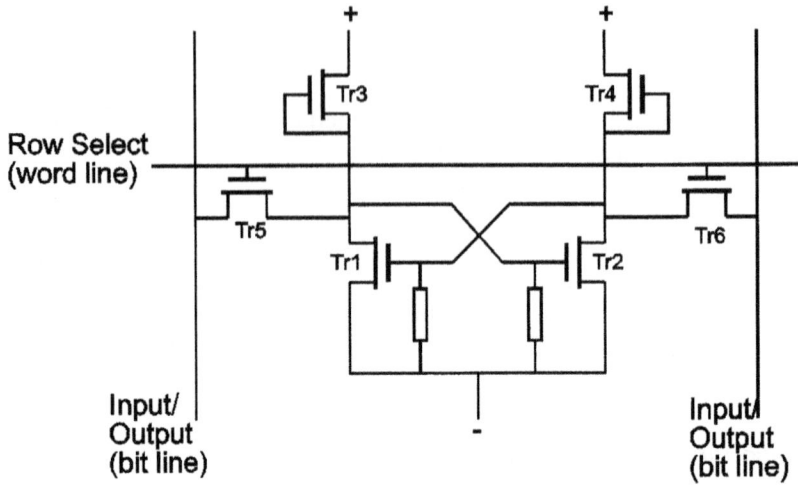

Fig. 15.25. Circuit of a memory cell in a static RAM

If the supply to a store of bistables is removed, the stored information is lost. Such storage is termed *volatile* and is used in digital computers to store information which frequently needs to be changed such as the data and program for the operations on which the computer is currently engaged. In general each new problem necessitates a different input so the loss of information on switch-off is no disadvantage. Such volatile stores are called read/write stores, direct-access or random-access stores (RAMs) and those using bistables are known as static RAMs (SRAMs).

Dynamic random access memories (DRAMs)

In an alternative form of RAM the basic storage element is a capacitor which, in some stores, is the internal gate-source capacitance of a m.o.s. transistor. A charge placed on the capacitor can represent logic 1 and the absence of a charge logic 0. Cell circuits can be very simple, only one m.o.s. transistor being required in addition to the storage capacitor as shown in Fig. 15.26. The transistor acts as a switch, connecting the capacitor to the input/output line when turned on by a signal on the word line. A disadvantage of such a circuit is that the charge leaks away through the inevitable resistance present but this can be overcome by a process termed *refreshing* in which the charge on each capacitor is regularly 'topped up'. The charge is compared with a standard value accepted as representing logic 1 and, if necessary, charge is added to restore the stored charge to this value. The simplicity of the cell circuit

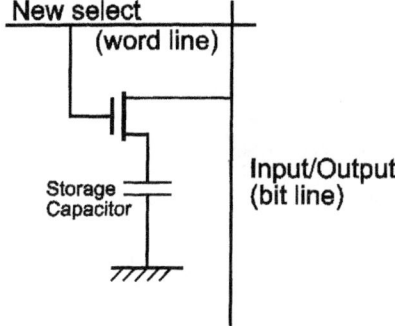

Fig. 15.26. Circuit of a memory cell in a dynamic RAM

gives this arrangement an enormous advantage over the static RAM even though refresh circuits are needed and most RAMs in modern computers are of this type.

Read-only memories (ROMs)

A computer needs a store to hold the information which enables it to carry out routine tasks such as the generation and display of alphanumeric characters. This information is required irrespective of the task currently being worked on by the computer and must not be lost when the computer is switched off: in other words the store for this essential information must be *non-volatile*. Storage of this kind is termed fixed, permanent or – more commonly – read-only memory (ROM).

Some ROMs are programmed by masking operations during manufacture of the i.c. In others (programmable ROMs or PROMs) the program is put in by the user in a once-and-for-all operation: each cell in the i.c. contains a fusible link which can be destroyed (to give a logic 0, say) by the application of a higher-than-normal voltage or left intact (to give a logic 1) so enabling the i.c. to be programmed. Yet another type (electrically-programmable ROM or EPROM) can be repeatedly reprogrammed by the user by use of high voltages but erasure of existing programs involves irradiation by u.v. light and the inconvenient removal of the i.c. from the equipment. More recent types of ROM (electrically-erasable and programmable ROM or EEPROM) can be erased and reprogrammed electrically in whole or part (even a single byte can be changed) with the i.c. in situ.

The need for ROMs to retain their information in the absence of a power supply has led to the introduction of the floating-gate transistor and the construction of one type of such a transistor is shown in

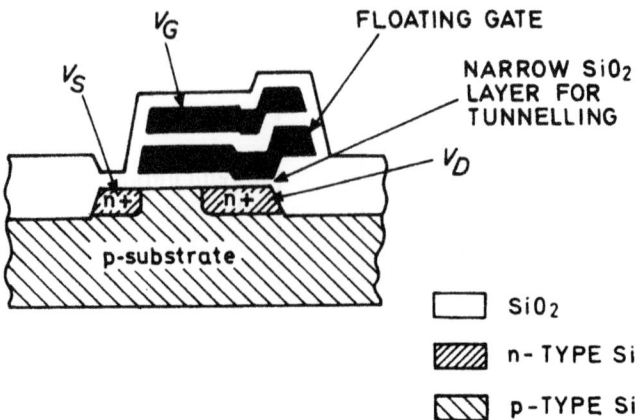

Fig. 15.27. Construction of one type of floating-gate transistor

Fig. 15.27. It is similar to the n-channel fet. of Fig. 2.17 but has two gates, that nearest the channel being floating, i.e. completely surrounded by silicon dioxide and thus highly insulated. Before the transistor can be programmed (written) it must first be erased. To do this a voltage of, say, 20 is applied to the outer gate. This causes electrons to be forced through the thin silicon dioxide layer to charge the floating gate. This has the effect of rendering the transistor immune to outer-gate voltages of about 6: it will not now respond to normal operating voltages of 3.5. But if the 20 V between outer gate and drain is reversed, most of the electrons on the floating gate are forced to return to the drain and the transistor can now be switched by normal operating voltages on the outer gate.

Charging and discharging of the floating gate is achieved by virtue of an effect known as Fowler-Nordheim tunnelling whereby certain electrons subjected to an electric field can cross the forbidden gap of an insulator to enter the conduction band and can thus flow a short distance to a positively charged electrode. To make use of this effect using voltages as low as 20 requires the insulating layer to be very thin: hence the narrowing of the silicon dioxide layer near the drain connection in Fig. 15.27.

In a typical EEPROM each memory cell comprises a floating-gate (storage) transistor and a select transistor connected in series to form a cascode as shown in Fig. 15.28. To erase the cell the select line and the word line are biased to +20 V whilst the column is grounded. To write, the select line is held at +20 V, the word line is grounded and the column is raised to +20 V. For reading, the select and word lines have normal operating voltages of 3.5 and the column is grounded. The read pulse

Column

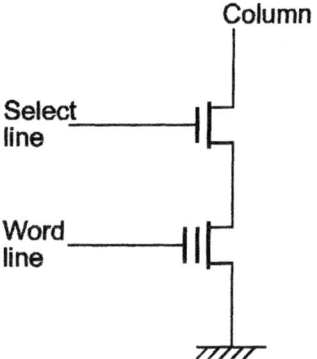

Select line

Word line

Fig. 15.28. Circuit of EEPROM memory cell

will switch a written cell so sending a current pulse to the output via the column but it has no effect on an erased cell.

The preferred word 'store' has been used throughout this section but the alternative 'memory' is likely to persist because of its use in the acronyms RAM, ROM etc.

Digital transmission of analogue signals

We have shown how digital techniques can be used to transmit information for display in alphanumeric form on a v.d.u. but there are other ways in which these techniques can be used for communication. Any analogue signal can be converted into a stream of constant-amplitude pulses and transmission of such pulses has a number of advantages over transmission of the analogue signal itself. For example, pulses received over a long link can, provided they exceed the noise level, be used to trigger a pulse generator which will deliver a replica of the received signal but with clean pulses, i.e. with a greatly enhanced signal-to-noise ratio. Secondly a signal in pulse form can readily be held in a static store thus effectively allowing the analogue signal to be delayed. Moreover a stored video signal can be read out at any desired speed so permitting a number of effects in television picture display. On the other hand certain types of signal processing, e.g. emphasis of higher frequencies to add crispness to speech or to enhance detail in television pictures, is easier to accomplish with the signal in analogue form.

One method of converting an analogue signal to digital form is by the use of pulse code modulation (p.c.m.). Briefly the technique is to sample the analogue waveform at a frequency high enough to resolve the highest-frequency component. The instantaneous amplitude of the waveform at the sampling instant is stored on a capacitor in a sample-

and-hold circuit. For high-quality sound transmission these amplitudes vary over an enormous range. Only a limited number of different amplitudes can be accepted and the amplitude nearest the stored value is selected for transmission. This value is expressed as a binary number and the stream of binary numbers is transmitted as the digital representation of the analogue signal. Synchronising signals are included in the bit stream to enable receiving equipment to identify the pulses constituting each binary number.

This technique is used by the broadcasting authorities to distribute high-quality stereo sound signals over the transmitter network. It is also used for the distribution of television sound, the digital signals being included in the video signal during the period of sync level within the line sync signals (sound-in-syncs). Compact disc recording also uses p.c.m. and more recently the system has been extended to domestic audio tape recorders and the NICAM system of television stereo sound.

Sample-and-hold circuit

The purpose of a sample-and-hold circuit is to take 'snapshots' of the instantaneous values of an analogue signal at regular intervals. One technique for doing this is to connect a capacitor to the input-signals source via a switch which closes for the duration of the sampling periods. To ensure that the capacitor charges rapidly to the signal value the signal source needs a low output resistance and is represented in Fig. 15.29 by an emitter follower. It is also essential to minimise charge loss in the intervals between sampling periods and the amplifier which reads the voltages on the capacitor requires a high input resistance. Another emitter follower is therefore shown in Fig. 15.29, which

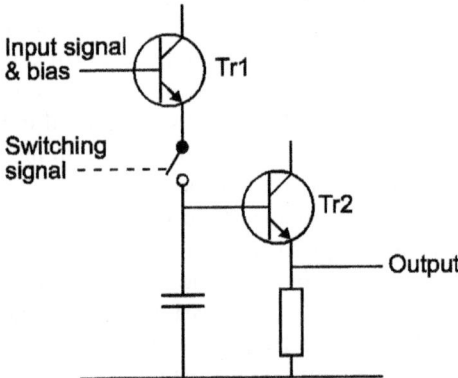

Fig. 15.29. Basic principle of a sample-and-hold circuit

illustrates the essential features of a sample-and-hold circuit. To avoid negative amplitudes a steady bias can be added to the analogue signal before its application to TR1.

In a practical circuit TR1 and TR2 are replaced by operational amplifiers with a large degree of negative feedback giving a much lower output resistance and input resistance than is possible from simple emitter followers.

Chapter 16

Further applications of transistors and other semiconductor devices

Introduction

In this chapter we shall describe a number of applications of junction diodes and transistors which do not properly belong in earlier chapters.

Supply-voltage stabilising circuits

As we have seen, many of the properties of transistors depend on the current in them and hence on the supply voltage. To ensure consistency in the performance of transistor equipment it is therefore common practice to stabilise the supply voltage.

Simple circuit using a voltage reference diode

Fig. 16.1 gives the diagram of a simple circuit suitable for use where only a small current is required from the stable voltage source.

An example of such a requirement may occur in a transistor car radio where the supply for the oscillator may require stabilising against changes in car battery voltage to secure good frequency stability and hence stable tuning. A suitable circuit is illustrated in Fig. 16.1. The junction diode must have a breakdown voltage equal to the value of the stabilised voltage required and the value of R_1 must be chosen to give an operating point on the nearly vertical part of the diode characteristic (Fig. 16.2). The diagram represents conditions in the circuit. The load line AB meets the axis at A at a voltage equal to the supply voltage, say 12 V. AB meets the diode characteristic at B and this point corresponds with the stabilised output voltage. OC represents the voltage drop across

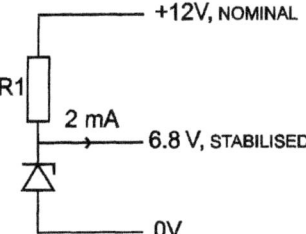

Fig. 16.1. Simple junction-diode voltage-stabilising circuit suitable for small currents

the diode and AC the voltage drop across R_1. In the chosen example the stabilised voltage is 6.8 V.

The slope of AB corresponds to the resistance R_1 and this clearly may vary within limits (as suggested by the dotted lines AD and AE) without much effect on the value of the stabilised voltage but it is preferable to choose a value for R_1 which keeps the dissipation in the diode well within the maximum value prescribed by the makers. For example, if the maximum dissipation is 50 mW we can choose to dissipate half of this, 25 mW, at a battery voltage of 12 V. The voltage across the diode is 6.8 V and the diode current must be 25/6.8, i.e. 3.7 mA. This current is supplied via R_1 together with the current for the load (the oscillator transistor). If the load current is 2 mA, the total current in R_1 is 5.7 mA.

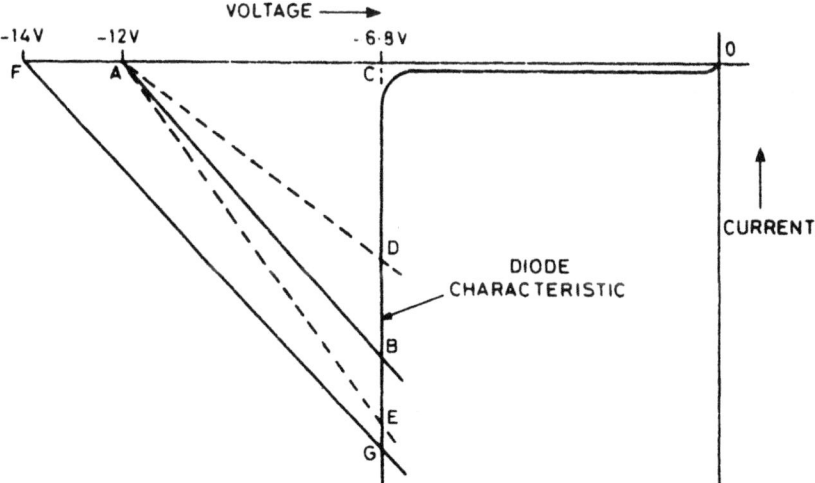

Fig. 16.2. Illustrating the operation of a simple voltage-stabilising circuit

The voltage across R_1 is 5.2 V and the required value of R_1 is given by

$$R_1 = \frac{5.2}{5.7 \times 10^{-3}}$$

$$= 900\,\Omega \text{ approximately}$$

The supply voltage may easily rise to 14 V when the car alternator is running. The effect such a voltage rise has on the stabilised voltage is illustrated by the load line FG which is parallel to AB (thus representing the same value of resistance R_1) but meets the axis at F corresponding to 14 V. FG meets the diode characteristic at G, representing a greater diode current than before (point B). The new stabilised voltage corresponds to point G which, because of the extreme steepness of the diode breakdown characteristic, is at almost the same voltage as before (point B). The dissipation in the diode is now greater than before the increase in supply voltage and care must be taken to see that the maximum safe dissipation is not exceeded when the supply voltage is at its maximum.

The effectiveness of the circuit depends on the steepness of the diode characteristic which is usually expressed as a slope resistance. This may be as low as 5 Ω, showing that a change of diode current of 10 mA gives an alteration in breakdown voltage of only $5 \times 10 = 50$ mV.

High-stability stabiliser IC

A Zener-like device has been developed for the special application of voltage supply for varicap tuners for radio and television. It is a multi-semiconductor combination in the form of an i.c. with two terminals, and has a closely stabilised reverse breakdown voltage of 33 V, chosen to suit the characteristics of the varicap diodes used for tuning. It has a slope resistance of typically $<5\,\Omega$ and a very low temperature coefficient of -10 mV per °C. It is used in a circuit similar to that of Fig. 16.1.

Series voltage stabilisers

More complex power-supply circuits are used to supply transistor equipments with current. For example, portable television receivers designed for operation from the mains or from car batteries incorporate a constant-voltage supply circuit which must be capable of operating satisfactorily with an input as low as 12 V.

Such a supply circuit has two distinct functions. Firstly it must maintain the output voltage at a particular wanted value (which can be predetermined) despite variations in input voltage whether from the mains or batteries. This is known as *stabilisation* of the output voltage. Secondly it must maintain that particular value of output voltage despite variations in the current drawn by the equipment. This latter quality is termed *regulation* and good regulation is achieved by giving the power supply circuit a low output impedance. This has the incidental advantage of minimising any tendency to instability in the equipment caused by the common impedance of the power-supply circuits. Regulation is an inherent feature of stabilisation.

In one commonly used stabiliser circuit the stabilised output is derived from the unstabilised input via a series element, the resistance of which is controlled so as to maintain a constant output voltage. The signal controlling this element is obtained from a comparator stage which compares a sample of the output voltage with a constant reference voltage. The sample of output voltage is usually obtained from a preset potentiometer which is adjusted to give the desired value of output voltage. The reference voltage is normally obtained from a Zener diode which is fed via a series resistor from the stabilised or unstabilised supply.

The comparator stage can be a single npn transistor, as shown in Fig. 16.3, in which the sample voltage is applied to the base and the reference voltage to the emitter. The Zener diode effectively presents the emitter with a very low impedance so that the full gain of a common-

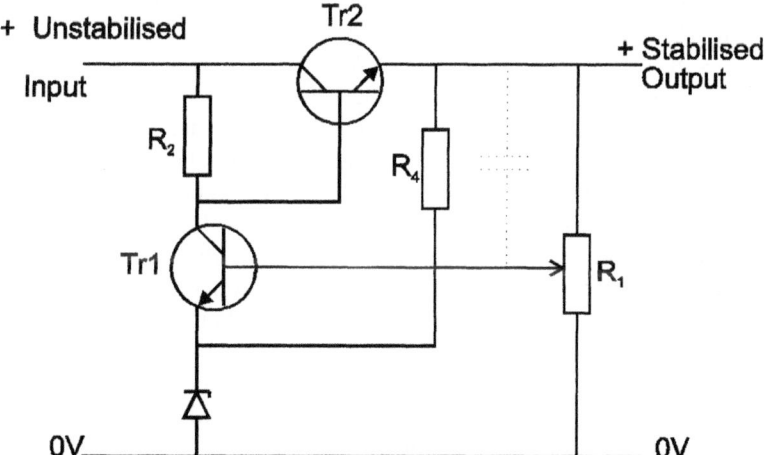

Fig. 16.3. Series stabiliser with zener diode in emitter circuit of the comparator transistor

emitter amplifier is available from the comparator transistor. If there is an increase in the current drawn from the output of the unit there is a tendency for the output voltage to fall. This causes a fall in the base voltage of the comparator transistor and its collector voltage therefore rises. This positive voltage change is applied to the base of the stabiliser transistor to enable it to supply the additional current required. The stabiliser transistor must hence be an npn type. A second requirement of the stabiliser transistor is that it must not invert the input signal: the positive voltage step applied to the base must give a positive step in output voltage. An emitter follower is therefore the obvious choice for the stabiliser stage and the circuit so deduced has the form shown in Fig. 16.3.

The potentiometer R_1 controls the fraction of the output voltage fed back, after amplification by TR1, to the base of TR2. As R_1 slider is moved upwards the degree of negative feedback is increased and variations in output voltage are decreased: such slider movement also decreases the stabilised voltage. If, in Fig. 16.3, the Zener diode has a reference voltage of 6.3, and if TR1 is a silicon transistor, then its base voltage must be approximately 7 for TR1 to conduct. If the potentiometer is at its midpoint, the stabilised output voltage must be 14. If the slider is lowered to give, for example, a 3:1 potential division ratio the output voltage rises to 21. If the upper arm of R_1 is shunted by a capacitor (as shown dashed in Fig. 16.3) the degree of negative feedback is increased for any alternating voltage components at the output terminals. If the capacitance is made large enough for its reactance at 50 Hz or 100 Hz to be negligible compared with the resistance of the lower arm of R_1, any ripple on the stabilised supply can be minimised.

Suppose we interchange the positions of the Zener diode and its feed resistor R_4 as shown in Fig. 16.4. By doing this we ensure that any changes in output voltage are fed back, without loss, to TR1 emitter by the Zener diode. They are also fed back, attenuated by R_1, to TR1 base. The effects of these two inputs on TR1 collector current are, of course, opposite, but the emitter input, being the larger of the two, dictates the phase of collector-current changes. The potentiometer R_4 is, however, still necessary to enable the output voltage to be adjusted to the desired value. In this circuit a sudden increase in output current causes a decrease in output voltage and thus in TR1 emitter voltage. This results in a decrease in TR1 collector voltage and therefore in TR2 base voltage. This input to the regulator stage must give increased current output and TR2 must hence be a pnp type. And because a decrease in base voltage must give increased output voltage TR2 must be connected as a common-emitter stage as shown in Fig. 16.4.

This circuit also minimises ripple on the output because the Zener diode has very low impedance at ripple frequencies. It also has another

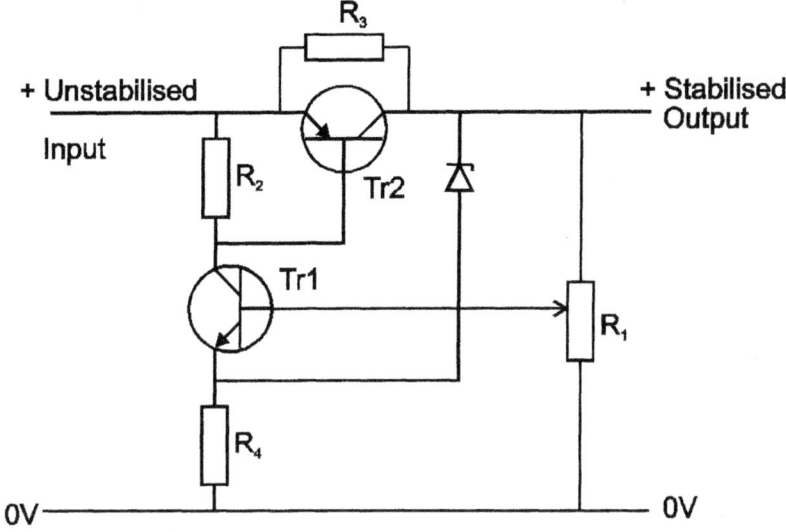

Fig. 16.4. Improved performance can be obtained by interchanging the positions of Zener diode and feed resistor R_4, as shown here

advantage over the circuit of Fig. 16.3 and this concerns the value of R_2, which must be large compared with the input resistance of the regulator stage in order to direct all the changes in current from the comparator stage into it – a normal requirement, of course, in any current amplifier. Now if TR2 is an emitter follower supplying, say, 100 mA its input resistance could be as high as 1 kΩ, and R_2 should ideally be large compared with this. Thus if TR1 has a mean current of 5 mA the voltage drop across R_2 could be considerable: indeed it could be comparable with the unstabilised input voltage! If, however, TR2 is a common-emitter amplifier its input resistance could be as low as 25 ohms and R_2 need only be 100 ohms or so, so that the voltage drop across it can be less than 1 V. This circuit is, therefore, better suited for use with low values of unstabilised input voltage such as the nominal 12 V from a car battery. In such stabilisers it is also desirable to minimise the voltage drop across the regulator transistor and a common technique is to connect a low-value resistor, e.g. 10 ohms across TR2 as shown as R_3 in Fig. 16.4. This necessarily reduces the degree of stabilisation achievable and emphasises the need for high gain in TR1 and TR2.

To determine the type of circuit required in a stabiliser we have so far discussed the circuits in terms of voltage changes but the comparator and regulator stages should, in fact, be designed as a current amplifier. The overall current gain is independent of whether TR2 is an emitter follower or a common-emitter amplifier because both have approximately equal values of current gain. From a knowledge of the current gain the output resistance of the stabiliser can be assessed in the following manner. Suppose the mean collector current of TR1 is 5 mA. Its mutual conductance is thus 200 mA/V and a voltage change of 5 mV at the base (or emitter) changes the collector current by 1 mA. Suppose this is handed on to TR2 without loss and that TR2 has a current gain of 50. Then TR2 current changes by 50 mA. We must now make an assumption about the transfer of changes in output voltage to the input of TR1. This depends on the setting of the potentiometer but assume that half the changes are effectively transferred to the comparator stage. Then for a change in output voltage of 10 mV, the output current can change by 50 mA, representing an output impedance of 0.2 ohm. Even lower values can be achieved by adding extra stages of current amplification between comparator and regulator.

Shunt voltage stabilisers

There is an alternative form of voltage stabiliser in which the stabilised supply is derived from the unstabilised supply via a fixed series resistance, control being achieved by an element across the output, the resistance of which is varied automatically so as to give a constant output voltage. The circuit of Fig. 16.1 is, of course, a simple example of such a stabiliser.

Switch-mode power supplies

The type of power supplier just described has a number of disadvantages. For example, the stabilised output voltage is necessarily less than the unstabilised input voltage. Moreover the regulating transistor is always conducting, and dissipation in it, and in any parallel-connected resistor, can be considerable. Both disadvantages can be overcome by arranging for the regulating transistor to be regularly switched between saturation and cut-off. This minimises dissipation in the transistor, in the on-state because of the low voltage across it and in the off-state because of the absence of current in it. Moreover regulation can be achieved by controlling the duration of the on-periods, an example of pulse-width modulation. As the regulating-transistor current is in pulse form the output can be obtained from a transformer-rectifier combination, so enabling output voltages greater than the input voltage to be obtained. The transformer need not be a bulky component: by arranging for the regulating transistor (known usually as a chopper) to be switched at a high frequency such as 15 kHz its inductance need only be 1/300th that of a 50 Hz component. The smoothing components are also correspondingly smaller.

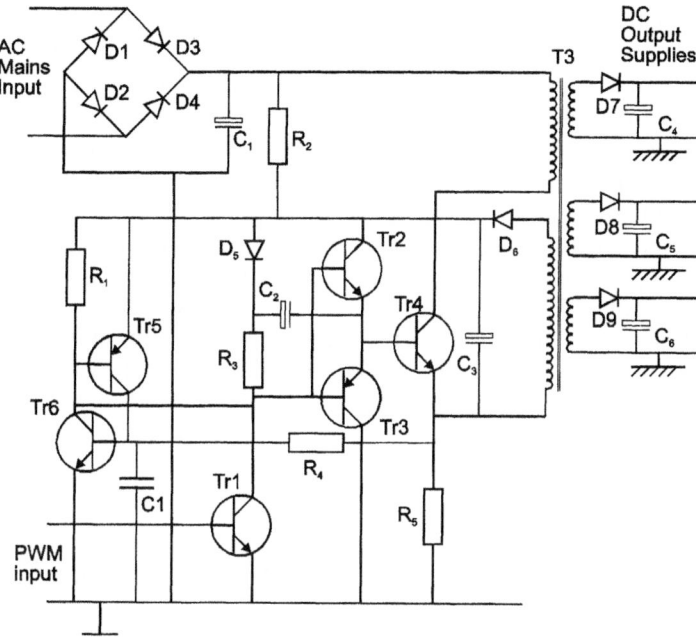

Fig. 16.5. Simplified switch-mode power supply for a television receiver

A simplified circuit diagram of a switch-mode power supplier typical of those used in television receivers is given in Fig. 16.5. The mains input is rectified by a full-wave rectifier to produce a supply at about 320 V which feeds the chopper transistor TR4 and its transformer T3. The chopper is switched at 15 kHz (sometimes synchronised at television line frequency) and is driven by the complementary stage TR2, TR3 which, in turn, is driven by the common-emitter stage TR1. TR1 input is a rectangular-wave signal pulse-width modulated by a sample of the output voltage of the power supplier. T3 has a number of secondary windings proportioned to supply the various steady voltages required by the receiver, including the chopper amplifier TR1 to TR4 itself. This poses a problem: the supply for the chopper amplifier is not available until the chopper is working but the chopper cannot work until it has a supply! This impasse is solved by providing an alternative supply for the chopper from the rectified mains supply via R_2 and C_3. This supply is available immediately after switch-on and is overtaken, as soon as the chopper is operating, by the supply from T3 and D6.

TR5 and TR6 constitute a direct-coupled complementary bistable (similar to that described on page 260) which is normally in the state in which both transistors are off. If, for any reason, the chopper transistor passes excess current, the voltage across R_5 turns TR6 (and hence TR5)

on and both transistors remain on. TR6 now short-circuits the output of TR1, so switching off the chopper and protecting the power supplier from possible damage. The charging of C_7 plays an essential role in this protective circuit.

One way in which the pulse-width modulation can be achieved is illustrated in the block diagram of Fig. 16.6 and the associated waveforms in Fig. 16.7. A square-wave signal at chopper frequency is integrated to give the symmetrical sawtooth (b) which is applied to one input of operational amplifier A2. A sample of the output voltage of the power supplier is compared with a reference voltage in operational amplifier A1 and its output forms the second input for amplifier A2. If this input increases it rises further up the sawtooth waveform, so

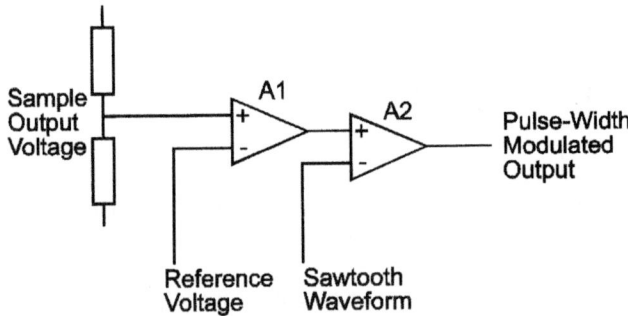

Fig. 16.6. Two operational amplifiers used to give a pulse-width-modulated output for a switch-mode power supplier

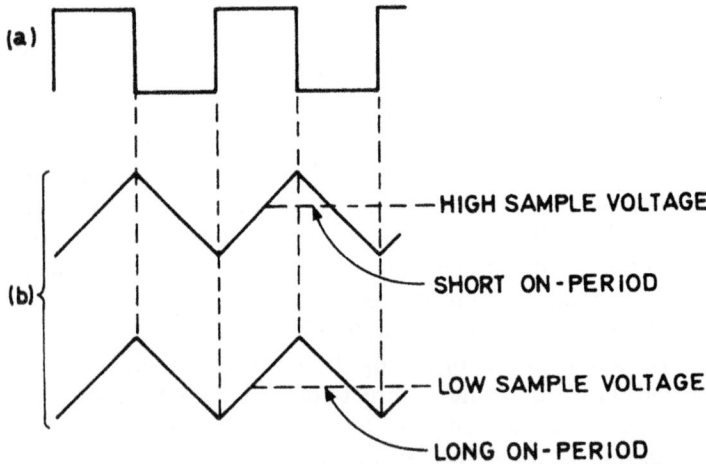

Fig. 16.7. Waveforms illustrating the operation of the circuit of Fig. 16.6

increasing the output of A2 for a shorter period in each cycle. This output signal is the input for the chopper amplifier TR1. The diagram shows a high sample voltage leading to a short on-period and a lower voltage giving a longer period.

There are many variations of mains-driven switch-mode power-supply circuits. That shown in Fig. 16.8 is based on the astable relaxation oscillator described on page 273 in which chopper transistor TR1 and switching transformer T1 form the basic oscillator. The energy in T1 is tapped off by the rectifier-diodes at the top left of the diagram.

When mains voltage is first applied, base-emitter current flows in TR1 via the resistor R_1. As a result the transistor is turned on and collector current flows through T1 primary winding L_1. This current induces a voltage in feedback winding L_2 which is passed on to TR1 base via C_1 and R_2, turning the transistor hard on. As soon as the current in L_1 stabilises, the positive pulse across L_2 falls to zero and then reverses rapidly, turning TR1 off. When the magnetic field of T1 has collapsed, the process repeats itself, giving continuous oscillation and feeding energy to the loads on the secondary windings.

Regulation is achieved by monitoring the voltage developed across feedback winding L_3 on T1. The alternating voltage it produces is rectified by diode D1 and smoothed by C_2 to produce a negative voltage proportional to that on the secondary supply lines. It is fed to TR2 emitter via Zener diode D2 and to TR2 base via potential divider $R_3R_4R_5$. The

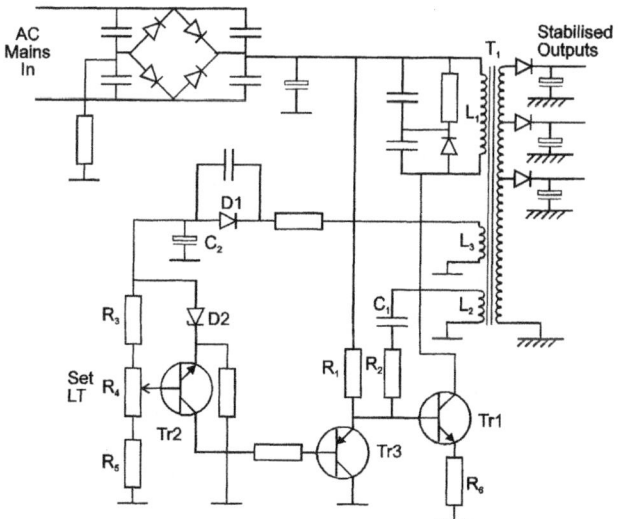

Fig. 16.8. Self-oscillating switch-mode power-supply circuit. Note the different 'ground' symbols used for isolated and mains-live sections

result is that conduction in TR2 is governed by the negative voltage on C_2: the greater the voltage, the greater the conduction. The basic operating point is set by R_4, the *set H.T.* control.

Conduction in TR2 governs the base-emitter current of transistor TR3 which shunts the base-emitter junction of the oscillator TR1. The effect is to control the turn-on time of TR1 and hence the duty cycle of the oscillator. A complete control loop is thus set up to regulate and stabilise all three output voltages.

The resistor R_6 develops a voltage proportional to the current flowing in T1 and TR1 and forms part of an overload protection circuit (not shown here) which removes the base drive from TR1 to shut down the oscillator in the event of a current overload. Over-voltage protection is effected by a 'crowbar' device (not shown here) at the secondary side of transformer T1. Its operation is described on page 351.

Many types of switch-mode power-supply circuits use purpose-designed i.cs incorporating 'soft-start', overload protection and standby facilities, the last-mentioned being governed by an infra-red remote-control system. One very efficient i.c.-based system directly drives the chopper transistor at frequencies between 22 kHz and 35 kHz, depending on input voltage and load conditions. It forms a self-oscillating, non-synchronous blocking converter, the frequency and duty cycle of which automatically adjust to variations in mains supply voltage and loading of the secondary windings. The secondary voltages are stabilised to ±0.5 per cent over a mains input voltage range of 185 V to 265 V, while for load variations between 30 W and 100 W the output voltages are held to within 1 per cent variation.

Power mosfets

The higher the operating frequency of a switch-mode power supply unit, the more efficient it becomes. The two limitations on operating frequency are hysteresis in the core of the chopper transformer and the switching speed of bipolar power transistors of the type required in the circuits described above. These difficulties have led to the development of ferrite transformer cores designed to meet the needs of high-power switching and of power mosfets capable of operating at frequencies of 100 kHz or more. Such fets have the advantages over bipolar transistors of lower losses, greater immunity to current overload and reduced drive power. On the debit side they are more expensive than similarly rated conventional transistors.

For a high current to flow in an fet the channel area must be large, making it difficult to create a depletion region wide enough to cut off the current flow between the source and drain terminals. In a power mosfet this problem is overcome by fabricating a large number of separate fets on a single substrate and connecting all their source, drain and gate terminals in parallel. Over 20,000 separate transistors may typically be formed inside a single package. With such construction the total current may be as high as 10A and the 'on' resistance as low as $10\,\text{m}\Omega$, depending on ambient temperature. With a drain source rating of 1 kV during cut-off, these devices are ideal for switch-mode power-supply applications at high switching speeds.

A power mosfet has a larger safe operating area (s.o.a.r.) than a similarly-rated bipolar device. The s.o.a.r. includes all the possible combinations of drain-source (collector-emitter) voltages and drain (collector) current values with which the device can operate without risk of breakdown. Outside the s.o.a.r. there is a likelihood that a destructive hot-spot may develop in a bipolar transistor, resulting in breakdown of the junction. The channels of a power fet have a positive temperature coefficient and the formation of hot-spots is less likely. If local overheating occurs the resistance in the region increases, diverting current to surrounding areas, so inhibiting further rise in hot-spot temperature and preventing breakdown.

The use of power mosfets in switch-mode power-supply units began in computer applications where reliability, efficiency and minimum bulk are required. Their use has now spread to include domestic television receivers. The circuits are similar to those using conventional transistors as shown in Fig. 16.9. The main difference is that no drive current is

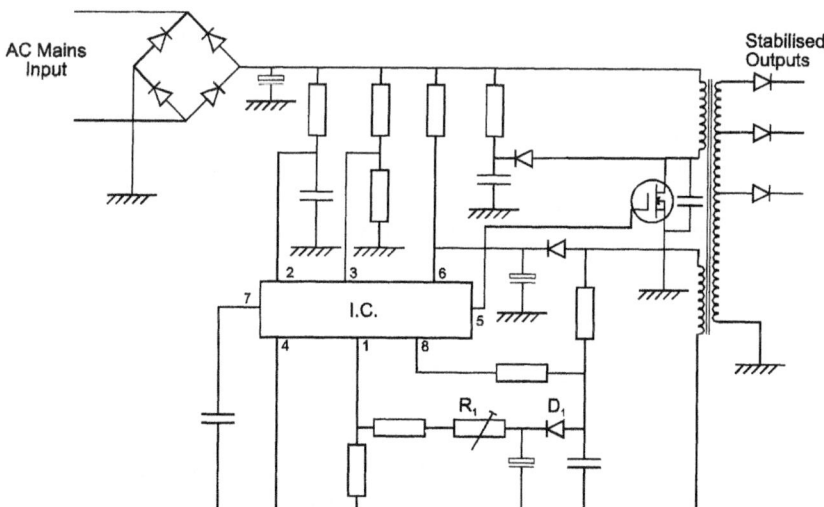

Fig. 16.9. Television power supply using a purpose-designed i.c. and a power-fet switching element TR1

required by the power fet. The drive device i.c. generates at pin 5 a square wave with a mark-space ratio (duty cycle) governed by the voltage applied to pin 1: this is generated by a secondary winding on the chopper transformer and preset to the required level by the trimmer resistor R_1 at D1 cathode, so setting up a control loop.

Switch-mode power supplies have the advantage of isolating the load (e.g. television receiver) circuits from the mains supply at less expense and with greater efficiency than circuits using heavy iron-cored 50-Hz mains transformers. For this reason they are widely used in home-entertainment equipment of all kinds as well as in industrial, instrumentation and test-gear applications.

D.C. to D.C. converters

Akin to switch-mode power supplies but smaller, lighter and generally working from very low voltage supplies are d.c.–d.c. converters. Typical applications are in domestic video recorders to generate medium-voltage auxiliary supplies and memory over-write voltages, and in video cameras to supply the operating voltages required by image-sensing devices.

An example from a home video recorder is given in Fig. 16.10, where transistors TR1 and TR2 form a 30-kHz oscillator with a feedback via C_5. A square-wave voltage is developed across the primary winding of transformer T1 and the secondary winding develops a 4-V heater supply for the fluorescent display panel (f.d.p.) via the voltage-doubler rectifier D1D2C_1C_2. A second voltage-doubler rectifier D3D4C_3C_4 is fed from TR1 collector to provide a supply at -35 V for use as accelerating voltage within the f.d.p. The square waveform at the collector of TR1 is also rectified by D5 and smoothed by C_6 to provide a 37-V tuning-voltage source for the channel-selecting section of the receiver within the video recorder. This 37-V source also provides the collector supply for TR2.

D.C. power converters

Transistors are particularly useful in the construction of d.c. converters, units which can be made remarkably compact and which can convert power from a low-voltage source (e.g. 6 V) to a higher voltage (e.g. 120 V) with an efficiency which can approach 85 per cent and is seldom less than 60 per cent.

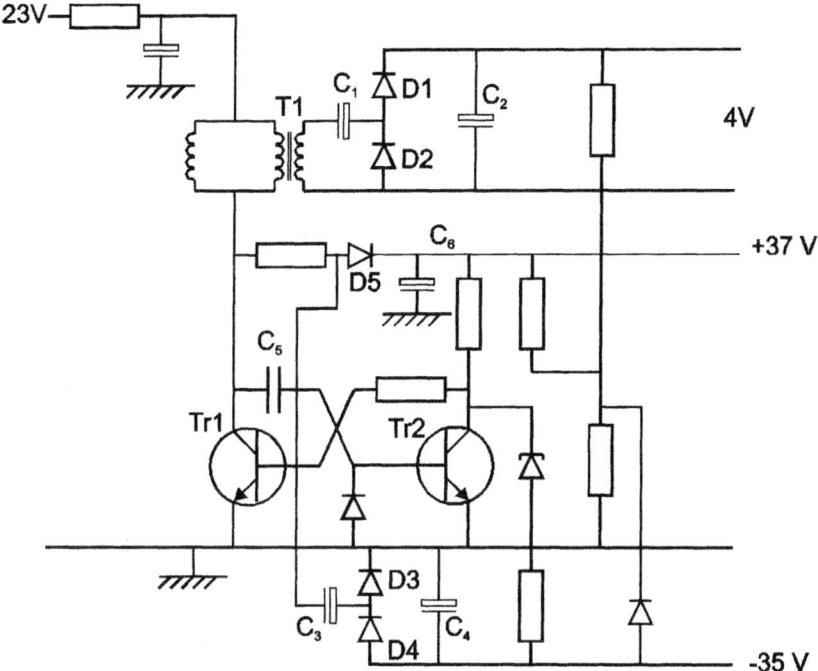

Fig. 16.10. Low-power d.c.–d.c. converter used for auxiliary power supply in a domestic video recorder

The transistor in such a converter is used as a switch which interrupts the d.c. supply from the low-voltage source to produce an alternating current. This is stepped up in voltage by a transformer or resonant circuit to give a high-voltage supply which is rectified and smoothed to obtain the high-voltage output. For high efficiency the power dissipated in the transistor itself must be small. The power dissipated in a transistor is of course given by the product of the collector current and the collector-emitter voltage. The power is therefore small when the collector current is nearly zero, that is to say when the transistor is cut off and also when the collector-emitter voltage is nearly zero, i.e. when the transistor is fully conducting. Thus the design of the d.c. converter must be such that the transistor is always either fully conducting or cut off. This is achieved by using the transistor as an astable relaxation oscillator which generates rectangular waves.

The circuit of one type of d.c. converter is given in Fig. 16.11. The collector and base circuits of the transistor are coupled to give positive feedback and consequent oscillation. Considerable feedback is necessary to drive the transistor hard into conduction and cut-off. The ratio of

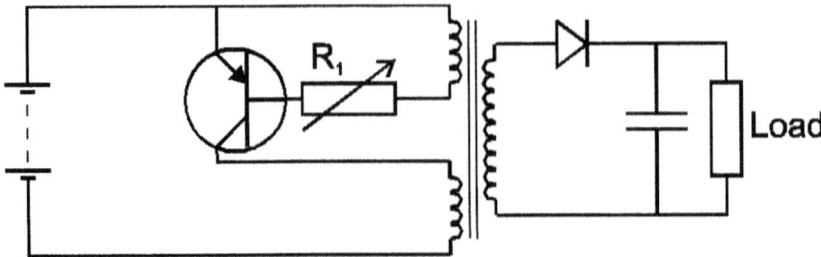

Fig. 16.11. Complete circuit of transistor d.c. converter

the periods of conduction and non-conduction can be controlled by adjustment of the value of the resistor R_1, which determines the base bias of the transistor. Frequently d.c. converters employ two transistors operating in push-pull.

Oscillation frequencies in d.c. converters may lie between 500 Hz and 10 kHz. If a converter is required to be particularly compact the transformer and smoothing capacitor must be small. This is practicable, provided the working frequency is high, and the tendency is therefore to have high working frequencies in compact converters.

Photo-diode

It is shown in Chapter 1 that the current which flows across a reverse-biased pn junction is carried by minority carriers, i.e. by the electrons and holes liberated by breakdown of the convalent bonds of the intrinsic semiconducting material. This current is substantially independent of the reverse-bias voltage, provided this exceeds approximately 1 V, but can be increased by heating the material or by allowing light to fall on it: both give the semiconductor atoms more energy and cause more covalent bonds to break. Where sensitivity to light is undesirable junction diodes and bipolar transistors are sealed in opaque containers: if sensitivity to light is required a transparent container is employed.

A junction diode in a transparent container is known as a photodiode and can be used to indicate the presence of light. An obvious form of circuit is that illustrated in Fig. 16.12. The current which flows in such a circuit when the diode is in darkness is due entirely to thermal dissociation of covalent bonds and increases rapidly as temperature rises. It was known as the *reverse current* in Chapter 1 but in photodiodes is usually known as the *dark current*. The ratio of light to dark current thus decreases as temperature rises. For a germanium photodiode the dark current is typically a few μA and the light current

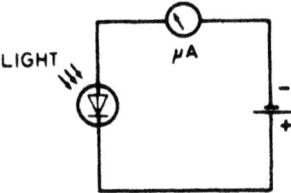

Fig. 16.12. Simple light-meter circuit using a photo-diode

150 μA for an incident illumination of 150 lumens per sq. ft. For a silicon photo-diode the dark current is unlikely to exceed 0.1 μA and the light current is similar in value to that for the germanium photo-diode.

The output power from a photo-diode is limited and amplification is essential if greater power is required, e.g. to operate a milliammeter or a relay. Amplification can be provided by a transistor direct-coupled to the photo-diode as shown in Fig. 16.13. In this circuit two photo-diodes and two transistors are used in a balanced circuit which largely eliminates the effects of temperature changes and gives a meter reading which depends only on the illumination falling on one of the photo-diodes.

To set up the circuit the two photo-diodes are screened from light and the potentiometer is adjusted to give zero meter reading. When one of the photo-diodes is now exposed to light the meter gives an indication proportional to the illumination and the meter can, in fact, be calibrated in terms of illumination.

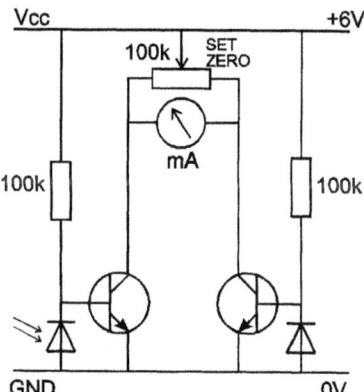

Fig. 16.13. Light-meter circuit employing two photo-diodes and two transistors

Photo-transistors

In a common-emitter amplifier the collector-base junction is, of course, reverse-biased and acts as a photo-diode if illuminated. The reverse current of this junction (i.e. the light current) flows through the forward-biased base-emitter junction and is amplified by normal transistor action, giving a light output from the collector β times that from a simple photo-diode. The dark current is also amplified and, for a germanium photo-transistor, can amount to 0.5 mA but is unlikely to exceed 10 μA for a silicon photo-transistor. The collector current-collector voltage characteristics for a photo-transistor with incident illumination as the parameter are plotted in Fig. 16.14. The curves are similar in shape to those for a common-emitter amplifier. They show that a change in collector current of 0.5 mA can be produced by a change in illumination of 20 lumens per sq. ft.

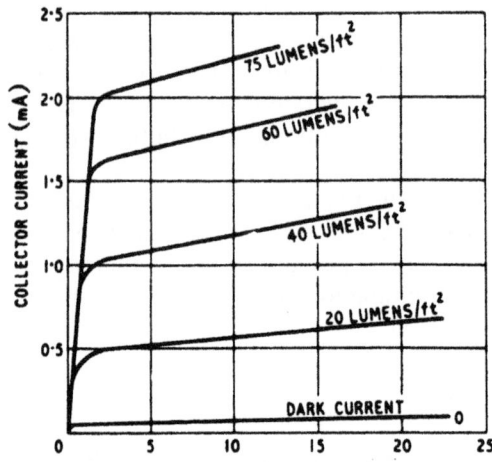

Fig. 16.14. Characteristics of a photo-transistor

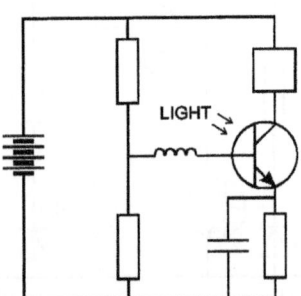

Fig. 16.15. Circuit using a photo-transistor with an interrupted light input

A typical circuit using a photo-transistor is given in Fig. 16.15. This employs the potential-divider method of dark-current stabilisation (see Chapter 6). Such a circuit might be employed as a punched paper tape reader where an electrical response is required from light passing through the holes in the tape.

Optocouplers

Where signals or commands are passed between two isolated electrical circuits an inexpensive and reliable alternative to transformers or relays is the optocoupler, an encapsulated combination of spectrally-matched LED and photo-diode as shown in Fig. 16.16(a). Complete electrical isolation between the two halves of the device permits them to operate in circuits 1 kV or more apart in potential. Amongst their applications are feedback couplers between live and 'earthy' sections of power supplies, e.g. in television receivers.

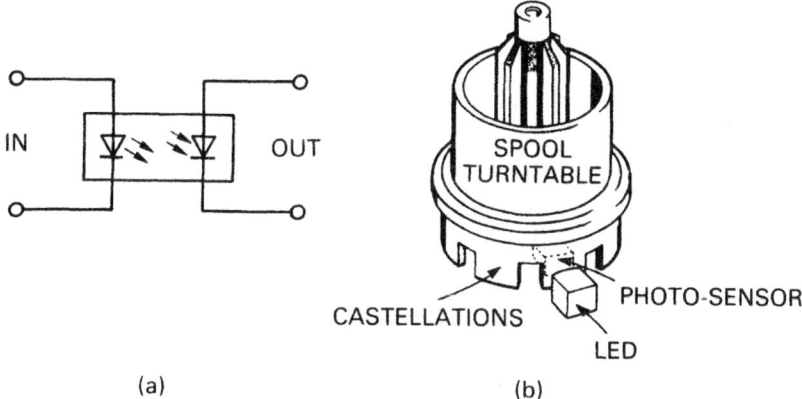

(a)   (b)

Fig. 16.16. (a) Optocoupler circuit symbol. (b) Used as rotation sensor. An alternative version of (b) has both emitter and sensor side by side, scanning alternate black and reflective segments of the underside of the spool turntable

Split optocouplers also find applications as movement sensors; when the beam is broken the photo-diode cuts off to signal the fact. Fig. 16.16(b) shows the use of such an optocoupler to detect rotation in the cassette spool of a video recorder. If the output from the photo-sensor ceases to alternate, the system-control sensor shuts down the deck to prevent tape spillage.

Laser diode in a CD player

The operating principle of a solid-state laser was described in Chapter 1. For use in audio compact disc players the laser diode is a miniature type with an output power of a few mW. As Fig. 1.18 shows, its current must be closely controlled for reliability, longevity and constant light output. In Fig. 16.17 the radiation from the laser diode LD (top left) is monitored by photo-diode PD which controls the current in the laser via amplifiers and regulator TR1. This closed loop ensures constant emission under all circumstances of temperature and ageing. LD and PD share the same encapsulation and have a common cathode connection.

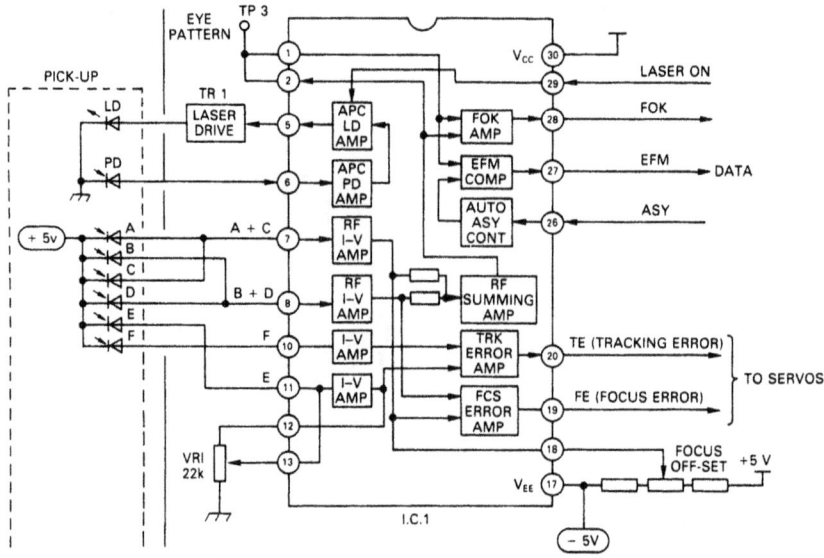

Fig. 16.17. Opto-electronics inside an audio compact disc player: the components inside the dotted box scan across the rotating disc

The other photo-diodes, A–F, are at the opposite end of the optical path, and read the data from the reflective surface of the disc. The bottom-most pair, E and F, are concerned only with providing tracking information to guide the optical assembly across the disc surface, while photo-diodes A–D provide signal data and focus-feedback information.

Unijunction transistor (double-base diode)

This device has a filament of, say, n-type silicon with ohmic contacts at each end and a p-type junction near the centre. If the junction is reverse-biased the filament may have a resistance of, say, $10\,k\Omega$ but this can be substantially reduced by forward biasing the junction. The device makes possible the simple pulse-generating circuit shown in Fig. 16.18.

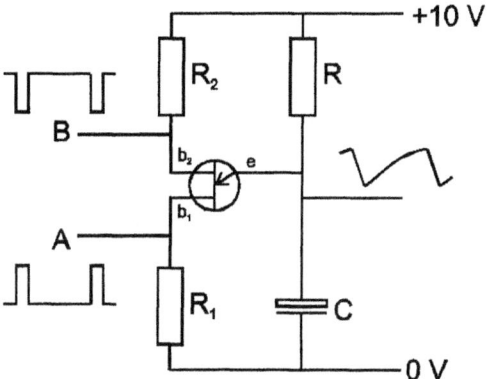

Fig. 16.18. Simple pulse-generating circuit using a unijunction transistor, i.e. double-base diode

For simplicity we will assume that the filament contacts are connected to a 10-V source and that the pn junction is at the centre of the filament. Thus the base at the junction has a bias of +5 V. Initially the capacitor C is uncharged and the emitter potential is 0 V. The pn junction is thus reverse-biased and a steady current of 1 mA flows through the filament via R_1 and R_2 (both assumed small compared with $10\,k\Omega$).

As C charges through R the potential of the emitter moves positively, following an exponential law, until it reaches the value +5.7 V. Any further rise in voltage across C causes the junction to conduct. The charge carriers are chiefly holes which are injected by the emitter p-region across the junction into the filament. Here the holes tend to move towards the most negative part of the filament and in doing so reduce the effective resistance of its lower half. This in turn increases the current crossing the junction and encourages the injection of further holes. Thus a regenerative process is set up which culminates in the rapid discharge of C by the current flowing across the junction and through the lower half of the filament and R_1. The burst of current in R_1 generates a positive-going pulse at terminal A. The momentary reduction in filament resistance due to hole injection causes a transient increase in current in R_2 and thus generates a negative-going pulse at terminal B.

The collapse in voltage across C causes the pn junction to become reverse-biased again and the circuit is back in the original state. The cycle then restarts and continues at a rate dependent on the time constant RC. This simple circuit thus gives pulses of both polarities and an approximation to a sawtooth output.

Inverter using thyristors

An account of the construction and principal properties of thyristors is given in Chapter 2. One application for such devices is in inverters, i.e. equipments which generate alternating supplies from a direct-current

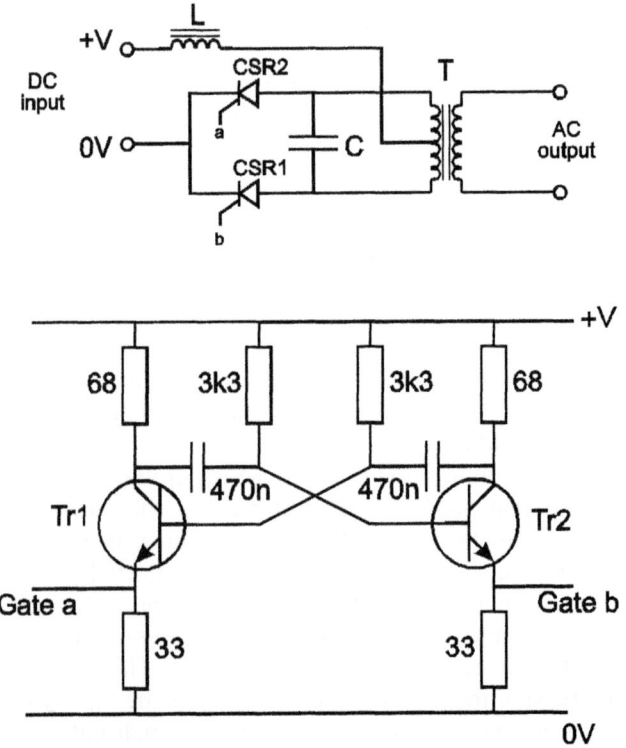

Fig. 16.19. (a) An inverter using thyristors; (b) the associated multivibrator switching circuit. Component values shown are suitable for an operating frequency of about 1 kHz

supply. Inverters enable mains-voltage apparatus, e.g. fluorescent lamps, to be operated where only a low-voltage d.c. supply is available.

The circuit diagram for a push-pull inverter is given in Fig. 16.19(a). This has two thyristors which conduct alternately to interrupt a d.c. supply at a rate determined by an astable multivibrator. The interrupted supply is fed to the primary winding of a step-up transformer T, from the secondary of which the high-voltage a.c. supply is obtained.

The multivibrator is of the collector-coupled type illustrated in Fig. 13.8 and the outputs, taken from resistors in the emitter circuits of TR1 and TR2, are directly applied to the gates of the thyristors; Fig. 16.19(b). Consider conditions in the circuit during the period when CSR1 is on. The voltage across CSR1 is negligibly low and the full battery voltage V appears across the lower half of T primary winding. By auto-transformer action an equal voltage appears across the upper half of this winding and thus the capacitor C is charged to $2V$. When CSR2 is abruptly switched on by the multivibrator, the voltage across it suddenly falls to zero. The voltage across a capacitor cannot change instantaneously and the effect is that the anode of CSR1 is immediately driven to $-2V$ volts, so switching the thyristor off. Thus the states of CSR1 and CSR2 are now interchanged and the voltage across T primary winding therefore reverses. C discharges and is recharged to $2V$ in the opposite direction. Half a cycle later CSR1 is turned on by the multivibrator, its voltage falls to zero and the capacitor C ensures that CSR2 is turned off. Thus the cycle continues. There is a brief period in each cycle when both thyristors are conducting and the choke L is included to limit the current taken from the battery during this period.

The multivibrator is usually designed to run at a frequency around 1 kHz, thus enabling much smaller transformers to be used than are necessary at 50 Hz. For operating fluorescent lamps or other mains equipment a sinusoidal output is desirable: capacitance C and the primary inductance of T are thus chosen to resonate at the operating frequency. If it is intended to rectify the inverter output, a square-wave output is preferable and can be obtained by choosing C and the primary inductance of T so as to avoid resonance.

Thyristor control for small electric motors

Thyristors can be used to control small electric motors such as those used in hand drills and food mixers where there is a need for several operating speeds. Using a simple circuit a thyristor can determine the motor speed by control of the fraction of each cycle during which mains current is allowed to flow (known as the *conduction angle*) and the speed can be maintained, in spite of varying mains voltage or changing

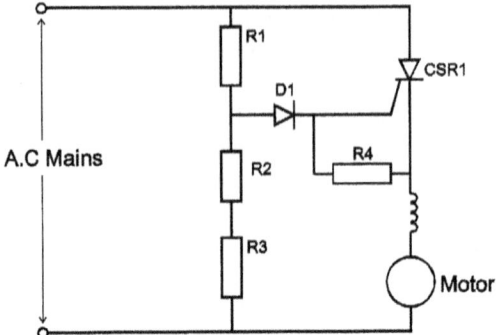

Fig. 16.20. Basic circuit for thyristor control of electric-motor speed

mechanical load on the motor, by feedback from the back-e.m.f. from the motor.

One possible circuit is shown in Fig. 16.20. When the thyristor anode is negative with respect to the cathode, the thyristor is non-conductive: thus for one-half of each mains cycle the motor receives no power. Nevertheless the motor is rotating as a result of the power received during the other half-cycle and generates a back-e.m.f. proportional to speed (and residual flux) which biases the thyristor cathode positively. When the thyristor anode is positive with respect to the cathode, the thyristor conducts if the gate is suitably biased: thus for this half of each mains cycle the motor receives power controlled by the gate circuit. The diode D1 conducts positive-going trigger voltages to the gate but when the thyristor conducts, the thyristor cathode potential takes up the anode value, so reverse-biasing D1 and isolating the thyristor from the trigger circuit. The gate voltage must exceed the cathode bias by a fixed amount to turn the thyristor on and the gate voltage is determined by the setting of R_2. The more positive the gate is made, the larger is the conduction angle and the higher the motor speed.

This simple circuit has disadvantages: the chief are that the range of conduction angle achievable by adjustment of R_2 is limited and that dissipation in the potential divider is high.

A better circuit is illustrated in Fig. 16.21. The gate circuit operates as a diode detector (see page 230). On positive half-cycles D1 conducts and charges C_1. On negative half-cycles D1 is non-conductive and C_1 discharges through $(R_2 + R_3)$. On the next positive half-cycle D1 conducts again and the charge lost from C_1 is restored. The extent of the discharge of C_1 depends on the value of $(R_2 + R_3)$ and can be controlled over a wide range by adjustment of R_2. The smaller R_2 is made, the more complete is the discharge of C_1 during negative half-cycles and the longer is the charging period during positive half-cycles. In fact the

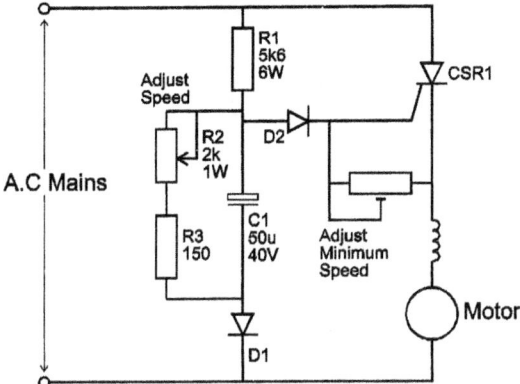

Fig. 16.21. Practical circuit for thyristor control of motor speed

charging of C_1 begins early in the positive half-cycle and by adjustment of R_2 can be made to last from a small fraction to three-quarters of the half-cycle. The charging current flows through R_1 and generates across it a negative-going pulse which reverse-biases D2, isolating the thyristor gate and preventing the thyristor from conducting. Thus the thyristor becomes conductive as the charging current ceases and is cut off when the anode voltage swings to zero at the end of the positive half-cycle. R_2 controls the duration of the charging process and hence the conduction angle of the thyristor.

Thyristor as crowbar

An effective (if crude) way to protect a circuit from voltage overload and consequent damage is to short-circuit its power supply line. An example of this 'crowbar' technique is given in Fig. 16.22(a), where Zener diode ZD monitors the 125-V line produced by a switch-mode power supply unit. If, due to a regulator fault, the line voltage rises to 130 V, the Zener diode breaks down and passes current into the thyristor gate. The resulting short-circuit through the thyristor shuts down the p.s.u., removing the risk of damage. The protection circuit is reset by switching the power off and on again.

There are two-terminal devices made to do this job, consisting internally of a thyristor-Zener diode or a form of diac. Their circuit symbol is shown in Fig. 16.22(b).

Thyristor field timebase

The field timebase section of a television receiver or monitor is required to produce a linear sawtooth current in the scanning yoke. In many

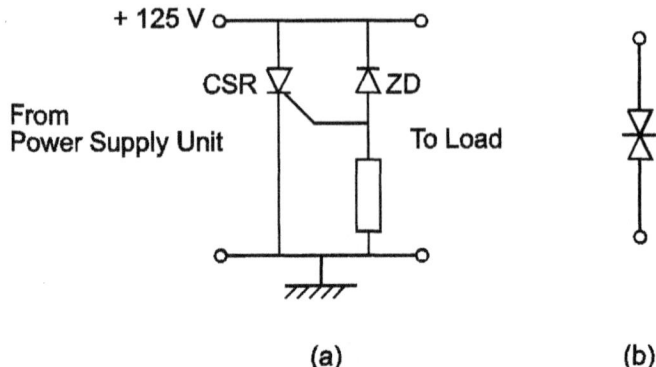

(a)   (b)

Fig. 16.22. (a) 'Crowbar' protection circuit with thyristor and Zener diode. (b) Breakover diode for crowbar protection

designs a complementary Class-B output stage is used for this purpose; the circuits are similar to those described in Chapter 8.

An alternative and more efficient circuit using a thyristor is depicted in block form in Fig. 16.23. The vertical scan coils have one end (A) permanently connected to a source of +21 V d.c., so that by varying the charge in C_1 (point B) between, say, 10 V and 30 V, current can be made to flow in either direction in the scan coils. The necessary energy comes from coil X–Y, which is a secondary winding on the line output transformer, so phased that during each line flyback point X goes 190 V positive of point Y. The active element is thyristor TH1, connected between point Y and ground. If TH1 remains permanently off, diode D1 rectifies the flyback pulses to charge C_1 to the peak voltage of +190 V. If TH1 stays permanently on, point Y is grounded, with the result that the pulses at point X position themselves symmetrically about ground potential: the charge on C_1 is zero under these circumstances.

If the conduction period of TH1 is varied throughout the field period, the charge on C_1 varies likewise and so does the scan-coil current. At the beginning of each field scan the thyristor is pulsed on relatively late in each line scan period so that it has a short duty cycle before being cut off at line flyback by the negative pulse at point Y. As field scan continues, the thyristor trigger point is progressively advanced so as to reduce the charge on C_1 linearly; at the end of the field scan it has dropped to around 10 V and maximum current flows in the scan coils. Now the thyristor gate pulses are suddenly retarded so that C_1 charge reverts to a high voltage, a large reverse current flows in the scan coils and field flyback takes place. As Fig. 16.23 shows, the thyristor trigger pulses are generated inside a purpose-designed i.c. which also generates line-scan drive pulses.

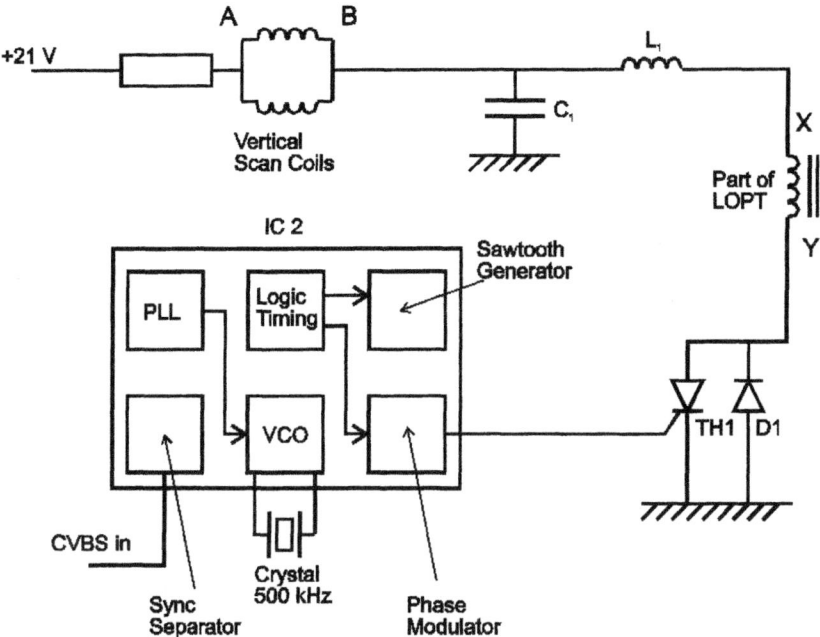

Fig. 16.23. Principle of thyristor-controlled field-scan current generator

Triac control of a.c. energy

The advantage of a bidirectional thyristor (triac) is that it works on both half-cycles of an alternating waveform so that all the energy of, for example, the mains supply is available to the load if required. Triacs can be used, then, as contactless switches for on-off control of inductive and resistive loads and, by varying the phase of their triggering pulses, as a continuously variable motor-speed or lamp-dimmer control. The gate-drive circuit is the same in principle as those given in Figs. 16.20 and 16.21.

Resistance-to-voltage converter

The principle of constant-current generators was described in Chapter 14 and Fig. 14.6. Another of their many possible applications is given in Fig. 16.24, where the collector load is a thermistor, a component with a resistance proportional to temperature. With a constant current flowing

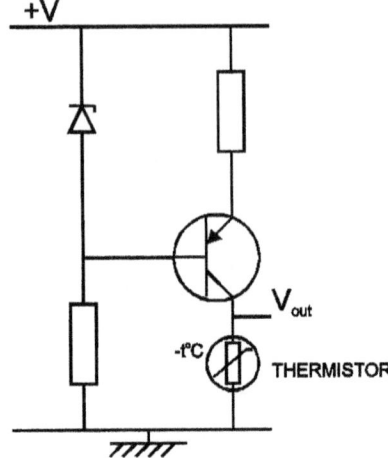

Fig. 16.24. Constant-current source used in thermometer circuit

in it, the thermistor generates a voltage across it proportional to its resistance and hence to temperature, facilitating direct readout of temperature by a voltmeter.

Motor-drive circuits

A common application of bipolar transistors is in control of small d.c. electric motors, usually in closed-loop servo systems. An example is given in Fig. 16.25. OM1 is a miniature brush-type motor driving a long spiral worm mounted along the rear of a disc-reproducing deck. It drives a tangential pickup arm across the audio disc surface and must be capable of moving it at varying speeds in either direction.

The drive circuit is in the form of a bridge made up of two complementary pairs of medium-power silicon transistors, TR3TR4 and TR5TR6, arranged as emitter followers. Each pair is controlled by two further complementary emitter followers, TR1TR2 and TR7TR8. The voltages at the bases of TR1 and TR8 are driven in opposite directions by the control signals. The polarity of the drive signals determines the direction of rotation of the motor whilst their amplitude controls the motor speed and hence the lateral speed of movement of the audio pickup arm.

There is no standing current through the output transistor pairs in Fig. 16.25 and as a result there is a little lag in operation of the motor, not detrimental in this application. Fig. 16.26 shows a more precise motor-drive circuit used in the auto-focus section of a camcorder. The

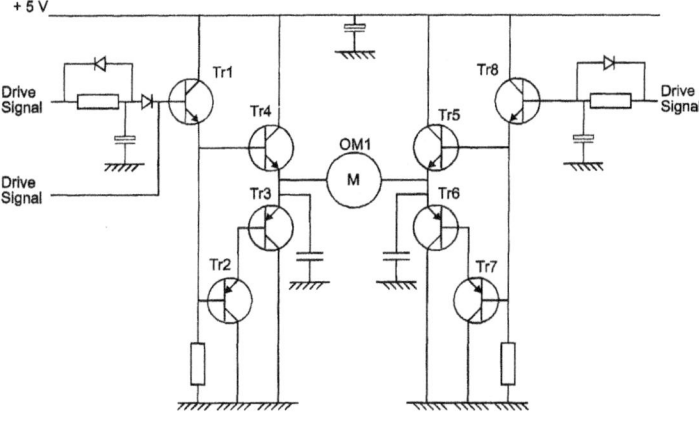

Fig. 16.25. Motor-drive circuit using a transistor bridge

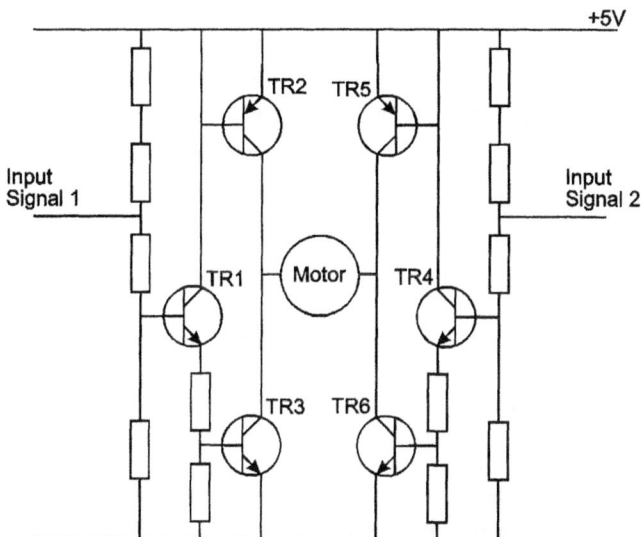

Fig. 16.26. Drive circuit for a camcorder auto-focus motor: TR1 to TR6 control speed and direction of the focus lens

drive motor is smaller than a thimble and rotates the focus barrel of the lens via a slipping clutch. Again the bridge configuration is used, this time with each of the complementary pairs TR2/TR3 and TR5/TR6 having a common collector connection, the emitters being connected directly across a 5-V supply derived from a stabiliser i.c. The pnp transistor of each pair is driven directly by the auto-focus control signal whilst the npn 'bottom legs' TR3 and TR6 are fed from driver transistors TR1 and TR4 respectively. Battery power is at a premium in camcorders and no power is therefore wasted in quiescent current in the drive transistors: TR2 and TR5 are biased just into conduction but TR3 and TR6 are held just below the threshold of conduction whenever the motor is stationary. The complementary drive-control voltages are derived from focus-feedback signals from a piezo-sensor or a through-the-lens (TTL) sharpness sensor circuit so that the motor and its drive circuit form part of a servo-loop embracing the subject or object at which the camera is pointed.

Other motor-drive transistor circuits involve simple speed-stabiliser configurations in which a single transistor controls the armature current in a small brush motor with back-e.m.f. control of speed to cope with varying mechanical loads. These are often used for capstan drive in cassette tape recorders.

Television picture enhancement

Fig. 16.27(a) illustrates one circuit for vertical contour enhancement of a television picture; the associated waveforms are given at (b). The input video signal (1) at the left is buffered in npn emitter follower TR1 and passes through a 100-ns delay line DL1, emerging as waveform (2). From TR2, a second emitter follower (pnp), it enters a second 100-ns delay line DL2, arriving at TR3 base as waveform (3). TR3 is a third emitter follower (npn) and to the doubly delayed signal at its emitter a sample of the original input waveform (1) is added via R_3. The result, at TR4 base, is waveform (4), a combination of waveforms (1) and (3). In TR4 (a common-emitter amplifier) waveform (4) is subtracted from waveform (2), which arrives at its emitter via R_7. TR4 collector waveform thus has the form shown at (5), which is added at TR5 base, to waveform (2) derived from TR2 emitter via R_6. Thus the output from TR5 (the final emitter follower) has the waveform shown at (6). It has overshoots and negative preshoots, the subjective effect of which is to sharpen or crispen the displayed picture. Circuits of this type are much used in television and video equipment, particularly in the replay circuits of domestic video recorders where they help to compensate for the slow rise and fall times due to the restricted bandwidth of the reproduced video signal.

Recording head drive

In a video recorder the spinning video heads are driven, via a rotary transformer, with a frequency-modulated luminance writing signal to

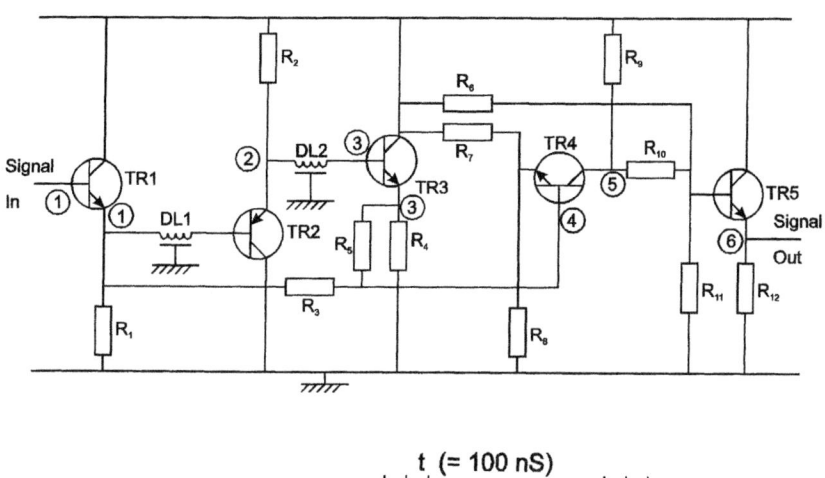

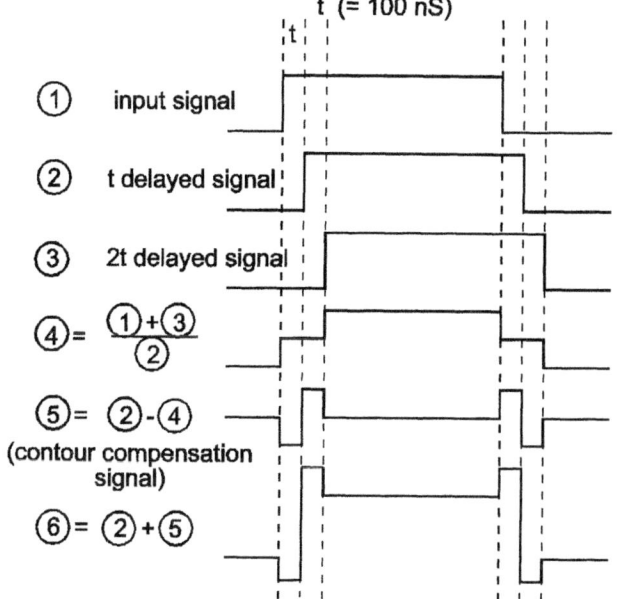

Fig. 16.27. (a) Vertical contour enhancer using two delay lines; (b) circuit waveforms and timings

which is added a relatively low-frequency chrominance signal. In Fig. 16.28 the luminance carrier enters as REC-Y to preset potentiometer R_1, from which the correct level of signal is tapped off for passage through a filter to the base of amplifier TR1. Here the chrominance signal REC-C (applied to the emitter) is superimposed upon the luminance signal. The response of TR1 is tailored by

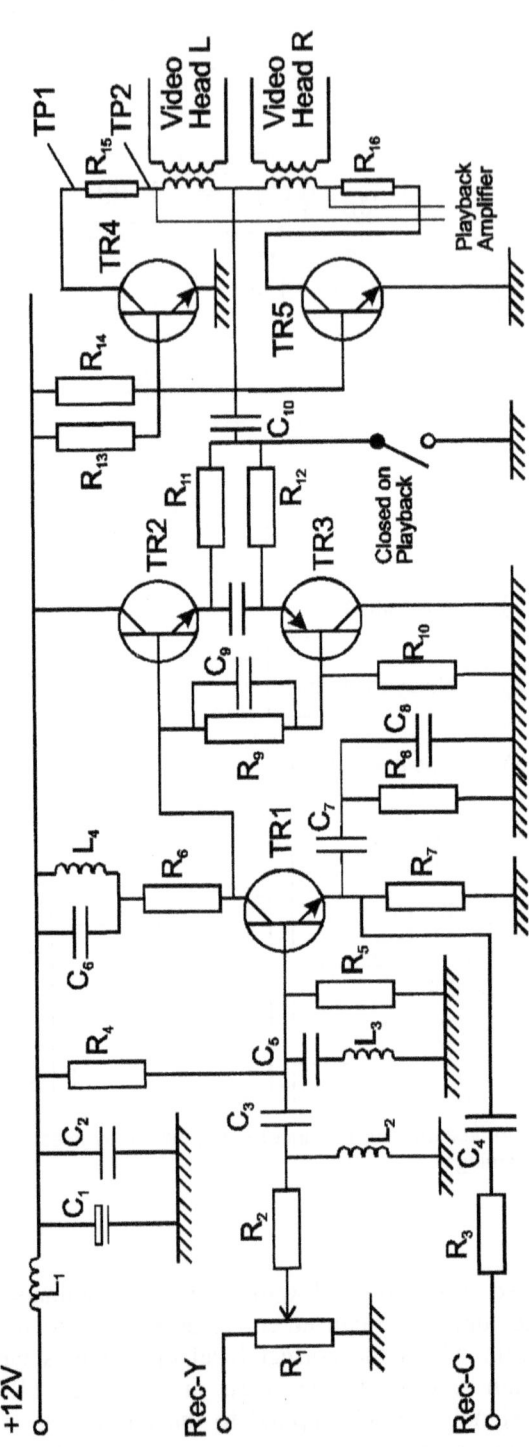

Fig. 16.28. Recording-head drive circuit for a video recorder

frequency-discriminating networks in emitter and collector circuits. The output is passed to complementary output pair TR2, TR3, the quiescent current being set by R_6, R_9 and R_{10}. The output passes via C_{10} into the recording heads, which are effectively paralleled during recording by switches TR4 and TR5, both turned on by the + 12-V line applied to the bases. The recording current passes through R_{15} and R_{16}, developing a proportional voltage across them which enables the recording current to be set by monitoring the voltage between test points TP1 and TP2 with an oscilloscope or millivoltmeter.

Transistors as switches

Transistors are widely used in electronic equipment as switches, examples of which are TR4 and TR5 in Fig. 16.28. Very common are sub-miniature devices with built-in resistors for direct connection to, for example, a 5-V switching line from a microprocessor control i.c. Fig. 16.29 gives a typical application using both inverting and direct switch transistors which close on high or low inputs respectively. They are used for function and route switching through the video recorder, of which this diagram shows a small section of the systems-control circuit.

Video clamp

Analogue television transmissions from satellites have an energy dispersal waveform superimposed on the baseband video signal to prevent spot frequencies appearing in the f.m. carrier signal. It takes the form of a triangular waveform at 25 Hz, half the field frequency. After demodulation it is removed to prevent flicker on the picture. The action is carried out by a clamping circuit as shown in Fig. 16.30. Diagram (a) depicts a peak clamp or d.c. restorer, in which the negative peaks of the incoming signal are held at a fixed potential by the action of clamp diode D and a source of fixed negative voltage –E. Fig. 16.30(b) illustrates a driven clamp in which a pulse, timed to coincide with some particular level in the video waveform (e.g. front porch, black level or sync tip), restores the charge on the right-hand plate of coupling capacitor C to a reference potential once per television line. These clamp circuits are useful wherever the d.c. level of a video waveform is liable to vary: they prevent changes in black level and brightness in reproduced pictures.

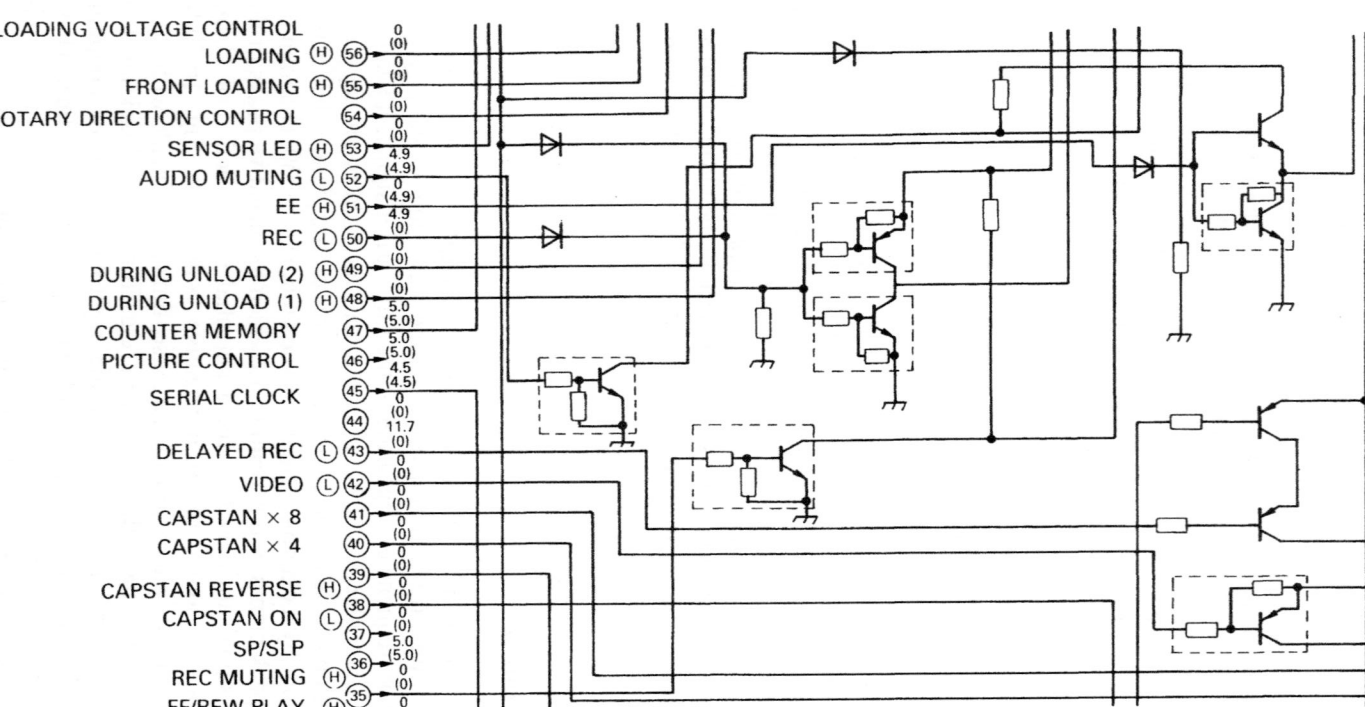

Fig. 16.29. System-switching by transistors. The circled numbers at left are the output pins of the control microprocessor

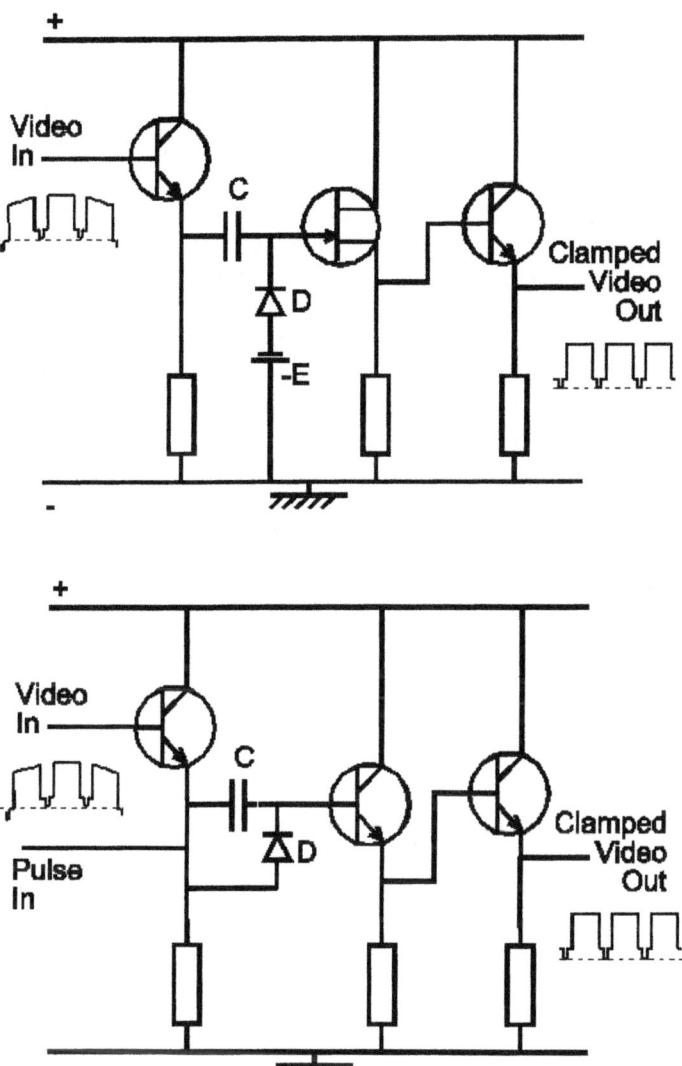

Fig. 16.30. Video clamping circuits: (a) d.c. restorer; (b) driven clamp

U.H.F. modulator

Video recorders, computers and games for use with home television receivers are often provided with r.f. modulators, the output of which can be directly connected to the aerial socket of the receiver. Their output power is only a few microwatts and the carrier frequency is adjustable over a small range centred on u.h.f. channel 36 (591 MHz). A circuit diagram of such a modulator is given in Fig. 16.31.

TR1 is a common-base oscillator, base-biased by R_{50} and R_{51}, the collector supply being via feed choke L_{50}. The resonant circuit is formed by stripline L_{51}, series-resonant with C_{51} and VC, a carrier-frequency preset trimmer. Feedback coupling is by C_{52} to the emitter to form a stable low-power oscillator. The output is taken from the adjacent stripline via pins 1, 2 and 3 on the screened box housing the oscillator. Pin 2 is effectively grounded and provides a bias-injection point (from RV1) for amplitude modulator pin diodes D2 and D3, through which video signal flows via R_8 from the amplifier i.c. IC1.*
The modulated u.h.f. carrier is passed via C_{17} to load resistor R_9 and to the aerial output circuit. Broadcast television transmitters produce a vestigial sideband signal but this miniature modulator generates a double-sideband signal, not detrimental in this application because the i.f. filter of the television receiver rejects the lower sideband.

The method of f.m. sound modulation is unlike that used in a conventional television transmitter. Here the audio signal is fre-quency-modulated onto a 6-MHz carrier generated by tuned circuit $T_1 C_{12}$ and the resulting carrier is simply added to the baseband video signal via R_7, C_{11} and L_1. It is seen by the modulator merely as a component of the video signal for which sidebands are generated, a characteristic of an a.m. modulating system. The sideband is spaced 6 MHz from the vision carrier and appears to the television receiver as a separate carrier: it is processed as such.

IC1 in Fig. 16.31 also contains a simple pattern generator con-trolled by the TSG switch connected to pin 9. An oscillator, based on the crystal at pin 11, generates simple line sync pulses and a white bar video pulse for injection into the r.f. modulator. This provides a clearly identifiable pattern for easy tuning of a television receiver to the output channel of the modulator. Supply voltage for the circuit is provided via R_1, stabilised by Zener diode D1 and decoupled by C_{10}.

---

* Reference is made to the characteristics of pin diodes in Chapter 1.

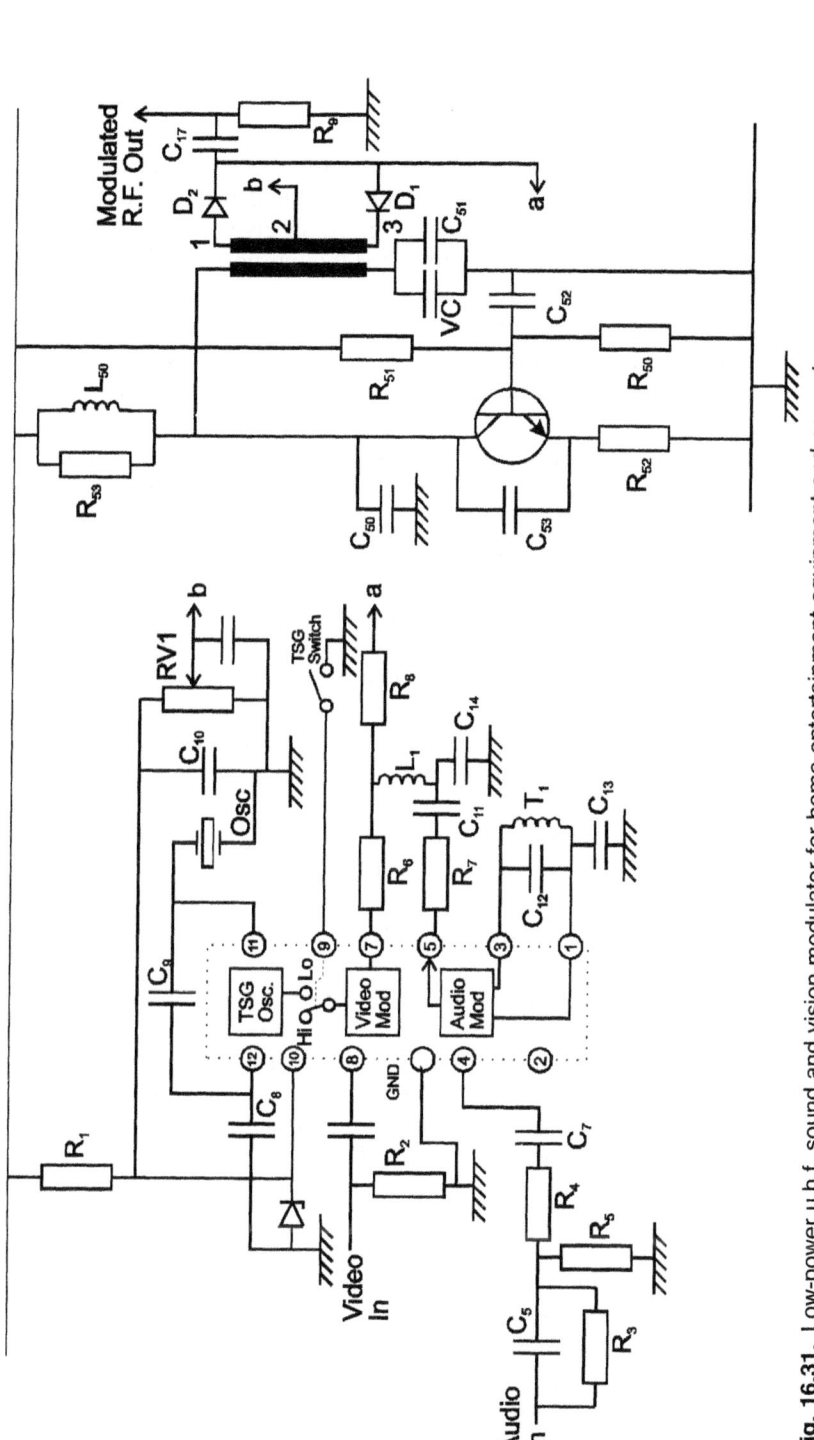

Fig. 16.31. Low-power u.h.f. sound and vision modulator for home entertainment equipment and computers

Capacitance-diode a.f.c. circuit

As mentioned in Chapter 1 the capacitance of a reverse-biased junction diode varies with the bias voltage. Such a diode can therefore be used for automatic frequency control and Fig. 16.32 gives a circuit diagram which can be used in an f.m. receiver for this purpose. Not all junction diodes are suitable for this application: for some types the damping due to the resistive component of the diode impedance may be sufficient to reduce the oscillation amplitude to a low value or even to prevent oscillation altogether. Diodes with very low damping have been developed for use in a.f.c. circuits.

$L_1 C_1$ is the oscillator tuned circuit and the junction diode is connected across the circuit via the isolating capacitors C_2 and C_3. The diode is reverse-biased from the supply by the potential divider $R_1 R_2$ and provided that R_1 is reasonably high in value, say, more than $40 \, \text{k}\Omega$, the damping of the oscillator circuit by this resistor should not seriously reduce the oscillation amplitude. The capacitance of the diode is effectively in parallel with C_1 and alteration in diode bias causes an alteration in oscillator frequency. To obtain a.f.c. the diode bias must be controlled automatically by the degree of mistuning and this can be achieved by returning R_2 to the d.c. output of a suitable f.m. detector, i.e. to its output *before* the capacitor which couples the detector output to the following a.f. amplifier.

The a.f.c. voltage required from the detector is one which represents by its polarity the direction of any mistuning and by its magnitude the extent of the mistuning, the voltage being zero at the correct tuning point. The most popular form of f.m. detector is the ratio detector described on page 238 but the circuit diagram of Fig. 12.10 must be modified to give an output of this form. The difficulty is caused by the earthing of one end of the long-time-constant combination $R_1 C_3$. As tuning is swept through the correct tuning point the voltage across $R_1 C_3$ rises to a maximum and this voltage also appears in the detector output, masking the wanted a.f.c. voltage. However, the voltage across $R_1 C_3$ can be eliminated from the detector output by transferring the earth connection to the midpoint of the resistor of the long-time-constant combination as shown in Fig. 16.32, in which the equal resistors R_3 and R_4 replace R_1 of Fig. 12.10. This change has no effect on the a.f. output of the detector but it turns the detector into a balanced circuit, the two diodes acting as equal impedances connected in series across C_4. The a.f.c. voltage must not be affected by a.f. signals in the detector output and these are therefore prevented from reaching the diode by the capacitor C_3, which forms with R_2 a potential divider which considerably attenuates all audio frequency signals.

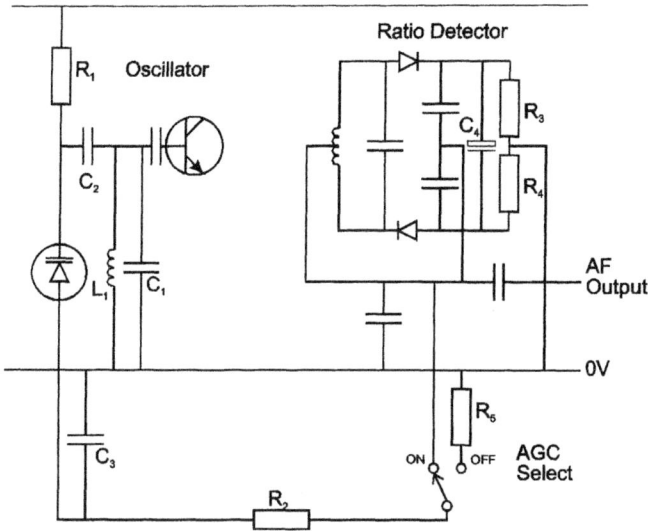

Fig. 16.32. Circuit illustrating the use of a capacitance diode to give a.f.c.

A.f.c. circuits of this type can be extremely effective, reducing mistuning effects by a factor of as much as 10:1. Manual tuning can be very difficult with a.f.c. and it is desirable to have some means of switching a.f.c. off whilst tuning is being carried out. As soon as the wanted signal is tuned in, a.f.c. is switched on to minimise subsequent tuning drift. Fig. 16.32 indicates one method of switching a.f.c. off. The resistor R_5 is approximately equal to the d.c. resistance of the ratio detector. Such a resistor is necessary to enable the a.f.c. to be switched off and on when the receiver is accurately in tune, without alteration of the bias across the capacitance diode.

Attenuator using an fet

It was pointed out on page 43 that a fet can be used as a voltage-controlled resistor with an enormous range of variation. The resistance is that of the drain-source path and the control voltage is applied to the gate. Fig. 16.33 gives the circuit diagram of an a.f. or r.f. attenuator based on this principle. The jfet forms with R_1 a potential divider, the attenuation of which can be controlled (remotely if desired) by adjustment of the negative voltage applied to the gate. $R_2 C_1$ are decoupling components. Because of the very high input resistance of the jfet, negligible power is needed to adjust the attenuator. The transistor behaves as a near-linear resistor provided the voltages applied between

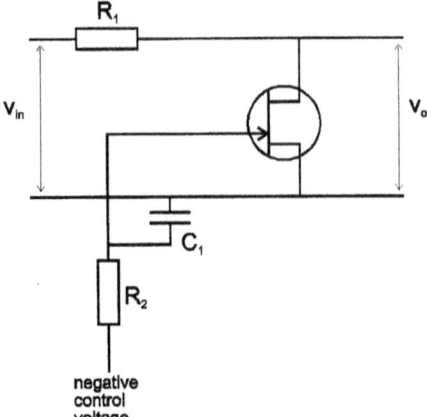

Fig. 16.33. A fet used as a variable resistance in a voltage-controlled attenuator

drain and source are small compared with 1 V. Thus the attenuator is suitable only for use with small-amplitude signals.

Bias/erase oscillator

In magnetic tape recorders a medium-power oscillator is required to provide bias for the signal current in the recording head and to energise the erase head which, in a video recorder, must cover the entire

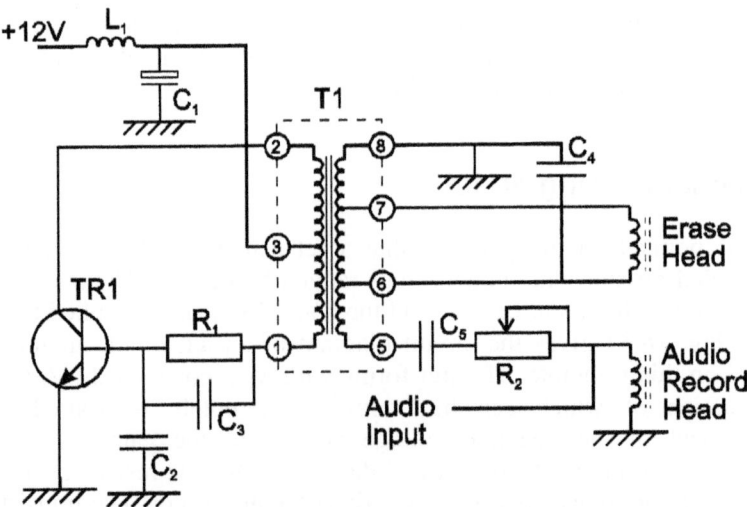

Fig. 16.34. Bias/erase oscillator used in a home video recorder deck

12.5-mm tape width with sufficient flux to wipe out completely any previous recording. Fig. 16.34 shows an arrangement commonly used in domestic video recorders. TR1 is a Hartley oscillator supplied with collector current via decoupling components L_1 and C_1 which feed tapping point 3 of the transformer T1. Positive feedback from winding 3–1 is applied to TR1 base via R_1 and C_3 to maintain oscillation at a frequency of about 60 kHz, determined mainly by the tuning capacitor C_4 across the secondary winding of the transformer. A sinusoidal voltage of 60 V peak-to-peak is generated for application to the ferrite-cored erase head connected between pins 6 and 7 of T1, while some 80 V peak-to-peak appears at pin 5 for application via C_5 and bias-level preset resistor R_2 to the audio recording head. The 12-V supply to L_1 is available only in the recording mode.

The manufacture of transistors and integrated circuits

This appendix surveys the various methods which have been used to manufacture transistors and semiconductor diodes, beginning with the 'dope-growing' and alloy-junction process used for the first germanium transistors and ending with the planar process now used to manufacture silicon transistors and i.cs.

Preparation of germanium for transistor manufacture

Germanium is obtained as a byproduct of metal refining, e.g. zinc-refining, in the U.S.A. and can also be obtained from the flue dust of certain coals in the U.K. After extraction by chemical methods, the germanium is usually marketed in the form of the dioxide GeO_2, a white powder.

The first step in the production of germanium for semiconductor manufacture is the reduction of the oxide to the element. This is achieved by heating the oxide to 650°C in a stream of pure hydrogen. The germanium powder so obtained is melted in an inert atmosphere and then cast into bars.

The bars are not pure enough for use in semiconductors: moreover, their crystalline form is unlikely to be suitable. Further purification is therefore carried out by the process known as *zone refining*. This relies for its success on the fact that, at the melting point of germanium, most of the impurities are more soluble in the liquid than in the solid form of the element. Thus, if a short length of a germanium bar is melted and if the molten portion is caused to move along the bar, say, from right to left, the impurities tend to follow the movement and concentrate at the left-hand end of the bar. The localised heating is usually carried out by r.f. induction, the germanium bar being supported in an inert atmosphere in a graphite boat which is slowly moved along a silica tube contained within the loops of the r.f. supply. By repeating the zone-refining process a number of times, the purity of the bar (except for the left-hand end, of course) can be raised to the degree required.

The single crystals required can be grown by dipping a small seed crystal into a bath of molten zone-refined germanium and slowly withdrawing, i.e. pulling the seed as the crystal grows and cools. Controlled amounts of p or n impurities can be added to the molten material to give the type of germanium required.

Preparation of silicon for transistor manufacture

Silicon forms approximately 25 per cent of the earth's crust but it is difficult to extract the element in a form pure enough for use in transistors, primarily because silicon melts at a very high temperature (1,400°C) and it is very reactive when molten, attacking most crucible materials. In particular, the removal of the last traces of boron is most troublesome.

Silicon occurs widely as the dioxide (sand) and the first stage in the extraction is the reduction of this oxide in an arc between carbon electrodes. The low-grade silicon so obtained is purified by a number of chemical processes and is then subjected to a zone-refining process similar to that used for germanium. To avoid contamination from containing vessels, however, the silicon, in the form of a bar, is supported vertically by its ends whilst heated locally by r.f. induction – a technique known as *vertical* or *floating zone refining*. Finally crystals of silicon can be obtained by the method of pulling or growing.

Grown transistors

The first method of manufacturing transistors was the so-called *dope-growing* method described by Shockley in 1951. In this method the molten semiconductor is first treated with sufficient n-type impurity to give the required collector resistivity (say, 1 to 2 Ω-cm) and an n-type ingot of the required crystalline form and resistivity is obtained by pulling or growing as described above. After a suitable length of crystal has been grown, a pellet of p-type impurity is added to the molten material and this is of sufficient mass to neutralise the n-type impurity and to give the p-type resistivity required in the base region (say, 1 to 2 Ω-cm). The ingot is now grown only a very short distance (equal to the thickness of base layer required – say, 0.001 in) and a second pellet, this time of n-type impurity, is added to the molten material to give the required resistivity of emitter region – say, 0.01 Ω-cm or less. After a

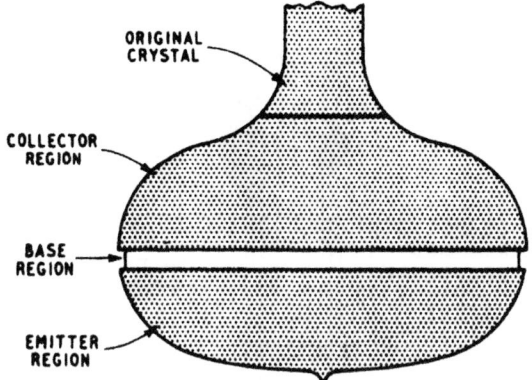

Fig. A.1. A grown crystal

suitable length of the emitter region has been grown the ingot is withdrawn from the molten material.

The ingot (Fig. A.1) is now cut in directions parallel to the direction of growth to produce transistors approximately 0.02 in square and 0.2 in long. Several hundred transistors can be obtained from one ingot. The emitter and collector ends can be distinguished by etching the transistors with a material which affects these ends at different rates. Connecting leads can then be soldered to the emitter and collector regions but the connection to the thin base region is difficult to make because, of course, this region cannot be seen. It can, however, be detected by exploring the surface of the transistor with a fine wire contact which is included in a circuit designed to give a signal when the base region is touched. An electric discharge can then be used to secure the wire to the base region.

Alloy-junction transistors

The starting point of this method of manufacture of germanium transistors is a wafer of crystalline semiconductor, say, n-type, of $5\,\Omega$-cm resistivity. The wafers are approximately 0.1 in square and 0.003 in thick. A pellet of a group-III element such as indium is placed at the centre of each face of the wafer which is then heated to a few hundred degrees Centigrade, well above the melting point of indium (155°C) but well below that of germanium (940°C). The indium melts and continues to dissolve germanium until a saturated solution is obtained. The wafer is now allowed to cool slowly and the dissolved germanium crystallises out. The recrystallised germanium retains some

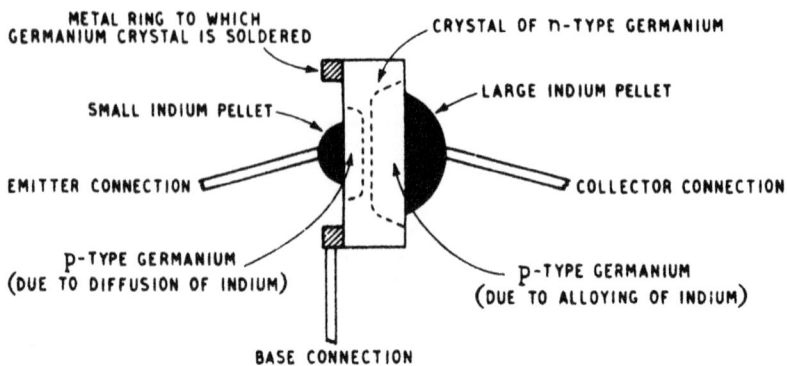

Fig. A.2. Construction of a germanium alloy-junction pnp transistor

indium and thus forms a region of p-type germanium on either face of the wafer as shown in Fig. A.2. These regions are separated by a region with the n-type conductivity of the original material, thus giving a pnp transistor. The final diameters of the indium regions are commonly of the order of 0.015 in (emitter) and 0.03 in (collector). Connections are soldered to both regions and the base connection is often made to a metal ring soldered to the wafer and closely surrounding the emitter region. After cleaning the assembly is hermetically sealed in a light-proof container.

A similar technique can be used to manufacture npn germanium transistors: a wafer of p-type germanium is used and pellets of a group-V element such as arsenic or antimony, carried in a neutral (group-IV) element such as lead, are alloyed to it.

Very large numbers of pnp germanium transistors were manufactured by this method and they can operate up to frequencies of the order of 5 to 7 MHz but to obtain reasonably consistent base thicknesses close control is required over the thickness of the wafers and the time and temperature of the processing.

Silicon transistors can also be made by this method, the usual alloying metal being aluminium but the very different coefficients of expansion of aluminium and silicon caused difficulties.

The principal factor limiting the performance of a transistor at high frequencies is the time taken for charge carriers to cross the base region and to obtain a good performance this time must be reduced to a minimum. The obvious way to improve the high-frequency performance is to reduce the thickness of the base region even though this may limit the maximum permitted collector voltage to a few volts. This is the technique employed in the surface-barrier transistor pioneered by Philco in 1953.

Surface-barrier transistor

The starting point of the process is a wafer of n-type germanium to which a ring is soldered to form the base connection. Both faces of the wafer are then etched electrolytically by jets of solution of precise cross-section which are directed against the surfaces, current being passed between the solution and the wafer so as to cause the germanium to dissolve. Etching is stopped when the required base thickness is reached: this can be determined by the transparency of the base region to light. The direction of the current is now reversed and the metal in the solution (commonly indium) is deposited on the faces of the wafer to form emitter and collector regions to which connections are made. This process is well suited to automatic operation and large numbers of transistors were produced. They were not so robust as those produced by other processes but operated at frequencies up to 50 MHz. The robustness of the transistors can be improved by lightly alloying the layers of indium to the base region by heating: the resulting transistors are termed *micro-alloy* types.

Another method of reducing transit time in a transistor is to vary the impurity concentration in the base region so as to produce an electric field which aids the passage of charge carriers across the region. To give this effect the impurity concentration must be a maximum near the emitter junction and a minimum near the collector junction; an exponential distribution or one approaching exponential form is desirable.

The technique of solid-state diffusion may be employed to produce such a graded base region. One method is to expose the semiconductor material to a vapour of the desired impurity in a furnace. This causes impurity atoms to diffuse into the crystal structure to give an impurity concentration which falls off as depth of penetration increases. If a germanium wafer with such a graded impurity concentration is used as the starting point in the manufacture of a micro-alloy transistor, a much-reduced transit time can be obtained and the transistor has a better high-frequency response than one with a uniform base region. In theory the cut-off frequency of a transistor can be improved by up to eight times by grading the impurity concentration in the base region but in practice the improvement is normally less than this. Graded-base micro-alloy transistors, known as *micro-alloy diffused transistors* (MADTs), operate at frequencies up to 300 MHz.

Drift transistors

Wafers of graded impurity concentration can also be used as the starting point in making alloy-junction transistors. The resulting transistors are known as drift types and operate at frequencies up to 100 MHz.

Diffused transistors

The technique of solid-state diffusion was mentioned above as a means of producing a semiconductor with a graded impurity concentration. This technique is also extensively used as a means of producing layers of p-type and n-type conductivity of very thin but controllable thickness which can be used as the emitter, base and sometimes collector regions of a transistor. There are many methods of manufacturing transistors using such layers: two will now be described.

In one method, introduced during 1956, the starting point is a wafer of p-type germanium which ultimately forms the collector region. To this, two pellets are alloyed in close proximity: one pellet is of a group-V element such as antimony and the other is of mixed group-III and group-V elements (e.g. antimony and aluminium). The wafer is now heated causing the elements in the pellets to diffuse into the germanium. Only n-type impurities diffuse from the first pellet but both n-type and p-type from the second. However, in p-type germanium, n-type impurities diffuse more rapidly than the p-type and thus, after diffusion, we obtain a structure such as that illustrated in Fig. A.3: the two diffused n-regions merge to form a base layer (to which the first pellet gives ohmic contact) whilst the recrystallised p-region acts as the emitter region (to which the second pellet gives ohmic contact). Finally the diffused n-type layer is etched away except around the two pellets which are masked during this operation. This leaves the active part of the transistor in the form of a mesa projecting from the collector material as shown in the diagram. This particular method of manufacture yields types of transistor known as *post-alloy diffused transistors* (PADTs).

The difficulty of manufacturing alloy-junction silicon transistors has already been mentioned. Many silicon transistors are now manufactured

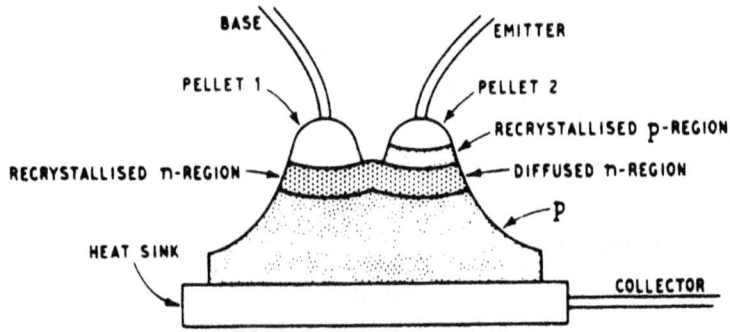

Fig. A.3. Construction of PADT pnp germanium transistor

by other techniques which include solid-state diffusion. In one method the starting point is a crystal of n-type silicon with a uniform resistivity of, say, 1 Ω-cm. This is cut into slices about 0.015 in thick each of which finally yields 100 or more transistors. The slices are heated to around 1,250°C in a furnace with a source of p-type impurity. The impurity volatilises to produce an atmosphere which diffuses into the surface of the slice to a small but controllable depth to give a p-region which is later used as the base areas of the transistors. The slice is now covered with a mask containing 100 or more apertures and is again heated, this time in an atmosphere of an n-type impurity. This diffuses to form a number of n-regions which are used as the emitter areas of the transistors. The slice is afterwards cut to separate the individual transistors. A sectional view of one transistor is given in Fig. A.4.

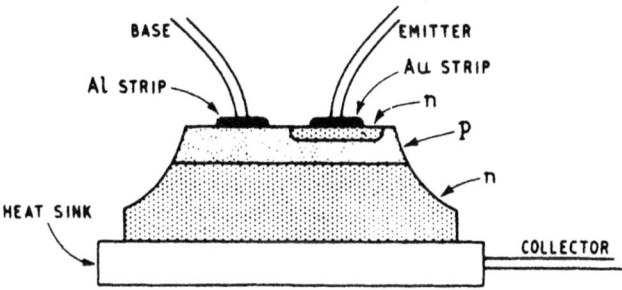

Fig. A.4. Construction of a diffused silicon transistor

Connections are now made to the three regions of each transistor. The body of the semiconductor is soldered to a heat sink which provides electrical connection to the collector. Contact to the emitter region is obtained by depositing a thin gold strip on the region by evaporation via a mask and finally lightly alloying it to the n-region by heating. The base connection is obtained by similarly depositing a thin strip of aluminium on the exposed p-region and once again lightly alloying it to provide an ohmic contact.

Only the base region between the emitter and collector junctions is vital to the action of the transistor and surplus base-collector junction area causes unnecessary collector capacitance. This capacitance is minimised by etching away the base material except that near the gold and aluminium, leaving the active part of the transistor standing up from the collector region in the form of a mesa. Such transistors operate at frequencies up to 500 MHz.

Epitaxial diffused transistors

The performance of silicon and germanium diffused transistors manufactured by the above processes is partly limited by the resistance of that part of the wafer forming the collector region. Only the upper crust of the wafer, containing the junctions, is effective in providing transistor action and the thickness of the collector region should be as thin as possible to minimise its resistance. However, it is difficult to work with a thickness less than about 0.003 in, yet such a thickness gives too high a resistance for some applications. If most of the semiconductor below the junctions could be replaced by low-resistivity material, the performance of the transistor could be greatly improved. This was effectively achieved around 1962 by using the technique known as *epitaxy*.

In a silicon epitaxial transistor the starting point is a wafer of very low resistivity, e.g. $0.002 \, \Omega$-cm. A thin skin of silicon of resistivity suitable for transistor action is deposited on the face of the wafer in such a way that it takes up the same crystalline structure as the wafer. The added layer is termed the *epitaxial layer* and is produced by heating the wafer to about $1,200°C$ in a furnace whilst it is exposed to an atmosphere of, for example, silicon tetrachloride and hydrogen. The hydrogen reduces the chloride to silicon which is deposited on the wafer. At the temperature used, the silicon atoms are mobile and take up their correct orientation relative to the crystalline structure of the wafer. A halide of a group-V impurity is also added to give the epitaxial layer the required n-type conductivity of about $1 \, \Omega$-cm. A typical thickness of an epitaxial layer is 10 microns.

The wafer, thus treated, can now be used for the manufacture of diffused transistors by the method described above, the diffusion occurring within the epitaxial layer. The characteristics of the transistors so produced are much better than those of a transistor without the epitaxial layer because of the much-reduced collector series resistance.

Planar transistors

When a silicon wafer is heated to about $1,200°C$ in an atmosphere of water vapour or oxygen a skin of silicon dioxide SiO_2 forms on the surface. This skin is a most effective seal against the ingress of moisture at room temperatures and has made possible the method of manufacture of planar transistors which is described below.

A crystal of n-type silicon, about 1 in diameter, is cut into slices about 0.008 in thick. The slices are lapped and etched to approximately

0.003 in thickness and, if required, an epitaxial layer can be formed on one surface. The slices are now heated in an oxidising atmosphere to acquire a protective coating of silicon dioxide. At this stage each slice has a sectional view similar to that shown in Fig. A.5(a). Each slice yields ultimately up to 1,000 transistors and the next stage is to mark off the individual transistors. This is achieved by a photo-lithographic process: each slice is coated in a dark room with a photo-sensitive material (known as photo-resist) and is then exposed to ultra-violet light via a mask containing an array of apertures corresponding to the base areas of the 1,000 transistors. The slice is now developed to remove the photo-resist from these regions thus exposing the silicon dioxide coating. Next the slice is treated with an etch which removes the silicon dioxide from the exposed regions. The remainder of the photo-resist is now dissolved: the cross-section of the slice now appears as in Fig. A.5(b) which shows a gap in the layer of silicon dioxide defining the base area for a single transistor.

The slice is now exposed at a high temperature to a boron-rich atmosphere. The silicon dioxide coating protects the slice against diffusion of boron except at the exposed areas and here boron diffuses isotropically, i.e. horizontally under the protective coating as well as vertically into the crystal, thus forming a p-type base region. Other more precise ways of forming such a region have been developed, for example by ion implantation. This involves a sharply defined bombardment of the substrate by a beam from an ion gun which enables the active base area to be closely controlled in area and shape, a process which can be compared with precision etching. The slice is now returned to the oxidising atmosphere and a coating of silicon dioxide is formed over the base areas (and the rest of the slice) to give a cross-section similar to that shown in Fig. A.5(c).

The emitter areas are now defined by a similar process of masking, photo-lithography, exposure to ultra-violet light, etching, etc., and the silicon dioxide is removed from the emitter areas to give a cross-section such as that shown in Fig. A.5(d). The slice is now heated whilst exposed to an atmosphere rich in phosphorus. This forms an n-type emitter region by diffusion and the exposed area is again sealed by heating the slice in an oxidising atmosphere to form a layer of silicon dioxide. See Fig. A.5(e).

Holes are now made in the silicon dioxide coating as shown in Fig. A.5(f) to permit ohmic contacts to be made to the base and emitter areas, the position of the holes being again determined by a mask. Contacts are then made to the transistors by a process of evaporation: the slice is placed in a vacuum chamber in which aluminium is evaporated, e.g. from a hot filament. This results in a deposition of a thin coating of aluminium over the entire face of the slice. Finally the

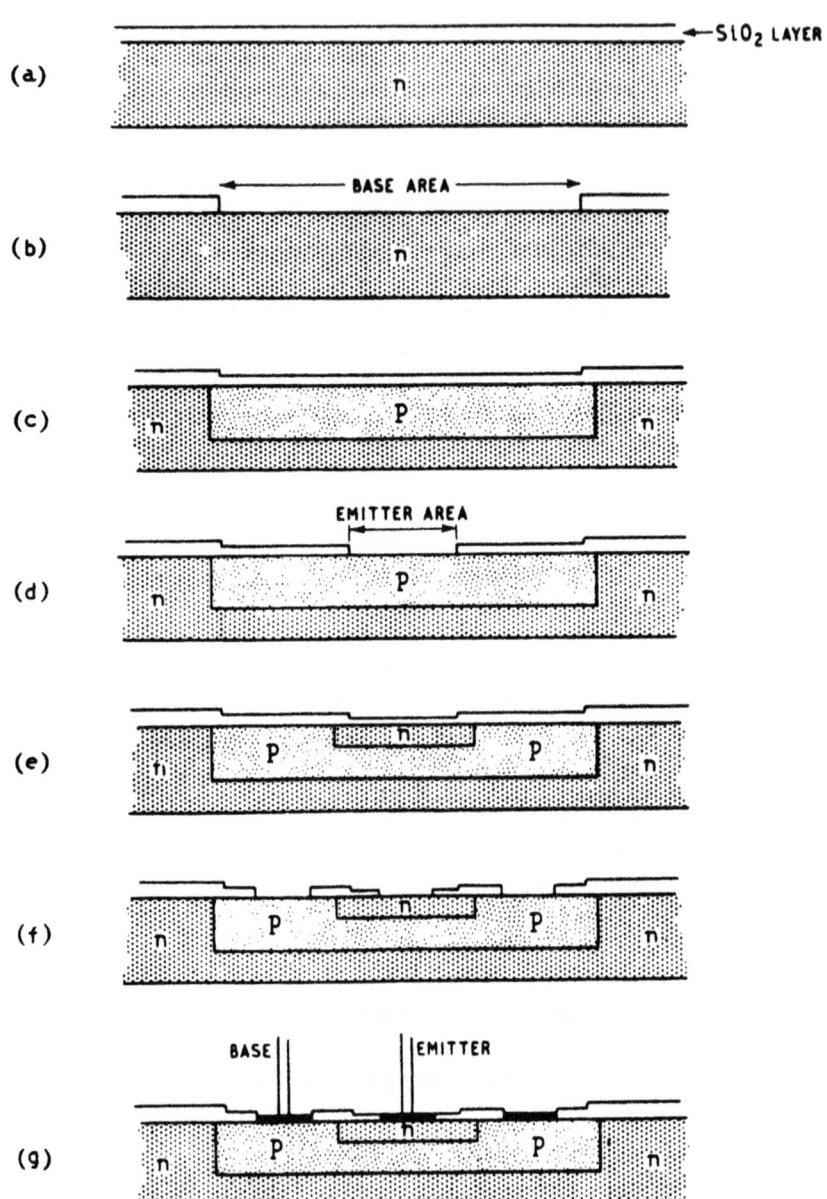

Fig. A.5. Stages in the manufacture of planar transistors

aluminium is removed from the areas in which it is not required by a masking and selective etching operation. The slice is now divided up into individual transistors and connections are made to the base and emitter regions of each transistor as shown in Fig. A.5(g). The base area of each transistor is sometimes of approximately annular shape surrounding a circular emitter area but in power transistors both base and emitter areas may be in the form of parallel strips.

The process described above produces planar transistors in large numbers and is well suited to mass production. Its introduction during the early 1960s revolutionised silicon transistor manufacture. The transistors are particularly robust and the protection of the silicon dioxide coating is such that even without sealing in cans the transistors will operate well under boiling water! Leakage currents are very low and the transistors can be designed to work at frequencies well over 1 GHz. In 1963 the process also made possible for the first time mass production of fets although the principle of this type of transistor had been described by Shockley 11 years earlier.

Integrated (monolithic) circuits

The method of manufacture of planar transistors lends itself well to the production of integrated circuits. These are circuits designed to carry out a particular function, or group of functions and may embody several transistors (bipolar or field-effect), diodes, resistors and all the necessary interconnections. All are produced on a single chip of silicon measuring perhaps less than $\frac{1}{4}$ in square by the photo-lithographic, masking, diffusion and evaporation processes described above. Resistors are areas of p-type or n-type silicon, the dimensions and impurity concentration being chosen to give the required value of resistance. Small capacitors, of the order of a few pF, can consist of reverse-biased pn junctions as suggested in Chapter 1. It is inconvenient, however, to make large capacitances and integrated circuits are usually designed with direct couplings, often using emitter followers, to avoid the need for such capacitances. This is illustrated in the examples of monolithic circuits described in earlier chapters of this book.

Most i.cs are embodied in a rectangular container known as a dual-in-line package with contacts spaced at 0.1 in intervals down the two longer sides, this spacing agreeing with that of the copper tracks on the printed wiring boards to which the i.cs are soldered.

Attempts were made to produce i.cs in the 1950s but it was not until the planar process was developed in 1960 that they became practical. Early i.cs were digital circuits (RTL, TTL, ECL etc.) composed largely of bipolar transistors and resistors. The number of components per chip was less than 100, a density known as small-scale integration (SSI).

The m.o.s. transistors introduced in 1966 required fewer diffusions in manufacture and occupied less area on the chip but were slower in operation than bipolar transistors. Using the m.o.s. technique enabled the component density to be increased to a few hundred per chip (medium-scale integration, MSI).

The computer industry demanded still greater component density and by 1969 an i.c. containing more than 10,000 m.o.s. transistors was available, this development being known as large-scale integration (LSI). By 1975 component densities were approaching 100,000 per chip.

Since 1980 further miniaturisation has made possible component densities of half a million components per chip (very-large-scale integration, VLSI). I.Cs now require more input and output leads than the 32 per side which can be accommodated on a reasonably sized dual-in-line package. Alternative methods of packaging are being introduced such as the chip carrier which has contacts on all four sides and smaller pin spacing.

There have been significant developments in i.cs for linear equipment. The first linear i.cs were operational amplifiers and towards the end of 1960 i.cs were developed for use in radio and television receivers. There is little limit to the complexity of i.cs which can therefore be developed to perform nearly all the functions required in receivers, high-fidelity equipments and video cassette recorders to improve their performance, their reliability and ease of operation.

Integrated circuits are now extensively used in linear and pulse equipment and their use is introducing a new design philosophy into electronics. Designers using monolithic devices are little concerned with components such as resistors, diodes and transistors: their main interest is in the function of the integrated circuits and their task is to choose devices with the functions necessary to carry out the purpose of the equipment and to combine them, ensuring that the impedances, signal levels, gains, bandwidths, etc., are correctly matched.

Appendix B

Transistor parameters

Introduction

There are three main systems of expressing the properties of a four-terminal network. In each the network is regarded as having an input voltage v_i, an input current i_i, an output voltage v_o and output current i_o as illustrated in Fig. B.1.

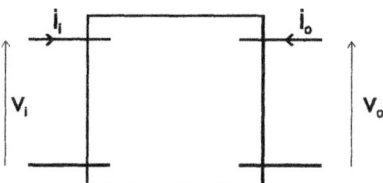

Fig. B.1.

Z parameters

In the first system the input and output voltages are expressed in terms of the input and output currents thus:

$$v_i = z_i i_i + z_r i_o \tag{B.1}$$

$$v_o = z_f i_i + z_o i_o \tag{B.2}$$

The factors z_i, z_r, etc., have the nature of impedances because, when multiplied by currents, they give voltages. The impedances can, in fact, be represented in the equivalent circuit for the transistor or shown in Fig. B. 2. z_i represents the input impedance of the transistor and z_o the

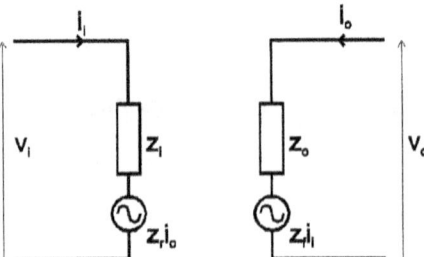

Fig. B.2.

output impedance. z_f can be called the forward impedance and z_r the reverse impedance.

At low frequencies the impedances can usually be replaced by resistances and the two fundamental equations become

$$v_i = r_i i_i + r_r i_o \tag{B.3}$$

$$v_o = r_f i_i + r_o i_o \tag{B.4}$$

in which the factors are known as r parameters.

z and r parameters are not greatly favoured largely because of the difficulty of measuring them. y and h parameters are generally preferred.

Y parameters

In this second system the input and output currents are expressed in terms of the input and output voltages thus:

$$i_i = y_i v_i + y_r v_o \tag{B.5}$$

$$i_o = y_f v_i + y_o v_o \tag{B.6}$$

The factors y_i, y_r, etc., have the nature of admittances because when multiplied by voltages, they give currents. The admittances can be represented in the equivalent circuit for the transistor shown in Fig. B.3.

The y parameters are more easily measured than the z parameters. If the output voltage v_o is made zero (to alternating signals) by connecting

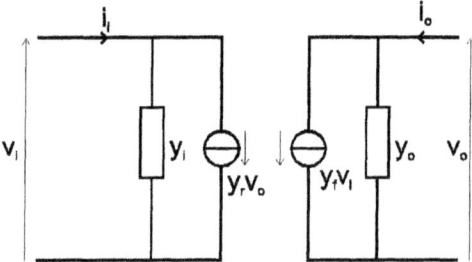

Fig. B.3.

a low-reactance capacitor across the output terminals, we then have, putting $v_0 = 0$ in Eqn B.5

$$i_i = y_i v_i$$

giving

$$y_i = \frac{i_i}{v_i}$$

i.e. y_i is the ratio of the input current to the input voltage.

y parameters are used in radio-frequency applications of transistors.

Hybrid parameters

These parameters combine some of the properties of the z and y parameters. In this system the input voltage and output current are both expressed in terms of input current and output voltage. The fundamental equations are as follows:

$$v_i = h_i i_i + h_r v_o \tag{B.7}$$

$$i_o = h_f i_i + h_o v_o \tag{B.8}$$

By inspection of these equations we can see that h_i has the dimensions of an impedance and h_o has the dimensions of an admittance. h_r and h_f are, however, both pure numbers. For this reason these parameters are known as *hybrid*. The parameters can be represented in an equivalent circuit for the transistor such as that shown in Fig. B.4.

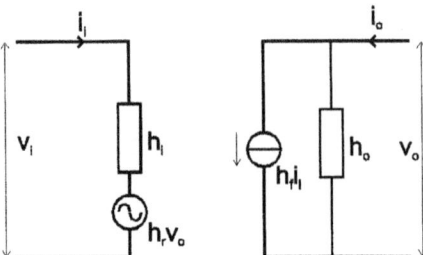

Fig. B.4.

The h parameters are quite simple to measure. For if the output terminals are short-circuited we have $v_o = 0$ and from Eqn B.7 we have

$$h_i = \frac{v_i}{i_i} \tag{B.9}$$

whilst from Eqn B.8 we have

$$h_f = \frac{i_o}{i_i} \tag{B.10}$$

If the input terminals are open-circuited $i_i = 0$ and from Eqn B.7 we have

$$h_r = \frac{v_i}{v_v} \tag{B.11}$$

whilst from Eqn B.8 we have

$$h_o = \frac{i_o}{v_o} \tag{B.12}$$

These parameters give useful information on the transistor performance. For example from Eqn B.9 h_i is equal to the transistor input impedance for short-circuited output. From Eqn B.12 the reciprocal of h_o is equal to the output impedance for open-circuited input. From Eqn B.10 h_f gives the current gain of the transistor. From Eqn B.11 h_r is equal to the ratio of the input voltage to the output voltage: this is known as the *voltage feedback ratio* of the transistor. These parameters are often quoted in transistor manufacturers' data sheets.

It is easily possible to convert from one set of parameters to another. For example if v_o is put equal to 0 in Eqns B.5 and B.7 we can easily show that

$$h_i = \frac{1}{y_i}$$

The parameters for a particular type of circuit configuration are generally distinguished by placing a suffix b for common-base, e for common-emitter and c for common-collector connection although the last-mentioned is not greatly used. For example h_{ib} is the input resistance (for short-circuited output) of a common-base circuit and h_{fe} is the current gain of a common-emitter circuit.

Relationship between hybrid parameters and the T-section equivalent circuit

From Eqn 3.7, we know that the input resistance of a common-base transistor circuit, for short-circuited output terminals, is given by

$$r_i = r_e + r_b/\beta$$

This, by definition, is the parameter h_{ib}.

$$\therefore h_{ib} = r_e + r_b/\beta \tag{B.13}$$

and, as we have seen, a typical value for this quantity is 31 Ω.

Also from Eqn 3.2 we know that the current gain for short-circuited output terminals is given by

$$\frac{i_c}{i_e} = \frac{r_b + \alpha r_c}{r_b + r_c}$$

This, by definition, is h_{fb}.

$$\therefore h_{fb} = \frac{r_b + \alpha r_c}{r_b + r_c}$$

Normally, of course, r_b is very small compared with αr_c and r_c. We can therefore write

$$h_{fb} = \alpha \tag{B.14}$$

and a typical value for h_{fb} is 0.98.

The output resistance for open-circuited input terminals is given by Eqn 3.14. Thus we have

$$r_o = r_b + r_c$$

This is, by definition, equal to the reciprocal of h_{ob}. Thus

$$h_{ob} = \frac{1}{r_b + r_c} \qquad (B.15)$$

As r_b is very small compared with r_c this may be simplified to

$$h_{ob} = \frac{1}{r_c}$$

A typical value for r_c is $1\,\text{M}\Omega$ and thus h_{ob} is equal to 10^{-6} mho or $1\,\mu\text{mho}$.

Finally the ratio of input voltage to output voltage for open-circuited input terminals is given by

$$h_{rb} = \frac{r_b}{r_b + r_c} \qquad (B.16)$$

This result can be obtained by inspection from Fig. 2.10 on page 33. Normally r_b may be neglected in comparison with r_c and this result can thus be simplified to

$$h_{rb} = \frac{r_b}{r_c}$$

Typical values of r_b and r_c are $300\,\Omega$ and $1\,\text{M}\Omega$, giving h_{rb} as 3×10^{-4}.

Appendix C

The stability of a transistor tuned amplifier

Fig. C.1 represents a transistor with input and output tuned circuits damped by parallel-connected resistances. As explained on page 182 feedback occurs via the internal collector-base capacitance c_{re} and this can cause oscillation at a frequency below the resonance value for the tuned circuits. Oscillation is most likely when the phase shift introduced by the tuned circuits totals 90°: if the circuits are similar and resonant at the same frequency oscillation occurs at the frequency for which each circuit gives 45° phase shift. From this information it is possible to arrive at a simple expression for the stability of the amplifier.

Fig. C.1 may be redrawn as in Fig. C.2 in which Z_f is the impedance of C_{re}. An alternating base input voltage v_b gives rise to a collector current of $g_m v_b$. This current flows through the transistor load which is

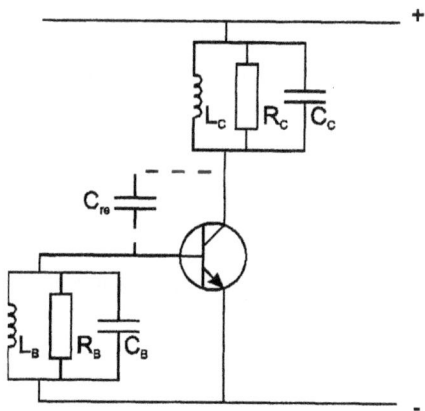

Fig. C.1. Essential features of a transistor tuned amplifier

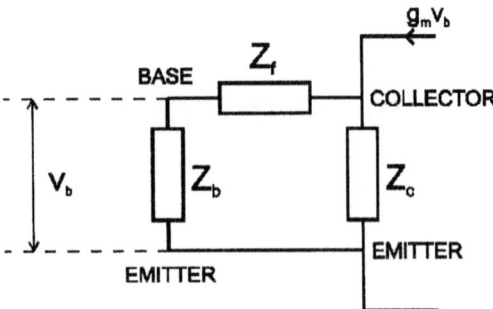

Fig. C.2. Equivalent circuit for Fig. C.1

composed of Z_c in parallel with the series combination of Z_f and Z_b. Thus the voltage generated across Z_c is given by

$$g_m v_b \cdot \frac{Z_c(Z_f + Z_b)}{Z_c + Z_f + Z_b}$$

Z_f and Z_b constitute a potential divider across Z_c and the voltage appearing across Z_b is equal to

$$g_m v_b \cdot \frac{Z_c(Z_f + Z_b)}{Z_c + Z_f + Z_b} \cdot \frac{Z_b}{Z_f + Z_b}$$

$$= g_m v_b \cdot \frac{Z_c Z_b}{Z_c + Z_f + Z_b}$$

Now C_{re} is normally a very small capacitance and its reactance is large compared with Z_c and Z_b. Thus the voltage across Z_b is given approximately by

$$g_m v_b \cdot \frac{Z_c Z_b}{Z_f}$$

If this is equal to v_b, oscillation can occur. The condition for oscillation is thus

$$\frac{g_m Z_c Z_b}{Z_f} = 1 \qquad\qquad\qquad (\text{C.1})$$

Now Z_c is composed of R_c in parallel with X_c the net reactance of L_c and C_c. We can thus say

$$Z_c = \frac{R_c j X_c}{R_c + j X_c}$$

To give 45° phase shift between the current in Z_a and the voltage across it, X_c, must equal R_c and we have

$$Z_c = \frac{jR_c}{1 + j} = \frac{j(1 - j)R_c}{2}$$

Similarly

$$Z_b = \frac{j(1 - j)R_b}{2}$$

$Z_f = 1/j\omega C_{re}$. Substituting in Eqn C.1 we have

$$\frac{j\omega C_{re} gm j(1 - j)R_c j(1 - j)R_b}{4} = 1$$

which on simplification gives

$$\omega C_{re} gm R_b R_c = 2$$

The condition for stability is thus

$$\omega C_{re} gm R_b R_c < 2$$

Semiconductor letter symbols

Bipolar

C_{cb}, C_{ce}, C_{eb} Inter-terminal capacitance (collector-to-base, collector-to-emitter, emitter-to-base).

C_{ibo}, C_{ieo} Open-circuit input capacitance (common-base, common-emitter).

C_{ibs}, C_{ieo} Short-circuit input capacitance (common-base, common-emitter).

C_{obo}, C_{oeo} Open-circuit output capacitance (common-base, common-emitter).

C_{obs}, C_{oes} Short-circuit output capacitance (common-base, common-emitter).

C_{rbs}, C_{res} Short-circuit reverse transfer capacitance (common-base, common-emitter).

C_{tc}, C_{te} Depletion-layer capacitance (collector, emitter).

f_{hfb}, h_{fe} Small-signal short-circuit forward current transfer ratio cut-off frequency (common-base, common-emitter).

f_{max} Maximum frequency of oscillation.

f_T Transition frequency or frequency at which small-signal forward current transfer ratio (common-emitter) extrapolates to unity.

f_1 Frequency of unity current transfer ratio.

g_{PB}, g_{PE} Large-signal insertion power gain (common-base, common-emitter).

g_{pb}, g_{pe} Small-signal insertion power gain (common-base, common-emitter).

g_{TB}, g_{TE} Large-signal transducer power gain (common-base, common-emitter).

g_{tb}, g_{te} Small-signal transducer power gain (common-base, common-emitter).

h_{FB}, h_{FE} Static forward current transfer ratio (common-base, common-emitter).

h_{fb}, h_{fe} Small-signal short-circuit forward current transfer ratio (common-base, common-emitter).

h_{ib}, h_{ie} Small-signal short-circuit input impedance (common-base, common-emitter).

$h_{ie(imag)}$ or $Im(h_{ie})$ Imaginary part of the small-signal short-circuit input impedance (common-emitter).

$h_{ie(real)}$ or $Re(h_{ie})$ Real part of the small-signal short-circuit input impedance (common-emitter).

h_{ob}, h_{oe} Small-signal open-circuit output admittance (common-base, common-emitter).

$h_{oe(imag)}$ or $Im(h_{oe})$ Imaginary part of the small-signal open-circuit output admittance (common-emitter).

$h_{oe(real)}$ or $Re(h_{oe})$ Real part of the small-signal open-circuit output admittance (common-emitter).

h_{rb}, h_{re} Small-signal open-circuit reverse voltage transfer ratio (common-base, common-emitter).

I_B, I_C, I_E Current, d.c. (base-terminal, collector-terminal, emitter-terminal).

I_b, I_c, I_e Current, r.m.s. value of alternating component (base-terminal, collector-terminal, emitter-terminal).

i_B, i_C, i_E, Current, instantaneous total value (base-terminal, collector-terminal, emitter-terminal).

I_{BEV} Base cut-off current, d.c.

I_{CBO} Collector cut-off current, d.c., emitter open.

$I_{E1E2(off)}$ Emitter cut-off current.

I_{EBO} Emitter cut-off current, d.c., collector open.

$I_{Ec(ofs)}$ Emitter-collector offset current.

I_{ECS} Emitter cut-off current, d.c., base-short-circuited to collector.

P_{IB}, P_{IE} Large-signal input power (common-base, common-emitter).

P_{ib}, P_{ie} Small-signal input power (common-base, common-emitter).

P_{OB}, P_{OE} Large-signal output power (common-base, common-emitter).

P_{ob}, P_{oe} Small-signal output power (common-base, common-emitter).

P_T Total non-reactive power input to all terminals.

$P_{tot\ max}$ Maximum total dissipation.

$r_b{'}C_c$ Collector-base time constant.

$r_{ce(sat)}$ Saturation resistance, collector-to-emitter.

$Re\ (y_{ie})$

$Re\ (y_{oe})$

$r_{e1e2(on)}$ Small-signal emitter-emitter on-state resistance.

R_θ Thermal resistance.

T_{amb} Ambient temperature.

T_j Junction temperature.

t_d Delay time.

t_f Fall time.
t_{off} Turn-off time.
t_{on} Turn-on time.
t_p Pulse time.
t_r Rise time.
t_s Storage time.
t_w Pulse average time.
V_{BB}, V_{CC}, V_{EE} Supply voltage, d.c. (base, collector, emitter).
V_{BC}, V_{BE}, V_{CB}, V_{CE}, V_{EB}, V_{EC} Voltage, d.c. or average (base-to-collector, base-to-emitter, collector-to-base, collector-to-emitter, emitter-to-base, emitter-to-collector).
v_{bc}, v_{be}, v_{cb}, v_{ce}, v_{eb}, v_{ec} Voltage, instantaneous value of alternating component (base-to-collector, base-to-emitter, collector-to-base, collector-to-emitter, emitter-to-base, emitter-to-collector).
$V_{(BR)CBO}$ (formerly BV_{CBO}) Breakdown voltage, collector-to-base, emitter open.
V_{RT} Reach-through (punch-through) voltage.
Y_{fb}, Y_{fe} Small-signal short-circuit forward-transfer admittance (common-base, common-emitter).
Y_{ib}, Y_{ie} Small-signal short-circuit input admittance (common-base, common-emitter).
$Y_{ie(imag)}$ or $Im(Y_{ie})$ Imaginary part of the small-signal short-circuit input admittance (common-emitter).
$Y_{ie(real)}$ or $Re(Y_{ie})$ Real part of the small-signal short-circuit input admittance (common-emitter).
Y_{ob}, Y_{oe} Small-signal short-circuit output admittance (common-base, common-emitter).
$Y_{oe(imag)}$ or $Im(Y_{oe})$ Imaginary part of the small-signal short-circuit output admittance (common-emitter).
$Y_{oe(real)}$ or $Re(Y_{oe})$ Real part of the small-signal short-circuit output admittance (common-emitter).
Y_{rb}, Y_{re} Small-signal short-circuit reverse transfer admittance (common-base, common-emitter).

Unijunction

η Intrinsic stand-off ratio.
$I_{B_2(mod)}$ Interbase modulated current.
I_{EB_2O} Emitter reverse current.
I_p Peak-point current.
I_V Valley-point current.
r_{BB} Interbase resistance.
T_j Junction temperature.

t_p Pulse time.
t_w Pulse average time.
$V_{B_2B_1}$ Interbase voltage.
$V_{EB_1(sat)}$ Emitter saturation voltage.
V_{OB_1} Base-1 peak voltage.
V_p Peak-point voltage.
V_v Valley-point voltage.

Field effect

b_{fs}, b_{is}, b_{os}, b_{rs} Common-source small-signal (forward transfer, input, output, reverse transfer) susceptance.
C_{ds} Drain-source capacitance.
C_{du} Drain-substrate capacitance.
C_{iss} Short-circuit input capacitance, common-source.
C_{oss} Short-circuit output capacitance, common-source.
c_{rss} Short-circuit reverse transfer capacitance, common-source.
$\overline{F}$ or F Noise figure, average or spot.
g_{fs}, g_{is}, g_{os}, g_{rs} Signal (forward transfer, input, output, reverse transfer) conductance.
g_m Transconductance.
g_{pg}, g_{ps} Small-signal insertion power gain (common-gate, common-source).
g_{tg}, g_{ts} Small-signal transducer power gain (common-gate, common-source).
$I_{D(off)}$ Drain cut-off current.
$I_{D(on)}$ On-state drain current.
I_{DSS} Zero-gate-voltage drain current.
I_G Gate current, d.c.
I_{GF} Forward gate current.
I_{GR} Reverse gate current.
I_{GSS} Reverse gate current, drain short-circuited to source.
I_{GSSF} Forward gate current, drain short-circuited to source.
I_{GSSR} Reverse gate current, drain short-circuited to source.
I_n Noise current, equivalent input.
$Im(y_{rs})$, $Im(y_{is})$, $Im(y_{os})$, $Im(y_{rs})$.
I_s Source current, d.c.
$I_{s(off)}$ Source cut-off current.
I_{SDS} Zero-gate-voltage source current.
$r_{ds(on)}$ Small-signal drain-source on-state resistance.
$r_{DS(on)}$ Static drain-source on-state resistance.
$t_{d(on)}$ Turn-on delay time.
t_f Fall time.

t_{off} Turn-off time.
t_{on} Turn-on time.
t_p Pulse time.
t_r Rise time.
t_w Pulse average time.
$V_{(BR)GSS}$ Gate-source breakdown voltage.
$V_{(BR)GSSF}$ Forward gate-source breakdown voltage.
$V_{(BR)GSSR}$ Reverse gate-source breakdown voltage.
V_{DD}, V_{GG}, V_{SS} Supply voltage, d.c. (drain, gate, source).
V_{DG} Drain-gate voltage.
V_{DS} Drain-source voltage.
$V_{DS(on)}$ Drain-source on-state voltage.
V_{DU} Drain-substrate voltage.
V_{GS} Gate-source voltage.
V_{GSF} Forward gate-source voltage.
V_{GSR} Reverse gate-source voltage.
$V_{GS(off)}$ Gate-source cut-off voltage.
$V_{GS(th)}$ Gate-source threshold voltage.
V_{GU} Gate-substrate voltage.
V_n Noise voltage equivalent input.
V_{SU} Source-substrate voltage.
y_{fs} Common-source small-signal short-circuit forward transfer admittance.
y_{is} Common-source small-signal short-circuit input admittance.
y_{os} Common-source small-signal short-circuit admittance.

Diodes

Diss Dissipation, watts.
c_T Barrier-layer capacitance.
I_F Forward current.
$I_{F(AV)}$ Average forward current.
I_{FRM} Repetitive peak forward current.
I_{FM} Peak forward current.
I_{FSM} Surge (non-repetitive) forward current.
I_G Gate current.
I_{GT} Gate trigger current.
I_O Continuous average forward current (in half-wave rectification).
I_R Reverse current.
I_Z Max. current in breakdown region (Zener diodes).
T_K Temperature coefficient of Zener voltage.
U_F Forward voltage.
U_R Blocking (inverse) voltage.

U_{RM} Peak blocking voltage.
U_Z Breakdown voltage (Zener diodes).
V_F Forward voltage.
V_{IRM} Input repetitive peak voltage.
V_Z Breakdown voltage (Zener diodes).

N.B. *U* and *V* are generally interchangeable.

Index

Printed and bound by CPI Group (UK) Ltd, Croydon, CR0 4YY

03/10/2024

01040432-0019